Modifying the Electronics of Modern Classic Cars
– the complete guide for your 1990s to 2000s car

More from Veloce:

SpeedPro Series
4-Cylinder Engine Short Block High-Performance Manual – New Updated & Revised Edition (Hammill)
Aerodynamics of Your Road Car, Modifying the (Edgar and Barnard)
Alfa Romeo DOHC High-performance Manual (Kartalamakis)
Alfa Romeo V6 Engine High-performance Manual (Kartalamakis)
BMC 998cc A-series Engine, How to Power Tune (Hammill)
1275cc A-series High-performance Manual (Hammill)
Camshafts – How to Choose & Time Them For Maximum Power (Hammill)
Competition Car Datalogging Manual, The (Templeman)
Custom Air Suspension – How to install air suspension in your road car – on a budget! (Edgar)
Cylinder Heads, How to Build, Modify & Power Tune – Updated & Revised Edition (Burgess & Gollan)
Distributor-type Ignition Systems, How to Build & Power Tune – New 3rd Edition (Hammill)
Fast Road Car, How to Plan and Build – Revised & Updated Colour New Edition (Stapleton)
Ford SOHC 'Pinto' & Sierra Cosworth DOHC Engines, How to Power Tune – Updated & Enlarged Edition (Hammill)
Ford V8, How to Power Tune Small Block Engines (Hammill)
Harley-Davidson Evolution Engines, How to Build & Power Tune (Hammill)
Holley Carburetors, How to Build & Power Tune – Revised & Updated Edition (Hammill)
Honda Civic Type R High-Performance Manual, The (Cowland & Clifford)
Jaguar XK Engines, How to Power Tune – Revised & Updated Colour Edition (Hammill)
Land Rover Discovery, Defender & Range Rover – How to Modify Coil Sprung Models for High Performance & Off-Road Action (Hosier)
MG Midget & Austin-Healey Sprite, How to Power Tune – Enlarged & updated 4th Edition (Stapleton)
MGB 4-cylinder Engine, How to Power Tune (Burgess)
MGB V8 Power, How to Give Your – Third Colour Edition (Williams)
MGB, MGC & MGB V8, How to Improve – New 2nd Edition (Williams)
Mini Engines, How to Power Tune On a Small Budget – Colour Edition (Hammill)
Modifying the Aerodynamics of your Road Car (Edgar)
Motorcycle-engined Racing Cars, How to Build (Pashley)
Motorsport, Getting Started in (Collins)
Nissan GT-R High-performance Manual, The (Gorodji)
Nitrous Oxide High-performance Manual, The (Langfield)
Optimising Car Performance Modifications (Edgar)
Race & Trackday Driving Techniques (Hornsey)
Retro or classic car for high performance, How to modify your (Stapleton)
Rover V8 Engines, How to Power Tune (Hammill)
Secrets of Speed – Today's techniques for 4-stroke engine blueprinting & tuning (Swager)
Sportscar & Kitcar Suspension & Brakes, How to Build & Modify – Revised 3rd Edition (Hammill)
SU Carburettor High-performance Manual (Hammill)
Successful Low-Cost Rally Car, How to Build a (Young)
Suzuki 4x4, How to Modify For Serious Off-road Action (Richardson)
Tiger Avon Sportscar, How to Build Your Own – Updated & Revised 2nd Edition (Dudley)
Triumph TR2, 3 & TR4, How to Improve (Williams)
Triumph TR5, 250 & TR6, How to Improve (Williams)
Triumph TR7 & TR8, How to Improve (Williams)
V8 Engine, How to Build a Short Block For High Performance (Hammill)
Volkswagen Beetle Suspension, Brakes & Chassis, How to Modify For High Performance (Hale)
Volkswagen Bus Suspension, Brakes & Chassis for High Performance, How to Modify – Updated & Enlarged New Edition (Hale)
Weber DCOE, & Dellorto DHLA Carburetors, How to Build & Power Tune – 3rd Edition (Hammill)

WorkshopPro Series
Car electrical and electronic systems (Edgar)
Setting up a home car workshop (Edgar)

EARTHWORLD – the new imprint from Veloce
Discovering Engineering that Changed the World (Edgar)

www.veloce.co.uk

First published in December 2018 by Veloce Publishing Limited, Veloce House, Parkway Farm Business Park, Middle Farm Way, Poundbury, Dorchester DT1 3AR, England. Tel +44 (0)1305 260068 / Fax 01305 250479 / e-mail info@veloce.co.uk / web www.veloce.co.uk or www.velocebooks.com.
ISBN: 978-1-787113-93-0; UPC: 6-36847-01393-6.
© 2018 Julian Edgar and Veloce Publishing. All rights reserved. With the exception of quoting brief passages for the purpose of review, no part of this publication may be recorded, reproduced or transmitted by any means, including photocopying, without the written permission of Veloce Publishing Ltd. Throughout this book logos, model names and designations, etc, have been used for the purposes of identification, illustration and decoration. Such names are the property of the trademark holder as this is not an official publication. Readers with ideas for automotive books, or books on other transport or related hobby subjects, are invited to write to the editorial director of Veloce Publishing at the above address. British Library Cataloguing in Publication Data – A catalogue record for this book is available from the British Library.
Typesetting, design and page make-up all by Veloce Publishing Ltd on Apple Mac. Printed in India by Parksons Graphics.

Modifying the Electronics of Modern Classic Cars
– the complete guide for your 1990s to 2000s car

JULIAN EDGAR

WORKSHOP PRO MODIFYING THE ELECTRONICS OF MODERN CLASSIC CARS

INTRODUCTION 6

1 ELECTRONIC MODIFICATION 10

2 OVERVIEW OF ENGINE MANAGEMENT 20

3 ENGINE MANAGEMENT INTERCEPTORS 28

4 PROGRAMMABLE ENGINE MANAGEMENT. 62

5 OTHER ENGINE BAY MODIFICATIONS 92

CONTENTS

6 MODIFYING OTHER ELECTRONIC CONTROL SYSTEMS 118

7 MODIFYING SOUND SYSTEMS 154

8 UPGRADING LIGHTING 192

9 UPGRADING DASHBOARDS, INSTRUMENTS AND WARNINGS 210

10 IMPROVING CONVENIENCE AND SECURITY 232

INDEX 256

WORKSHOP PRO | MODIFYING THE ELECTRONICS OF MODERN CLASSIC CARS

Introduction

I have been mechanically modifying my road cars for about 35 years, doing it first as an enthusiast, and then as a journalist writing about those modifications. I've rebuilt a six-cylinder BMW engine; supercharged and turbocharged naturally-aspirated engines; upgraded intercoolers and turbos and exhausts and radiators. I've done full suspension makeovers, engine swaps and major aerodynamic modifications. In the distant past, I even internally modified an automatic transmission valve body to change shift behaviour, developing the whole modification myself from scratch. I've also got a home workshop with a lathe, mill, bandsaw, welders and lots of other equipment.

So? Well, the point is this: *I think that electronic modification of cars beats mechanical mods hands-down for cheapness, ease of work and results.*

With electronics, you can change turbo boost curves to get more performance at the bottom of the rev range – and the top as well. You can improve fuel economy and alter the way the car feels to steer. On one car with active torque-split four-wheel drive, you can, in 30 minutes and for very little money, make a greater improvement to handling than spending thousands on new suspension parts and tyres. Nothing – simply nothing – beats home electronic mods.

This book is about electronic modification of cars of the 1990s and 2000s – a golden era for making automotive electronic modifications. Why is that? Well, more recent cars often integrate electronic systems in a way that makes it hard to change one system without then negatively affecting others. Analog signals have been in many cases replaced with bus signals (eg CAN bus), in turn making intercepting and changing these signals impossible for the

The era covered by this book is a golden one – engines sufficiently sophisticated in their electronics to be highly capable, but not so complex as to make them difficult to modify. (Courtesy Saab)

average modifier. Engine management is now integrated with transmission control, stability control, and the dashboard – and usually other systems as well. And much older cars? These often used electronic systems that were quite basic, and so were limited in their capability. But for about that 20-year period of the 1990s and 2000s, cars had sufficient sophistication in their electronics to achieve great outcomes, but were not so complex as to make modification difficult – or even near-impossible.

But of course, electronic modification is not the answer to everything. You can't change the boost curve of a turbo that isn't there, or use electronics to improve the flow of a restrictive exhaust. So, as always, the final outcome also depends on mechanical factors – electronic modification doesn't get rid of those.

THIS BOOK

This book has ten chapters. Chapter 1 is an overview of the electronic modification approach that I suggest you take. It also talks about some of the essentials you need – a workshop manual for your car, and a multimeter.

In Chapter 2, I summarise some aspects of engine management systems. (Note that Chapter 2 is only a brief overview – for a more detailed look, I suggest that you read the companion book to this volume – *Car Electrical & Electronic Systems*, also published by Veloce.)

In Chapter 3 I look at electronic modifications that can be made to engine management systems using interceptors. The modification techniques covered in this chapter vary from those that cost literally just pocket change, through to much more expensive commercial interceptors.

In Chapter 4 I look at programmable engine management. If you want to have complete control over every aspect of how the engine is run – including power, response and fuel economy – this chapter is for you.

Chapter 5 covers other aspects of working in the engine bay: how to move the battery to the other end of the car to create more space, controlling the air-conditioner compressor in a smarter way to give better power and fuel economy, and installing a diagnostics socket in non-OBD cars, among other projects.

In Chapter 6 I turn to non-engine car systems that can be electronically modified. While modifying engine management is commonplace, very few people seem to realise that the same approach can be taken with almost any electronic systems in a car.

In Chapter 7 I look at modifying car sound systems. While in the above section I waxed lyrical about electronic car systems of the 1990s and 2000s, one area where major improvements have since occurred is in in-car entertainment. However, with cars of the described era usually having standard one-DIN or two-DIN dashboard

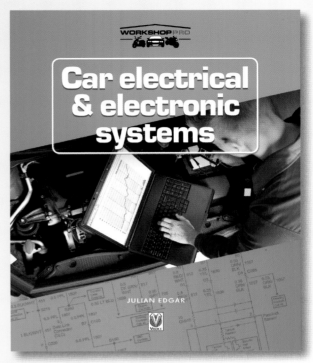

The basics of car circuits, switches and relays, analog and digital signals, oscilloscopes, and so on, are covered in the book *Car Electrical and Electronic Systems*, as seen here.

PERFORMANCE?

This book is unusual in that it looks at car performance in a wide context.

Most people think of 'performance' as being acceleration – going hard in a straight line. And techniques are covered here that will allow you to develop more engine power – reducing intake restriction by upsizing the airflow meter, the electronic modifications that will allow you to fit larger injectors, running higher turbo boost without fuel-cuts, and improving intercooler efficiency – among others.

But this book also looks at electronic modifications that can improve car handling – altering the influence of electronic stability control, changing front/rear torque split in an all-wheel-drive, and actively controlling air suspension.

And since I consider fuel economy (mileage) an aspect of automotive performance, it also looks at how you can improve that – fitting a dashboard indicator that will allow you to drive more fuel-efficiently, running lean cruise with programmable engine management, and improving regen braking on a hybrid.

The greater the number of electronic systems in a car, the greater the opportunity for modification!

Changing the mixtures from full rich to full lean can be as simple as turning an added knob. However, that's a powerful tool that, used unwisely, could kill your engine.

apertures, upgrading is easy. Much of this chapter I spend on speaker systems, an area where careful selection and good installation can make or break a car's sound system.

Chapter 8 covers upgrades to lighting – both external lights and those inside the car. If you drive fast on empty roads at night, you need good lights, and interior lighting upgrades simply make the car a much nicer place to be. Some of these upgrades are also cheap and easy: eg swapping filament bulbs to LED. Others, like using Tuned Mass Dampers to prevent driving light vibration, are unique.

In Chapter 9 I cover instrument and dashboard upgrades. I'm someone who has always liked having a lot of information on what the car is doing – and to do that, you need gauges and instruments. Upgrades in this chapter vary from the simple – eg adding a voltmeter or intake air temperature gauge – through to the complex and more expensive, such as installing a fully programmable digital instrument panel.

Finally, in Chapter 10, I look at items you can add to provide more convenience and comfort. Some of the projects covered in this chapter are easy and cheap – perfect for the person who is just starting out in this area.

SAFETY – OR, DON'T BLOW UP YOUR ENGINE

Almost every technique covered in this book I have used successfully on my own cars (if I haven't used a technique, but think it would work well, I say just that.) Over the years, I have lost only one engine – a tiny 660cc turbo engine screaming to 8500rpm and boosting at 21psi … still on tweaked original equipment engine management. That engine died because it ran lean at high boost. Considering that in those days, wide-band air/fuel ratio meters did not exist for amateur use, I think I did well! (The replacement engine also ran strongly for years.)

The key point is this: the techniques covered in this book give you enormous power over electronic car systems. Literally one turn of a knob at high load could destroy your engine. Modifying electronic systems that impact car handling could make your car a dangerous, unpredictable machine that could kill you.

Of course, none of this is new: give someone a screwdriver and spanner and they can tune a carby-and-points car so badly that it blows up. Let someone fit the wrong anti-roll (sway) bar to a car's suspension, and the car can become a lethal weapon in the wet.

But people unversed in electronic modification tend to think of it as being more like turning the knobs on a home sound system – you won't do any harm and it's all just a matter of tuning the system to your taste. *But modifying car electronic systems is not like that!*

It's also not something to launch into thoughtlessly. Wiring a programmable engine management system into place can be complex; even just adding a resistor to the intake air temperature sensor requires concentration and care. I find tuning programmable engine management both engrossing and exhausting … there are just so many things to think about simultaneously!

I don't want to make it sound like some boring, dreary work – it isn't. But as with any car modification, it requires concentration, effort and care if you're to get the best result. And it doesn't matter if you're modifying with a single electronic component that costs just loose change – or an engine management system worth thousands. In both cases, you still need that concentration, effort and care. But on the other side of the coin, the results that can be achieved are quite incredible.

(And of course I should add that not everything in this book is in the same category. Improving sound systems, upgrading lighting, and giving your car more convenience and security are all low-risk pursuits!)

CAR MODIFICATION WORKSHOPS

Many of the techniques covered in this book are ones that you'll never see in a car modification workshop. For example, using a simple pot as an interceptor (an approach with enormous power when used with lots of different car electronic systems) is a technique I have never seen a commercial workshop use. So, does that mean this approach is bad?

Not at all.

Workshops are businesses that exist to make money, and it's hard to charge the customer a lot when the component costs less than a can of soft drink, fitting it takes 10 minutes, and tuning it takes about the same length of time! Much better in that situation to use a commercial interceptor, where the total bill for the customer will be literally 10-20 times as much!

(That's not to say that commercial interceptors should never be used – far from it. But it's horses for courses: if a cheap component can achieve the desired outcome, great!)

The other reason that some of the approaches seen in this book are unknown in commercial workshops is that very few workshops modify electronic systems other than engine management. In this book, I modify lots of different car electronic systems – from regen braking in a front-wheel drive hybrid, to power steering weight in a V8 rear-wheel drive, to automatic transmission control in a front-wheel drive turbo V6, to torque split control in an all-wheel-drive twin-turbo six-cylinder.

For modern classic cars of the 1990s or 2000s, I doubt that there is even one electronic system that cannot be modified. And that leads me to the final point in this introduction: every car is different, and every car electronic system is different. There is therefore little point in covering numerous different modifications that are specific to individual cars. Instead, it's much better if modifications are shown and explained so that the *ideas* – the principles – can be applied in your situation.

So for example, Chapter 6 shows how I modified the power steering weight in a Prius – but also points out that just the same electronic approach could have been taken in altering the front-rear active torque split in a Skyline GT-R. In the same chapter, I show how to use a fully programmable ECU to control an air suspension system – but the same approach could be taken to controlling active aerodynamic elements like moveable wings, radiator shutters and rear diffusers. In Chapter 4 on programmable engine management, you can see the wiring diagrams for a particular brand and model of programmable engine management – not so that you can fit this unit, but so that you see the depth of information that you will need before fitting *any* programmable engine management system.

Therefore, think of this book as being a *tool kit of techniques*, where the illustrated step-by-step examples just show how the idea was applied to a particular car.

So, enough of the prelude – let's get into it.

All this disassembly was required to access just one wire – the signal from the speed sensor! Making electronic modifications requires concentration, effort and care – but the results can be fantastic.

WORKSHOP PRO: MODIFYING THE ELECTRONICS OF MODERN CLASSIC CARS

Chapter 1
Electronic modification

- Modifying electronic car systems
- Types of signals
- Selecting and using a multimeter
- Working with wiring looms

WORKSHOP PRO: MODIFYING THE ELECTRONICS OF MODERN CLASSIC CARS

MODIFYING COMPLEX CAR SYSTEMS
Let's start by looking at the approach that I recommend to modifying car electronic systems.

Workshop manual
Some people delve into their car's wiring harness, not knowing any of the colour codes, plug pin-outs or the functionality of the systems. Sure, this can be done – but it's also an easy way of blowing up expensive Electronic Control Units (ECUs), or getting a frustratingly ineffective outcome. To achieve success in almost all electronic modification, you need to have a good workshop manual – and, in nearly all cases, that's the factory's workshop publication.

For most cars of the 1990s-2000s, a full, manufacturer-produced workshop manual will be available. Even if it costs a quite a lot, I'd recommend that you buy it. (And if you can't buy it, see if you can beg or borrow one to do a major scanning exercise.) Most of these workshop manuals have detailed coverage of the electronic systems in a car, including the basics on how things work as well as how to diagnose faults, check trouble codes, etc. The more you can find out about how the electronic systems work in your car, the greater the chance of success when modifying them.

So you're now armed with the workshop manual, complete with circuit diagrams, wiring colour codes and

This engine in my Honda uses aftermarket programmable engine management, larger injectors, a new ignition system, an added turbo with water/air intercooling, and a new fabricated airbox. (An EGT probe was being installed when this photo was taken.) Inside the cabin there's a new digital dash. Without the workshop manual, wiring-in the new systems would have been almost impossible.

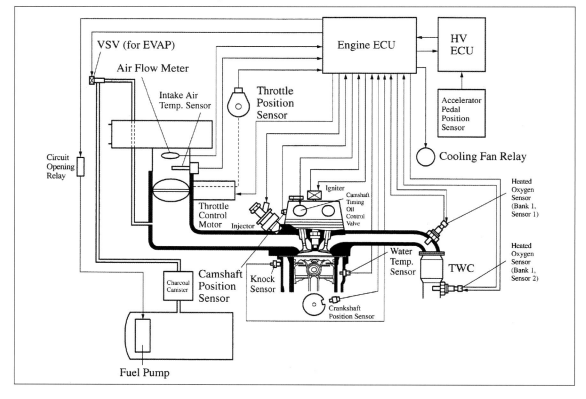

For the era of cars described in this book, the manufacturers' workshop manuals are usually excellent. This is from the manual for an NHW10 Prius. (Courtesy Toyota)

1. ELECTRONIC MODIFICATION

some coverage of how things work. What next? What you need to do now is to work out the best approach to achieve your desired modification outcome.

DECIDING ON THE BEST APPROACH

There are a lot of ways of carrying out electronic modifications, but let's start with the cheapest (and often easiest) of approaches – intercepting signals in an engine management system, so causing a change in the system's behaviour. In this approach, the signals going to (or, more rarely, coming from) the ECU are intercepted and altered. Doing this means that the ECU's internal software and logic works as it always did – it's just that the outcomes are different. Cost? Simple but powerful interceptors can be developed for almost nothing.

But before we can start the modification, we need to consider three questions:

1. What modification outcome do you want?
2. What parameter (factor) do you need to alter to achieve this outcome?
3. To alter this parameter, is it easiest to make modifications on the input or output side of the ECU?

Let's say that the outcome you desire is leaner full-load mixtures. (Many cars of the era ran full-load mixtures that were too rich for best power.) This means that the parameter that you want to alter is the *amount of fuel that the injectors flow at high loads*. (If you don't know much about engine management systems, you might want to read Chapter 2 first.)

Next question: do you make the intercepting modification on the *input* or *output* side of the ECU? If you make the modification on the output side of the ECU, you're going to need to alter the duty cycle with which the injectors are driven – that is, lessen the full-load duty cycle so that they pass less fuel.

Alternatively, you can make the modification on the input side of the ECU, for example changing the signal coming from the airflow meter so that the ECU thinks that less air is flowing into the engine than there actually is. This in turn will lean the air/fuel ratio, as the ECU will have the injectors on for less time than it did previously.

In nearly all cases, you will find a variable voltage signal coming out of the airflow meter. Variable voltage signals are a lot easier to modify than variable duty cycle signals (see the breakout box on the next page for more on signals), so in this case the approach is clear – you modify the input signal to the ECU. (And the 'full-load only' part of the requirements? That's easy – you just switch your modification in and out on the basis of throttle position. You could do this with a microswitch, or by using a voltage switch triggered by the throttle position sensor.)

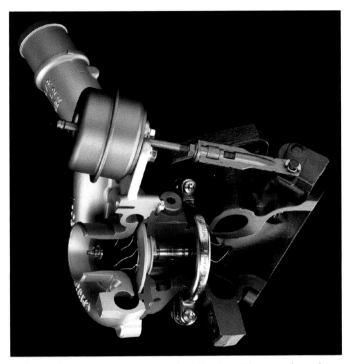

So you have a turbo car, and, when the boost is wound up, the ECU shuts down the fun. How do you disable such a boost cut? You can do this easily by limiting the measured boost that the ECU 'sees.'

OTHER APPROACHES?

Rather than modifying the input or output signals to the standard ECU, you can choose instead to replace the ECU with a brand-new controller, and then program ('map') the required behaviour. With engines, this is termed fitting programmable engine management, and is covered in Chapter 4.

The same 'programmable controller' approach can also be taken with any of the other electronic systems in the car. The new ECU will need to have sufficient inputs and outputs, and have internal logic that can be programmed to provide the required control strategies – and both are quite possible. (In Chapter 6, I look at doing this with a controller for an air suspension system.)

A completely different approach is to remap the software in the standard ECU. To be able to do this, you need aftermarket software that has been developed for that specific ECU. I don't cover this approach in this book, for two reasons. First, the vast majority of ECUs have never been 'cracked' in this way, and so do not have user-adjustable software available. Second, those that have had such packages developed tend to have the software priced at 'workshop' levels, and so are out of reach of the average modifier.

WORKSHOP PRO: MODIFYING THE ELECTRONICS OF MODERN CLASSIC CARS

Let's look at a different example. A turbo car has a modification problem – when the boost is raised, the ECU cuts fuel. (This is a common over-boost safety precaution.) The outcome that you want is to disable the fuel cut – you don't want it to occur. Modifying the output signal of the ECU is nearly impossible (doing this modification would require that you kept the injectors pulsing correctly, even after the ECU has switched them off), so you need to look instead at the input side of things. In this case it's the MAP sensor that is measuring the over-boost condition, therefore you need to modify the ECU input signal from this sensor, so that a ceiling isn't exceeded (How to do this is covered in Chapter 3).

Let's stay for a moment with the turbo car. The problem you now have is high-boost detonation – the outcome you want is more stable combustion, and the parameter you want to alter is the ignition timing at high loads. To alter ignition timing on the output side of the ECU is hard – again it's a pulsed signal – and on the input side it's scarcely less easy, because again the main timing signal (eg from a crank angle sensor) is pulsed. (To alter these you'd need a commercial interceptor, as described in Chapter 3.)

But is there a cheaper and easier way of doing it? Yes, because there's another input signal that can be altered to tweak timing – intake air temperature. This sensor uses a variable resistance, so it's easy to modify (also covered in Chapter 3). Again, the modification can be activated with a load switch – a boost pressure switch would be ideal in this case.

What about systems other than engine management? To show how the same three decisions need to be made with *any* car electronic system (the questions again: the desired outcome, the parameter to alter, and whether it's done on the input or output side of the ECU), let's now take a completely different example.

The outcome you want is heavier power steering in a car with electronically-controlled, variable-assistance steering. So, the parameter you want to alter is the behaviour of the power steering pressure control valve – but do you try to modify the power steering ECU inputs or outputs to achieve this?

Let's say that the workshop manual shows you that the system is very simple – there's a single power steering ECU input of vehicle speed, and a single output of a variable duty cycle that goes to the power steering

SIGNAL TYPES

There are two basic types of signal that you will find in car electronic systems – analog varying voltages, and pulsed signals. The first is the easier to understand and measure.

An analog voltage is one that steplessly varies as the parameter changes. For example, the airflow meter in most cars has a voltage output that alters with engine load. At idle the voltage output from the airflow meter might be 1.2V, at a light load 2.0V, at a heavier load 3.4V, and at full load 4.2V. At 'in between' loads, the voltages will be in between these figures. Sensors that have analog voltage outputs include: coolant, intake air and cylinder head temperatures; most airflow meters; most MAP sensors; and throttle position sensors. A normal multimeter can be used to measure these signals.

The other type of common signal is one that is pulsed – it turns on and off rapidly. For example, a signal from a road speed sensor might vary rapidly from 0-5V in a square wave. At any point in time the signal is either at 0 or 5 volts – there are no 'in betweens.' The way that the ECU makes sense of this signal is to look at its frequency – how quickly it turns on and off per second. This is measured in Hertz – abbreviated to Hz. The old name for Hertz is 'cycles per second' and in many ways this gives a better mental picture of what is happening – how many up/down cycles of the signal occur each second.

The shape of the waveform is also very important in many sensing applications. For example, a crankshaft position sensor needs to communicate with the ECU both the rpm of the engine (which the ECU works out from the frequency of the waveform coming from the sensor) and the position of the piston – normally the Top Dead Centre (TDC) position of No 1 cylinder. This extra information can be communicated by a change in the waveform occurring at that point. For example, if the waveform is being generated by a toothed cog passing a sensor, and at TDC there's a tooth on the cog missing, then the

A commercial interceptor that can alter both pulsed and analog voltage input signals. (Courtesy AEM)

1. ELECTRONIC MODIFICATION

assistance solenoid. Neither the input nor output is a constantly varying voltage (ie both signals are pulsed) so that is likely to make things harder. You then look at the diagrams in the workshop manual and realise that the speed input also connects to the cruise control ECU and engine management ECU. So if you attempt to modify the speed input, you might also upset these other systems – in this case it might be better to modify the variable duty cycle *output* signal (covered in Chapter 6).

Thinking about the modification outcome you want to achieve, the parameter you need to alter to achieve that, and whether that's best done on the input or output side of the ECU gives you a strategy that you can use when modifying any electronic control system in the car.

Backwards diagnostics

As car systems get more complex, it can be difficult working out what parameter to actually intercept and

ECU will be able to sense the missing pulse. Sensors that have pulsed outputs include some airflow meters and MAP sensors, and all crankshaft, camshaft and speed sensors.

But what about when the ECU is *sending out* the pulsed signal? When the ECU is controlling something by the use of a pulsed signal, there are two parameters which are critical. Firstly, the frequency. Just as with an input signal, how fast the output signal is being turned on and off is important. However, it's the second parameter which is more widely used as a control variable – the duty cycle of the signal.

Consider a square wave signal that is being used to open the injectors. When the current is switched on, the injector is open. When the current is switched off, so is the injector. But what proportion of the time is the injector on for? If the 'on' time is the same as the 'off' time, then the duty cycle is 50 per cent. If it is on for three-quarters of the available time, the duty cycle is 75 per cent. By varying the proportion of on and off times, the ECU can control the injector flow. Sometimes this approach is called 'pulse width modulation,' or PWM.

A pulsed output signal can vary in both frequency and duty cycle – and sometimes both simultaneously. For example, the frequency with which injectors squirt is tied to engine revs, so as rpm increase, so does the injector pulsing speed. However, as previously mentioned, the duty cycle of that signal will also vary with engine load.

While injectors vary in both frequency

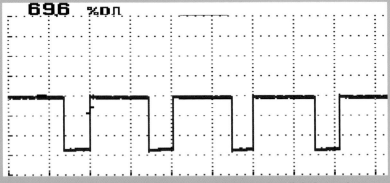

A digital multimeter can be used to measure analog signal voltages.

An oscilloscope screen grab showing a rectangular, digital waveform.

and duty cycle, many other pulsed actuators have a fixed frequency and vary only in duty cycle. For example, the flow control solenoid in an automatic transmission or power steering system is likely to have a fixed frequency but a variable duty cycle. These valves aren't 'open' and 'closed' like injectors; rather the valve pintle hovers at mid positions, giving a flow able to be continuously varied by ECU command. Seeing the shape of a pulsed signal waveform requires an oscilloscope – these days, cheap enough to purchase even just for amateur use.

WORKSHOP PRO: MODIFYING THE ELECTRONICS OF MODERN CLASSIC CARS

change. You know what performance outcome you want – but how do you get there? One approach is to *read the workshop manual backwards* – looking at what the manual describes as fault symptoms, but what you see as a good performance outcome.

For example, an electronically-controlled automatic transmission might have listed a fault condition termed 'harsh engagement (1st → 2nd),' with a variety of possible causes for this condition then listed. Look then at what is also shown as causing 'harsh engagement' for the 2nd → 3rd, 3rd → 4th and 4th → 5th gear changes – that is, all of the up-changes. In the case of the manual that I have open in front of me now, the only common cause of these conditions is the 'solenoid modulator valve.'

From this you can deduce that modifying the signal that goes to the solenoid modulator valve can be used to firm up the gear-shifts. Having come to this decision, you can look at the diagnosis tables to see what other conditions will occur if this signal is modified. In this case, the table also shows that it's likely that the car will also then have a harsh engagement when changing from Neutral → Drive, and Neutral → Reverse. If you didn't want the latter conditions to occur, you could switch the modification in only when the throttle is open past the idle position (eg by using a voltage switch monitoring the throttle position sensor).

Even very complex systems can be examined with this 'backwards diagnostics' approach – another excellent reason to have the complete workshop manual available.

WORKING ON CAR ELECTRONIC SYSTEMS

Before you start to delve into your car's wiring harnesses, intent on gaining a better performance outcome, there are some things you should know – how to use a multimeter (and what to look for when buying one if you don't already have one), and how to find the right wires and then tap into them.

Selecting a multimeter

The most important tool that you will use when making electronic modifications to a car is a multimeter.

A multimeter is a test tool which can measure a variety of

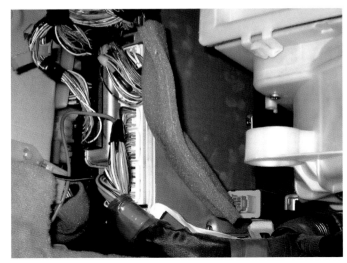

The control logic of the ECU is hidden inside this box, however by using the workshop manual in a 'backwards diagnostics' way, the correct signals to intercept can sometimes be found.

different electrical factors – voltage (volts), current (amps) and resistance (ohms) are the basics. However, while you might be able to pick up a basic volts-ohms-amps meter

Measuring the signal output of a vane-type airflow meter at idle.

very cheaply, it pays in the long run to dig deeper to get a meter with these extra functions:
- Frequency (Hz)
- Duty cycle (per cent)
- Temperature (degrees C or F)
- Continuity (on/off buzzer)

Multimeters are available in auto-ranging or manual-range types. An auto-ranging meter has much fewer selection positions on its main knob – just amps, volts, ohms and temperature, for example. When the meter probes are connected to whatever is being measured, the meter will automatically select the right range to show the measurement. Meters with manual selection must be set to the right range first. On a manual meter, the 'volts' settings might include 200mV, 2V, 20V, 200V and 500V. When measuring battery voltage in a car, the correct setting would be '20V,' with anything up to 20V then able to be measured.

While an auto-ranging meter looks much simpler to use – just set the knob to 'volts' and the meter does the rest – the meter can be much slower to read the measured value, because it needs to first work out what range to operate in. If the number dances around for a long time before settling on the right one, it can be annoying for quick measurements – and very difficult if the factor being measured is changing at the same time as well! To speed up readings, some auto-ranging meters also allow you the option of fixing the range.

Using a multimeter

So how do you go about measuring volts, amps, ohms and all the rest?

When measuring volts, the meter should be connected in parallel with the voltage source. Most commonly in a car you're trying to find 12V for a power supply, or measuring the voltage output of a sensor. In either of these cases, the meter would be set to its 20V (or 40V or auto-ranging DC volts scale, depending on the meter), and the meter probes inserted into the connected wiring. If the polarity is wrong (you've connected the negative multimeter probe to the positive supply line) then no damage will be done – the meter will simply show negative volts instead of positive volts. When measuring voltage, the circuit does not need to be broken – the meter is inserted in parallel. In most car measurements, the negative probe of the meter is connected to ground (the car body).

Measuring current (amps) requires that the circuit is broken and the meter placed into it, so that all the current flows through the meter. To do this, the meter will often require that the positive probe plug be inserted into another meter socket. Failure to do this will result at best in the blowing of an internal multimeter fuse, and at worst in damaging the meter.

Measuring injector duty cycle at idle.

Resistance measurements require that the device be isolated from its normal circuit, otherwise the measurement could be false. In the case of a sensor (such as throttle position), this means that it needs to be unplugged. Always check that the multimeter indicates zero resistance when its leads are touched together; if it doesn't, you need to press a 'delta' button or zero ohms reset.

Duty cycle should be measured with the meter in parallel with the working device. Injectors and the operation of other pulsed actuators should be measured in real operating conditions, with the best way of doing this being on the road with the multimeter remotely located inside the cabin.

Temperature is most often measured using a bead or probe-type thermocouple. The bead unit has very little mass, and so reacts to temperature changes quickly – but it is fragile and hard to handle. The probe type has a slower reaction time, but is easier to handle and more robust. Using the thermocouple feature of a multimeter is as easy as selecting that function and plugging in the probe. Some meters also have an internal sensor, which allows measurement of the ambient temperature of the day – useful when comparing different test days.

The continuity function causes an internal buzzer to

WORKSHOP PRO: MODIFYING THE ELECTRONICS OF MODERN CLASSIC CARS

When making electronic modifications, you must be certain that you have found the correct wires. Use wiring colour codes, count the pins, and then check with a multimeter or scope.

For protection, all new engine bay wiring should be wrapped in tape or covered in convoluted tube.

sound when the meter's probes are connected together. If the probes are inserted in different points in the wiring and the buzzer sounds, it indicates that there is a complete circuit between the probes. This function is useful for checking that you have a ground, or that there is no break in a wire.

Working with wiring looms

One of the very first steps when modifying a car electronic system is to find the right wires. That's harder than it sounds – some cars have ECUs with hundreds of conductors disappearing into plugs, while in other cars, even finding the right ECU itself can be a major issue. A fundamentally important step is therefore to have an accurate and clear guide to the wiring. This is normally gained from a workshop manual for the car.

Let's take an example – you're fitting a voltage switch that will monitor throttle opening. (That way, you could, for example, trigger an intercooler water spray at large throttle openings.)

The first step is to decide whether you're going to tap into the sensor output at the engine bay end of things, or at the ECU end. In one case, you'll be working near to the throttle position sensor itself (which is in the engine bay), and in the other case you'll be working near the ECU, which in most 1990s-2000s cars is inside the cabin.

There are advantages and disadvantages in each approach – since the voltage switch will probably be mounted inside the cabin, if you tap into the loom near to the sensor, you will need to then run the wire back into the cabin through a hole in the firewall – but on the other hand, locating the correct sensor wire will be easier. In this case, we'll assume that the connection will be made at the ECU.

Here are the steps you'd follow.

1. Finding the wire

Using the workshop manual, find out which wire carries the signal from the throttle position sensor to the ECU. You can either look on an overall ECU inputs/outputs diagram, in a table showing the same information, or under 'throttle position sensor' itself.

For this example, the workshop manual could show that the throttle position signal input occurs at ECU terminal 2D. The next step is to find where terminal 2D actually is on the ECU connector. For example, the diagram could show this to be on ECU connector #2, one position in from the right-hand, bottom end when viewing the plug from the ECU side, with the plug tabs uppermost. This step is very important – make sure that you check whether the plug is shown from the loom or ECU side, and how is it orientated in that view.

2. Checking it's the right one

To check that you've found the right wire, two more steps should be taken.

First, make sure that the wire's colour code matches the described plug location. In other words, if the throttle position sensor signal wire is supposed to be a blue with a red trace, make sure that the wire going into the designated connector placement actually *is* blue with a red trace!

Second, does the wire have the correct signal on it? In this case, the voltage of the sensor with the ignition turned on should be in the 0-5 volt range, and alter with throttle position. To find out if this is the case, connect the multimeter's red probe to this wire (either by using a thin

1. ELECTRONIC MODIFICATION

A temperature-controlled soldering iron like this makes excellent soldered connections on car wiring. The supple cord allows the handpiece to be easily controlled.

A soldered connection wrapped in electrical tape. Add a cable tie to hold the new green wire in place and this join is finished.

piece of stiff wire to push into the ECU connector from the back, or probing directly through the insulation of the signal wire) and grounding the other multimeter probe, eg by connecting it to the car's body.

3. Making the connection
In the case of fitting the voltage switch, the signal wire from the throttle position sensor to the ECU is not broken – the voltage switch simply taps into the signal wire. There are a few ways in which this connection can be made, including crimp-on clips. However, my preference is to do it like this:

- Use a razor blade or sharp utility knife to remove a section of insulation (easier to do if the insulation is sliced around in two circumferential cuts about 5mm (³⁄₁₆in) apart and then the separated insulation sleeve pulled off).
- Remove 10mm (⅜in) of insulation from the new wire and then firmly wrap the bared section around the original car loom wire at the point where the insulation has been removed.
- Solder the two together and make sure that the join is shiny all around (which indicates a well-soldered join).
- Wrap the join with high quality insulation tape and then use a cable tie to stress-relieve the new joining wire, so that a tug on the new wire doesn't pull on the new connection.

Taking this approach has a number of benefits – it doesn't weaken the original electrical connection, it can be reversed, and it gives excellent mechanical and electrical connection to the new wiring.

In this example of voltage switch wiring, you'll now also need to find ground and ignition-switched battery voltage connections to power the module. Again, these can be found on the ECU without too much trouble.

WORKSHOP PRO MODIFYING THE ELECTRONICS OF MODERN CLASSIC CARS

Chapter 2

Overview of engine management

- Electronic fuel injection (EFI)
- Ignition
- Input and output signals
- Closed and open loop
- Variable intake manifolds
- Variable valve timing
- Automatic transmission control
- Turbo boost control
- Electronic throttle control

WORKSHOP PRO: MODIFYING THE ELECTRONICS OF MODERN CLASSIC CARS

The basics of engine management are very easy to understand. Despite people talking about MAPs and MAFs and EGO sensors and all these sorts of weird things, getting a grasp of what's going on will take you only as long as it takes to read these pages.

Firstly, what is EFI? 'EFI' simply stands for *electronic fuel injection*. It's a system where the addition of fuel to the engine's intake airstream is controlled electronically, rather than occurring through a carburettor. *Engine management* is the term used when *both* the fuel and the timing of the spark are controlled electronically. In addition, management systems on many cars also control the automatic transmission, turbo boost, camshaft timing and the operation of the throttle.

Before we get into an overview on how engine management systems work, let's take a quick look at the layout of the fuel and ignition systems.

ELECTRONIC FUEL-INJECTION

EFI cars use a multipoint system of injection. That is, each cylinder has its own *injector* that opens to squirt a mist of fuel onto the back of the intake valves. When the valves next open, the fuel (and lots of air) are drawn into the combustion chamber. So what's an injector? An injector is simply a solenoid valve: when power is applied, the valve opens, allowing fuel to flow through it. When the power is removed, the valve shuts and the flow stops. When the engine is running, the injectors each open and briefly squirt once every two crankshaft revolutions (that is, once per intake stroke). Injectors are either fired sequentially (each squirts just a moment before its associated intake valves open), all together, or in one or two groups (sometimes called 'banks').

The amount of fuel that is supplied to the engine is dependent on how long each injector stays open for. If an injector was open and flowing for half of the available time, it would be said to have a 'duty cycle' of 50 per cent. If it was squirting for only 2 per cent of the time, the duty cycle is said to be 2 per cent. On a standard car, duty cycles are often around 2-4 per cent at idle and 80 or 90 per cent at full load, full rpm. When the duty cycle reaches 100 per cent, the injector is flat out – it can't flow any more fuel because it is already open continuously.

For fuel to squirt out in a fine spray whenever the injector opens, the fuel must be fed to the injector under pressure. This process of pressurisation starts at the other end of the car, in the fuel tank. Here a roller-type *pump* works at full speed all of the time – in most cars, it's flowing just as much fuel at light engine loads as at full load. The fuel leaves the pump, passes through a filter and then is fed into the *fuel rail* on the engine. The fuel rail is a long, thin reservoir that joins the injectors together. Mounted on the fuel rail is a *pressure regulator*, which allows a

In this engine cutaway, the direct-fire ignition coil can be seen between the two valves, and the fuel injector squirting into the intake port is on the right. (Courtesy GM)

proportion of the fuel to escape from the rail and flow back to the tank through a return line. The more fuel that the regulator lets out of the fuel rail, the lower the pressure in the rail will be.

Fuel pressure is automatically set by the regulator on the basis of manifold pressure. As manifold pressure rises, so does fuel pressure, so that the fuel pressure is always a fixed amount above the pressure in the intake. In this way, if the injector is open for 3 milliseconds, the same amount

The fuel pump in most systems runs at full speed all of the time, with much of the fuel coming back to the tank through the return line.

2. OVERVIEW OF ENGINE MANAGEMENT

A fuel pressure regulator, located on the end of the fuel rail.

of fuel will flow out of the injector irrespective of whether the manifold pressure is experiencing turbo boost – or is in vacuum.

The above description is typical of most systems, but there are some exceptions which should be mentioned. First, some cars run fuel systems that lack a fuel return line. On these cars, the fuel pressure regulator is at the tank-end of the system. Second, over the years some cars have been produced that use only one or two injectors mounted in a carburettor-looking arrangement. This system is called 'throttle body injection'. Third, some EFI systems squirt the injectors once each crankshaft rotation (that is, twice each intake stroke) rather than only once every two crank rotations. And finally, some cars electronically control fuel pump speed, meaning that in the cars equipped with this sort of system, the pump runs more slowly at light loads.

IGNITION

Most cars with engine management use multiple *ignition coils*. Sometimes there is a coil for each plug – with the coils often mounted directly on the plugs – while in other cars, double-ended coils are used. In the latter case, the number of coils is half the number of sparkplugs. Older cars use distributors, where the output of a single coil is distributed in turn to each sparkplug by a moving mechanical rotor arm. Each coil has an ignition module, which is a computer-controlled switching device that can handle the high current requirements of coils. The ignition modules (sometimes called igniters) can be built into the coils, but are sometimes contained within a separate box mounted nearby.

The key parameter that the engine management system varies is the timing of the spark, referenced against the rotation of the crankshaft and the position of the piston (ie the spark timing is said to be so many crankshaft degrees before piston Top Dead Centre).

INPUT AND OUTPUT SIGNALS

The best way of visualising an engine management system is to consider it on the basis of its inputs, outputs and decision-making. We've already covered the two major outputs – the fuel injectors and the ignition coils – but what about the inputs and the decision-making? The decisions on how long to open the injectors for, and when to fire the ignition coil(s), are made by the *Electronic Control Unit*, or ECU. If you like, it's the brain. ECUs are sometimes referred to by different abbreviations (eg ECM for engine control module) but their function is largely the same in all cases.

ECUs make decisions mostly on the basis of software that has been pre-programmed into them. This software determines that when the engine is operating under a particular condition, the correct fuelling requires that the injectors are triggered with (say) a 20 per cent duty cycle and the ignition timing requires that the sparkplugs are fired (say) 15 degrees before Top Dead Centre. However, for the ECU to make these sorts of decisions, a lot of information about the engine operating conditions needs to be continually fed to it. This information is provided to the ECU by input sensors.

The time at which the sparkplug is fired during the rotation of the engine is called ignition timing. It is measured in degrees before Top Dead Centre. (Courtesy Bosch)

The Electronic Control Unit (ECU) is the system's brain. It monitors input signals and then outputs appropriate signals to control the fuel injectors, ignition timing and other factors. (Courtesy GM)

WORKSHOP PRO: MODIFYING THE ELECTRONICS OF MODERN CLASSIC CARS

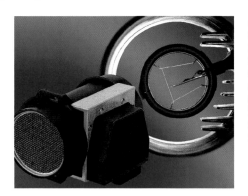

A hot-wire airflow meter that measures the mass of air being drawn into the engine. In engines so-equipped, this is the major load sensor. (Courtesy Bosch)

A vane airflow meter. As the name suggests, it uses a vane that opens in response to the pressure difference across it.

The most important aspects of engine operation for which the ECU needs to have accurate and timely information are:
- engine load
- crankshaft rotational position
- engine temperature
- air/fuel ratio

Engine load is most often determined by an *airflow meter* – a device that measures the mass of the air being drawn into the engine. If the ECU knows how much air is being drawn into the cylinders, then it can add the right amount of fuel to go with it. Airflow meter-based systems are sometimes referred to as MAF (mass airflow) systems. There are a number of designs of airflow meter available:

- *Hot-wire airflow meters* use a very thin, heated platinum wire. This wire is suspended in the intake air path or in a bypass passage and the temperature of the wire is electrically related to the mass of air passing it. Meters of this sort normally have a 1-5V analog output signal, although just a few have a frequency output.
- *Vane airflow meters* employ a pivoting flap which is placed across the inlet air path. As engine load increases, the flap is deflected to a greater and greater extent. The flap moves a potentiometer, which in turn alters the analog output voltage signal, which typically is 0-5V (although some meters use a 1-12V output range).
- *Karman Vortex* airflow meters generate vortices whose frequencies are measured by an ultrasonic transducer and receiver. They use a flow-straightening grid plate at the inlet to the meter. This type of meter has a variable frequency output.

Of the three meter types, the hot-wire design is by far the most common on cars of the period covered by this book, followed by the vane and then Karman Vortex – the latter used only by a few manufacturers (eg Mitsubishi and Hyundai).

The other way of measuring engine load is indirectly, through the monitoring of manifold pressure. These systems are called MAP (Manifold Absolute Pressure) systems. By measuring three factors – manifold pressure, engine rpm and the temperature of the intake air – the ECU can estimate the mass of air flowing into the engine.

Crankshaft (and, often additionally, camshaft) position sensors tell the ECU where the crank is in its rotation. This is vital if the spark is to be fired at the right time, and is also used to time the injectors in sequential injection engines. The ECU can also work out engine rpm from this sensor. Again, different types of sensor exist:

- An *optical position* sensor uses a circular plate with slots cut into it. The plate is attached to the end of the camshaft and is spun past an LED. A sensor on the other side of the disc registers when it sees the light shining through the slots, with the ECU then counting the light pulses.
- A *Hall Effect* position sensor uses a set of ferrous metal blades that pass between a permanent magnet and a sensing device. Every time the metal vane comes between a magnet and the Hall sensor, the Hall sensor switches off.
- An *inductive position* sensor reads from a toothed cog. It consists of a magnet and a coil of wire, and as a tooth of the cog passes, an output voltage pulse is produced in the coil.

2. OVERVIEW OF ENGINE MANAGEMENT

A MAP sensor mounted on the intake manifold, which measures pressure after the throttle body. In cars equipped with MAP sensors, the ECU uses this signal and engine rpm to calculate load.

All of these sensors have frequency outputs, with an alteration in the waveform used to indicate piston position.

Engine temperature is another important factor for the ECU to know, especially during cold start. Two engine temperatures are usually monitored: coolant temperature and intake air temperature. Invariably, these sensors use a design where resistance changes with temperature. The sensor is fed a regulated voltage by the ECU and the ECU then measures the changing voltage coming back from the sensor. Some cars use other temperature sensors to measure fuel, cylinder head and exhaust gas temperatures.

The *oxygen sensor* (sometimes called the EGO sensor) is located in the exhaust. It measures how much oxygen there is in the exhaust compared with the atmosphere, and by doing so indicates to the ECU whether the car is running rich or lean. Many cars use multiple oxygen sensors, for example before and after the cat converter(s).

Two different types of oxygen sensor are used. A *narrow band* sensor generates its own voltage output, just like a battery. When the air/fuel ratio is lean, the sensor emits a very low voltage output, eg 0.2V. When the mixture is rich, the voltage output is higher, eg 0.8V. A *wide-band* sensor is more complex and develops a voltage signal that is proportional to mixture strength. Unlike a narrow-band sensor, a wide-band sensor can indicate mixture strength to the ECU over a wide range of air/fuel ratios.

A number of other sensors are also common to most engine management systems. The *throttle position sensor* indicates to the ECU how far the throttle is open. Most throttle sensors use a variable pot giving a 0-5V analog output. The *vehicle speed sensor* lets the ECU know how fast the car is travelling; it can be mounted on the gearbox or in the speedometer, and has a variable frequency output. The *knock sensor* is like a microphone listening for the sounds of knocking (detonation). It's screwed into the engine block and works in conjunction with complex filtering and processing circuitry in the ECU to sense when knocking is occurring.

CLOSED AND OPEN LOOP

Two key operating conditions of the ECU need to be identified – closed and open loop.

Closed loop refers to the situation where the air/fuel ratio is being controlled primarily on the basis of feedback from the oxygen sensor. In these conditions, the ECU is programmed to most of the time keep the air/fuel ratio close to 14.7:1 – the air/fuel ratio at which the catalytic converter works best at cleaning the exhaust gases. The oxygen sensor sends a voltage signal back to the ECU, indicating to the ECU whether the car is running rich or lean. If the engine is running a little rich, the ECU will lean it out. If it's a bit lean, the ECU will enrich the mixtures. The oxygen sensor then checks the effect of the change.

Closed loop running in cars with narrow band oxygen sensors occurs primarily in cruise and idle conditions. In nearly all narrow-band systems, the oxygen sensor is ignored at full throttle – this is called open loop running. When in open loop, the ECU bases its fuelling decisions totally on the information that has been programmed into it. If the ECU senses a high load, it will open the injectors for a long time and so spray in large amounts of fuel.

On cars that use wideband sensors, the ECU may keep the air/fuel ratio around 14.7:1 at all loads, or may allow the mixtures to become richer at high loads. Because the ECU is using the output of the wideband sensor all the time, the car is in constant closed loop.

The ECU uses a software table of information (called a map) that tells it how long to open the injectors at all the different engine loads. In addition to closed loop running, the oxygen sensor is also used as part of the ECU's self-learning system, where changes in the mixtures that would otherwise occur over time, can be automatically corrected.

The oxygen sensor monitors exhaust gases, and indicates the mixture strength to the ECU. When the system is operating in closed loop, the ECU uses this feedback to adjust the mixtures so that they remain correct. (Courtesy Bosch)

VARIABLE INTAKE MANIFOLDS

Variable intake systems change the length of the intake manifold runner, or the volume of the plenum chamber. This allows the intake to have more than one tuned rpm – giving better cylinder filling at both peak torque and peak power, for example. The changeover is normally performed as a single step – the intake system is either in one configuration or the other.

The intake system can be variably tuned in a number of ways, including (especially on six-cylinder engines) connecting twin plenums at high rpm but having them remain separate smaller tuned volumes at lower revs. The introduction of a second plenum into the system at a particular rpm is another approach taken. However, the most common method is to have the induction air pass through long runners at low revs, and then swap to short runners at high rpm. Note that this doesn't mean that the long runners need to be positively closed – opening parallel short runners is sufficient to change the effective tuned length of the intake system.

The changeover is normally performed by a solenoid valve, which directs engine vacuum to a mechanical actuator that opens or closes the internal manifold changeover valves. The changeover point can be based on engine rpm (this is most common), engine load, or a combination of both.

VARIABLE VALVE TIMING

Variable valve timing systems alter the timing and/or lift of the valves. In most 1990s-2000s cars, variable camshaft timing is on only a single camshaft, and the camshaft timing varies in a single step. That is, when the engine reaches a certain rpm and/or load, the ECU moves the camshaft timing – so one cam is either in the advanced or retarded position. Depending on the engine and manufacturer, that variable cam can be either the intake or exhaust cam.

Steplessly variable cam timing has also been used by many manufacturers. This allows lots of 'in-between' camshaft timing positions to be used, giving a far better result than single-step cam timing variation. Steplessly variable cam timing is most commonly used on just one camshaft, but some manufacturers used steplessly variable cam timing on both the intake and exhaust camshafts.

Systems that additionally vary the valve lift as well as cam timing are also employed. Honda's VTEC system is probably the best known of this type of single-step system.

The techniques used to alter the camshaft timing and/or lift also vary. However, where the camshaft timing alters in one step, an off/off signal from the ECU is used to activate a solenoid that feeds oil pressure to the mechanism, causing the change to take place. Where camshaft timing varies steplessly, a pulsed solenoid is used to allow the cam phasers to vary in their position. Camshaft timing can be varied according to input signals including engine rpm, throttle position, engine coolant temperature and intake airflow.

AUTOMATIC TRANSMISSIONS

On many cars, the control of the automatic transmission is integrated into the engine management system. This allows the same input sensors (eg throttle position, intake airflow, engine temperature, etc) to be used in transmission control without the need for duplicate sensors. It also allows the engine operating conditions to be varied as required – eg the ignition timing to be retarded during the gear changes to momentarily drop engine power and so give smoother shifts.

The automatic transmission control is carried out by the actuation of a number of hydraulic valves within the transmission. These control the flow of hydraulic oil which apply and release the internal clutches and bands, causing the gearshifts to take place. The two main inputs used to determine both the internal clamping pressures and when gearshifts occur are throttle position and road speed. Throttle position can be signalled to the ECU by the output of the throttle position sensor, or the ECU may internally model the torque output of the engine (eg by looking at

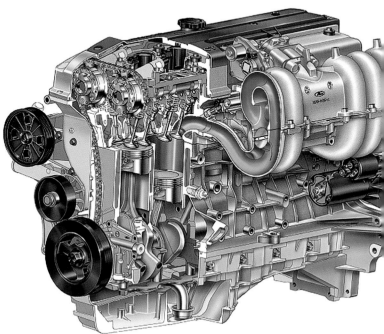

An inline six-cylinder engine with variable intake manifold. One of the changeover butterflies can be seen in the cutaway manifold. (Courtesy Ford)

throttle position, airflow, etc) and then use this information. Some transmissions that are otherwise electronic still use a throttle cable that mechanically connects the throttle to the transmission.

Line pressure is also varied within auto transmissions. This pressure controls the clamping forces and has a major influence on when gear changes occur; as engine power output increases, line pressure is increased. Transmissions also have a lock-up clutch in the torque converter, which stops any slip when engaged. This clutch is controlled on the basis of road speed and load, and may also be automatically disengaged when braking.

Transmission fluid control solenoids use two approaches – they're either turned on or off, or they are a variable flow design where the ECU steplessly alters their opening. The solenoids that control the gear change process are generally either on or off, whereas fluid pressure control and torque converter clutch engagement are achieved by steplessly varying the amount of fluid that flows through their respective solenoids. These variations in flow are achieved by varying their duty cycle.

TURBO BOOST CONTROL

Nearly all turbo cars use electronic boost control. The system of wastegate control is based on the older approach of using a wastegate moved by a sprung diaphragm. When boost pushing against the diaphragm overcomes the spring pressure, the diaphragm is deflected, in turn moving a lever that opens the wastegate and allows exhaust gases to bypass the turbo. This prevents the turbo from rotating any faster, and so limits the peak boost able to be developed. Electronic control adds a variable duty cycle solenoid that bleeds air from the wastegate hose, so altering the pressure that the wastegate actuator sees. Wastegate actuators in electronically controlled boost systems have quite weak springs – that is, if no boost is bled from the line by the solenoid, peak boost levels will be low.

Electronic turbo boost control systems can be open or

An electronically-controlled throttle. This design uses a DC motor and geartrain to open and close the throttle butterfly. (Courtesy GM)

closed loop. In open loop systems, the signal sent to the solenoid valve has been completely pre-mapped. That is, the system doesn't have any way of directly monitoring the resulting boost level. (Note, however, that many cars have an over-boost fuel-cut to shut the car down if something goes catastrophically wrong.) Other cars use a closed loop boost control system, where the boost level is monitored by a manifold pressure sensor which is able to adjust the solenoid valve pulsing to give the desired boost level, even at different altitudes and temperatures.

ELECTRONIC THROTTLE CONTROL

Electronic throttle control replaces a direct mechanical connection from the accelerator pedal to the throttle blade. Instead, pushing on the accelerator moves a position sensor which send this 'torque request' information to the ECU. The ECU then drives an electric motor which opens the throttle blade. The actual opening of the throttle is monitored by a throttle position sensor similar to those fitted to conventional engine management systems. Elaborate safeguards prevent the throttle operation from going mad if any faults develop in the system.

Electronic throttle has significant advantages in the integration of traction control, stability control and cruise control, and can also be programmed to reduce emissions. Note than in systems with electronic throttle, the terms 'accelerator position' and 'throttle position' are no longer synonymous – all electronic throttle systems at times use throttle blade openings that don't directly match the driver's request.

In systems where a DC motor is fitted, it is driven in either direction by a variable duty cycle, variable polarity current. Other systems use stepper motors, with control through the sequential pulsing of the windings.

A turbo boost control valve. It works by bleeding a variable amount of air from the wastegate control hose.

WORKSHOP PRO — MODIFYING THE ELECTRONICS OF MODERN CLASSIC CARS

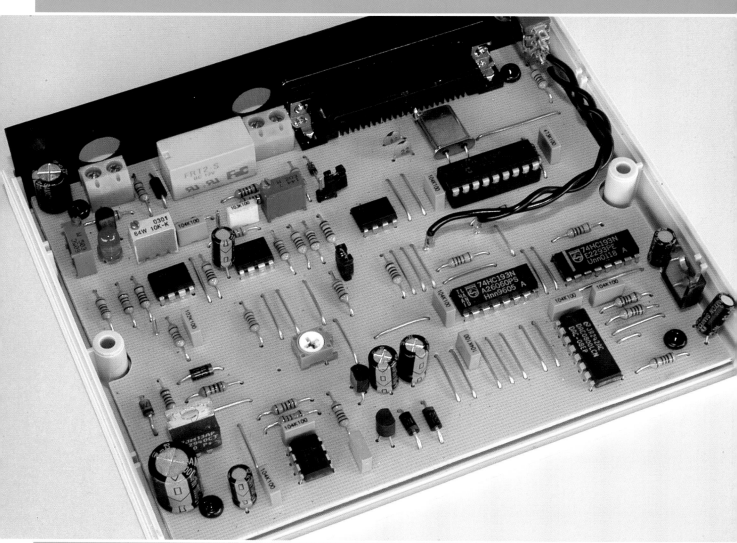
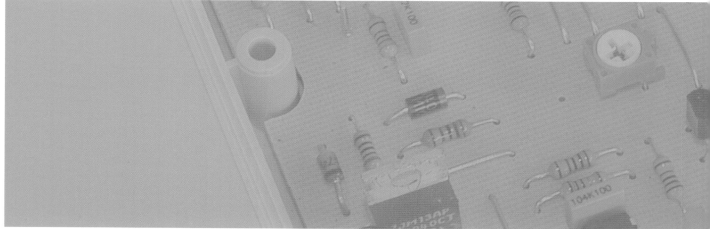

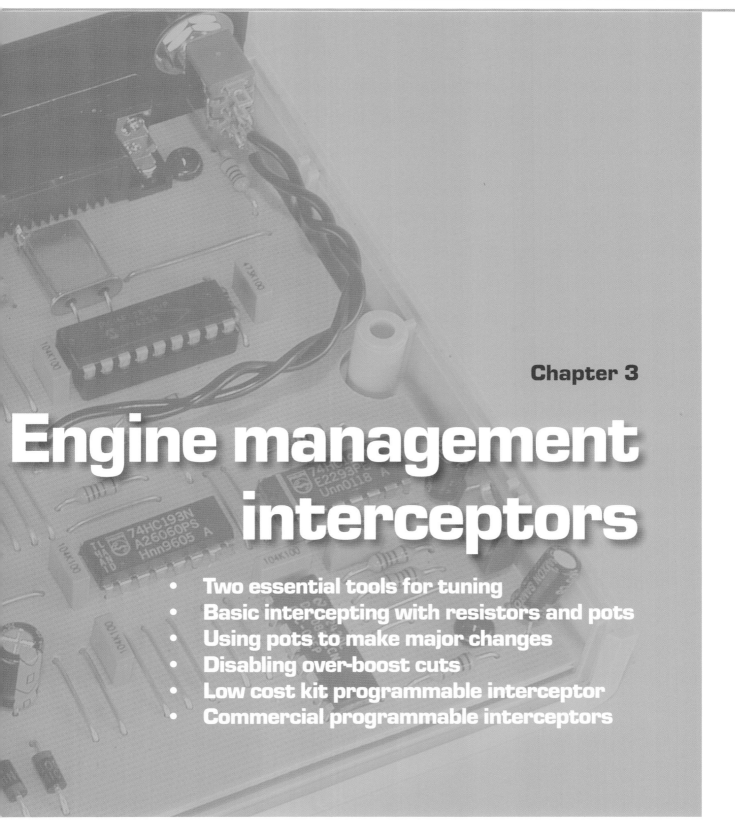

3. ENGINE MANAGEMENT INTERCEPTORS

Chapter 3

Engine management interceptors

- Two essential tools for tuning
- Basic intercepting with resistors and pots
- Using pots to make major changes
- Disabling over-boost cuts
- Low cost kit programmable interceptor
- Commercial programmable interceptors

WORKSHOP PRO: MODIFYING THE ELECTRONICS OF MODERN CLASSIC CARS

This chapter is divided into two major sections – simple and cheap intercepting techniques using pots and resistors, and then more expensive and sophisticated techniques using programmable interceptors.

Using resistors or potentiometers (pots) is a very powerful and extraordinarily cheap way of increasing or decreasing the output of any analog sensor. Examples of the technique that I use in this chapter include having the ability to run bigger injectors by decreasing the output of the MAP sensor, altering Exhaust Gas Recirculation (EGR) by changing the output of the EGR valve position sensor, and adding a little ignition timing by changing the output of the intake air temperature sensor. (Chapter 6 describes how I used similar techniques to alter the weight of power steering and to alter regen braking strength on a Toyota Prius.)

A 10-turn potentiometer (pot). These cheap electronic components are very powerful in modifying the output of analog voltage signals, including airflow meters and MAP sensors.

When using a pot in this way, the signal output of the sensor is proportionally changed, a very important but subtle point often overlooked by those who dismiss this technique. Using resistors and pots costs less than the price of a can of soft drink.

However, a pot or resistor can change the output only by a constant proportion through the signal range – it can't, for example, easily add more at the bottom of the range and less at the top, or vice versa. To do this, more sophisticated interceptors are needed, and these are covered in the second part of this chapter.

The cheapest of these interceptors is a kit device that uses 128 load sites – it's a one-dimensional interceptor programmed by four pushbuttons, and is ideal for altering airflow meter output to tune air/fuel ratios, for example. In cars that do not run closed loop feedback (ie they don't use an oxygen sensor), it can change mixtures throughout the rev and load range. For cars that use feedback (ie have an oxygen sensor) it can change mixtures only outside of closed loop operation – typically at high loads in cars with narrow-band oxygen sensors. In all cars, such an interceptor can get the system back to the point where fine-tuning will be carried out by the oxygen sensor in closed loop operation. An example of this requirement is when you fit a larger, less restrictive airflow meter. This type of interceptor costs about the same as buying four or five average takeaway pizzas.

The next level of sophistication is to use a commercial interceptor that is lap-top programmable. With such a device, ignition timing can be advanced or retarded at all loads and engine speeds (rpm), and with the qualification of being in or out of closed loop described above, air/fuel ratios can be tuned. Turbo boost can also be modified. A good interceptor can achieve outstanding results, within certain limitations. It can't, for example, make the knock sensor more sensitive, or increase the rev limit – that is, it cannot change the inbuilt logic of the ECU. Depending on the type of interceptor you buy, your budget has just jumped to 10 or 20 pizzas! But the capability is also now pretty good, too.

But before we look at any technique that can be used to alter engine tune (including programmable engine management, covered in the next chapter), two tools that you must have.

TWO ESSENTIAL TOOLS FOR TUNING

For *any* modification and tuning of engine air/fuel ratios and ignition timing, you need two essential tools. Don't even think of making engine management changes without these.

Wide-band air/fuel ratio meter

A wide-band AFR meter monitors the mixture strength real-time. It comprises a wide-band sensor and a dedicated controller and digital display. For occasional tuning, the sensor can be inserted up the exhaust pipe, and the display placed on the dash. For more serious tuning, the sensor should be mounted close to the engine. Good quality wide-band air/fuel ratio meters are now relatively cheap.

A wide-band air/fuel ratio meter is a requirement before you alter air/fuel ratios. (Courtesy Innovate)

If you are intending buying a high quality digital dash, the dash may have the ability to control and read a wideband sensor. If you have both programmable engine management and a matching digital dash, the programmable ECU can normally send air/fuel ratio data to the dash real time, allowing you to again see mixtures displayed.

Knock detection system

You need a way of detecting engine detonation. If you are using software that works with the factory engine management and can therefore monitor the factory knock sensor (and work with the standard ECU knock sensor logic), you should have a very effective system. Next most effective is programmable engine management, monitoring the standard knock sensor and allowing you, the tuner, to also monitor it. That monitoring might be through headphones plugged into the ECU, where you can actually hear the detonation, or it might be by means of a gauge or other display.

If neither of those apply in your situation, or you want to do it as cheaply as possible, use a remote-mounted microphone in the engine bay, connected to a small, adjustable in-cabin amplifier. The person listening to the engine should wear fully enclosed headphones – and that's especially the case if the car has a loud exhaust.

It's easy to make your own listening system, starting with a cheap eBay battery-operated 'personal listening' amplifier, and then extending the microphone wiring and adding good-quality headphones. (The step-by-step process for making this system can be found in my book *Optimising Car Performance Modifications*.)

Detonation sounds like 'ting ting' – after all, it's the sound of the hammer-blow of incorrect combustion hitting the cylinder walls and pistons, and that sounds pretty much as you'd expect it to! If you become familiar with the engine, you can usually recognise the harder edge that the combustion sound gains prior to actual detonation occurring. Testing the car briefly at very low revs and loads, with overly advanced timing, should cause detonation that you can hear and then later recognise. It is *not* a sound you want to hear at high loads!

Note that in real-world tuning, being able to hear the engine is very useful in other ways as well. You can hear when the radiator fan is running (useful when setting up idle speed control); in a turbo car, you can hear the operation of a plumbed-back blow-off valve; and you can hear slight engine staggers (eg when engaging the clutch and moving away from a standstill) that may not be obvious in normal circumstances.

MODIFYING WITH POTS AND RESISTORS

Two very cheap electronic components – resistors and potentiometers ('pots') can be used to achieve amazing outcomes on engine management systems. Let's take a look.

Tweaking ignition timing

It needs to be stated up-front that the following modification has gained a poor reputation through those claiming that it will revolutionise the performance of your car. It won't! For example, I've seen people claiming that on one particular car, you can gain 30hp – or, if you want, 20 per cent better economy! These figures are completely and utterly absurd. However, in terms of 'bang for your buck', the results on some cars can be absolutely outstanding. So, what modification is being discussed?

In most cars, the engine management systems uses an inlet air temperature sensor. As the name suggests, the ECU uses this to determine the temperature of the inlet air. But it's what it does with this information that's interesting. In addition to working out the density of the air (ie how heavy it is, which indicates how much oxygen there is in a given volume), the ECU uses inlet air temperature as an important input in determining what ignition timing to use.

If the inlet air is colder, the ECU will advance the ignition timing. If the inlet air is hotter, the ECU will retard the ignition timing.

Advanced ignition timing, especially when using higher octane fuel, will result in more power. In many cars, this will be felt as better light-throttle response – for example, the ability to use a higher gear in a given light-load cruise situation. This makes the car sweeter to drive and can improve fuel economy. So, if we can make the ECU think that the intake air is colder than it really is, the ECU will advance the ignition timing. And this can be achieved very simply!

Intake air temperature sensors use variable resistance

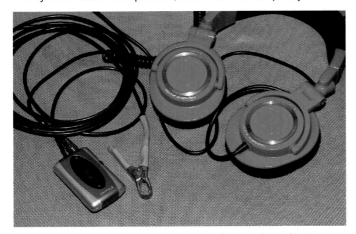

A detonation detection system made from a cheap battery-operated amplifier. The microphone is mounted in the metal clip, that is attached to the engine.

WORKSHOP PRO — MODIFYING THE ELECTRONICS OF MODERN CLASSIC CARS

An intake air temperature sensor. Modifying its output normally changes ignition timing by a small amount.

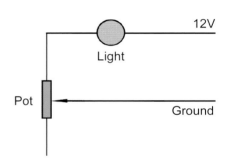

A pot shown in a circuit. The arrow shows the wiper (the moveable arm) that can go from one end of the resistance track (green rectangle) to the other. When the wiper is at the top, no resistance is introduced into the circuit, and the light would be bright. When the wiper is at the bottom, maximum resistance is introduced into the circuit and the light will be dim. At in-between pot positions, the light will vary in brightness. Note that in this circuit only two of the pot connections are used.

designs – that is, they vary in their resistance to electrical current flow with changes in temperature. Normally, they use a design where resistance (measured in ohms) increases as the temperature decreases. If you want the ECU to think that the temperature is colder, you add extra resistance in *series* with the sensor – this increases the resistance that the ECU sees. If you want the ECU to think the temperature is higher, you add extra resistance in *parallel* with the sensor (ie across the sensor) – this reduces the resistance that the ECU sees.

To change the resistance of the intake air temperature sensor, you need to use a resistor. This is an electronic component available in a range of resistance (ohms) and power (watts) values. (In this application the lowest power rating – ¼ watt – is fine.)

Resistors are available on-line for literally a few cents/pence each. For example, you might buy a ¼ watt, 3.2 kilo-ohm (kΩ) resistor (3.2kΩ = 3200Ω). But in this modification (and many others), it's easiest if you start off with a *variable resistor*. This allows you to adjust the modification until you get the results you're after. A variable resistor sounds expensive, but again the component is very cheap. In electronic terms, a variable resistor is called a potentiometer, or pot. At time of writing pots cost about $2/£1 each.

Let's have a closer look at these two components.

A resistor poses a resistance to the flow of electricity. It doesn't have any polarity (ie it can go into the circuit either way around) and its resistance is marked with colour-coded bands on the body of the resistor. Don't bother trying to learn the colours – just buy one with the correct resistance and then check it with your multimeter. The circuit diagrams throughout this book will show resistors as rectangles.

A pot has three connections. Two of the connections are to the ends of the internal resistor, while the third connection is for a wiper that can be moved along the resistance track. When you use a pot as a variable resistor, just two connections are used – one end of the track, and the wiper. On most pots, the connections are physically arranged as described – the ends of the track at either end of the pot, and the wiper connection between them.

When you get a pot – any pot – always check its resistance. Set the multimeter to 'resistance' (or 'ohms') and then connect the probes to the two outer terminals. The meter should show the maximum value of the pot. So a 10kΩ pot should have a value close to 10kΩ (10,000Ω). It doesn't matter if it isn't exact, but if the measurement shows (say) 4.8kΩs, it's a 5k pot not a 10k pot!

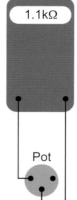

Next, connect the multimeter between the central terminal and one of the end terminals. Rotate the shaft of the pot and you will see the ohms read-out on the multimeter increase. (If it decreases, swap the outer connection to the other side of the pot – ie to the unused pin.)

You might be wondering why a pot has three connections, when here we're using only two.

The answer is that a pot can also be used in a completely different way to being just a variable resistor. The other use is as a *voltage divider*, which is also extremely useful in electronic car modification – more on this follows.

3. ENGINE MANAGEMENT INTERCEPTORS

USING A RESISTOR ON THE INTAKE AIR TEMPERATURE SENSOR

My Honda Insight is designed to run on 95 RON fuel but I normally use 98 RON. The RON value is purely a measure of the fuel's resistance to detonation; nothing else. Higher octane fuel therefore has a higher resistance to detonation, so can tolerate a higher engine compression ratio, and/or more ignition timing advance. To an extent, the ECU will automatically advance timing when running on higher octane fuel – but only to an extent. It expects 95 RON fuel, so it's never going to advance timing to the degree it would if it had been originally calibrated for 98 fuel.

The Honda workshop manual provides no real detail on the intake air temperature sensor, but it is easily removed and tested. At about 35°C (95°F), the resistance was 1600Ω, at about 20°C (68°F) it was 2000Ω, and when packed briefly in ice, it increased to 5000Ω. All measurements were done with a multimeter. So (to reiterate), higher resistances equal lower temperatures.

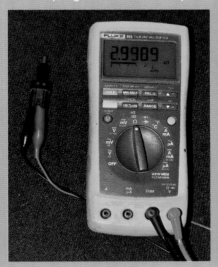

Measuring the resistance of the Honda intake air temperature sensor at ambient temperature.

The Honda Insight, on which the intake air temperature modification was made.

So what, I wondered, would happen if I altered the signal the ECU saw from the intake air temperature sensor? Since the lower the intake air temperature, the greater is the engine's resistance to detonation, if the ECU was convinced that the intake air temperature was actually lower than it really was, it could be expected to run more ignition timing advance. This could, in turn, well suit the higher octane fuel.

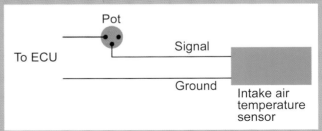

The pot inserted in series with the intake air temperature sensor. Note how only two of the terminals on the pot are used – the wiper (central terminal) and one end of the resistance track.

I snipped the signal feed near the sensor itself (this could have been done at the ECU but it was simpler to do it under the bonnet) and wired-in a 5kΩ (5000Ω) pot as a series variable resistor (ie using one end and one central terminal). (Note: if neither wire is connected to the sensor body ground, the pot can be inserted in either wire. If one side of the sensor wiring is earthed at the sensor, then the pot must go in the signal wire.)

I used a 10-turn pot so that changes could be made very gradually, but a normal pot could be used if care was taken with rotation. (10-turn pots are much rarer than normal rotation pots – I bought mine online.)

The temporary wiring for the pot at the intake air temperature sensor.

(continues overleaf)

WORKSHOP PRO: MODIFYING THE ELECTRONICS OF MODERN CLASSIC CARS

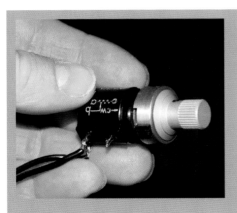

A 10-turn pot with a locking collar. This allows very fine adjustments to be made to the added resistance.

Initial results

By turning the pot, I could therefore make the ECU think the intake air temperature was colder than it really was. I turned the pot and noted that idle sped rose slightly before then falling. This is indicative of an advance in ignition timing (what was wanted) followed by an idle speed correction. (Note that this change in rpm didn't always occur – it depended on other parameters like engine coolant temperature.)

I wound in about 3000Ω of extra series resistance and went for a drive.

The greatest care should now be taken to listen for detonation. If you don't have an acute ear for it, build an amplified set of headphones with a remote microphone mounted on the engine. Of course, high octane fuel should also be used.

On the road, the Honda was clearly far more driveable. In light load driving, gear changes could be made earlier, a characteristic of increased light load torque. The earlier up-changes also suggest that, in urban driving, the fuel consumption might be a little improved. Specifically, 5th gear could be used up slight inclines at 60km/h (37mph), something the car was reluctant to do previously. However, on a highway fuel economy test, there was no discernible change. No detonation could be heard, and no check engine light appeared.

Final iteration

By using a pot (rather than a fixed resistor) in the initial configuration, you can adjust it to your heart's content. When you have decided on the value that is required, the pot can be removed and its resistance measured with a multimeter.

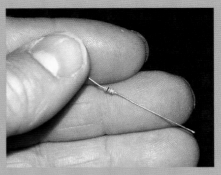

It's easy to incorporate a fixed value resistor into the wiring loom.

A fixed value resistor of the same value can then be bought, and wired in series with the temperature sensor, the lot then covered in heat-shrink or tape. And that's just what I did, using a 3.2kΩ resistor (which actually measured closer to 2.9kΩ!).

And the wonderful thing is that if you try it on your own car – and it doesn't work – you've spent only an hour and a few coins.

MAKING MAJOR CHANGES TO VOLTAGE-OUTPUTTING SENSORS

As I have already described, a pot can be wired as a variable resistor. A pot wired as a variable resistor is good when you want to add resistance to a sensor circuit. But that's all a variable resistor can do – add resistance, or, if the wiring is done differently (ie in parallel), subtract resistance. But most sensors in cars don't use a variable resistance designs. Instead, they have inbuilt electronics that cause them to output a varying *voltage*.

For example, most airflow meters, MAP sensors, accelerometers, yaw sensors and throttle position sensors have an output voltage that varies across the range of about 1-5 volts. If you just insert a series variable resistor in their circuit, things are liable not to work in the way you might expect! But by using a pot in a different way, very good modification results can be gained with voltage outputting sensors. Before we look at how to do it, let's examine the type of sensors that we'll probably be applying this modification to.

Many voltage-outputting sensors have three connections. For example, a MAP sensor is typically fed a regulated 5V supply from the ECU. (Regulated means it is held at a fixed voltage, irrespective of battery voltage.) There is also a ground wire from the sensor that somewhere is connected to the car's body – probably back at the ECU. Finally, there is another wire that is the signal output.

If you connect a multimeter to the signal and ground wires, and then drive the car, you'll be able to see on the multimeter how the signal varies in different driving conditions. For example, you might find that the lowest MAP sensor output is 0.8V, and the highest is 4.6V. You might also be able to see that the highest occurs at full throttle and the lowest occurs on the engine over-run. The ECU uses this signal to detect what is occurring – in this case, what the intake manifold pressure is at any given moment.

The highest that the MAP sensor output can go is 5V, and the lowest it can go is 0V. But of course this sort of sensor is rarely holding a steady signal output – it's up and down as the driver moves their foot and manifold pressure varies. So if we wanted the ECU to see a higher voltage coming from this sensor, we can't just cut off the sensor wire and connect it to 5V. Sure, that would mean that the signal voltage the ECU sees from the sensor is higher – but it's also now fixed, not varying with engine conditions. For the same reason, we can't just connect the signal wire to ground. If we want to raise the sensor output voltage (so for example, the ECU thinks that manifold pressure is higher than it really is), we need to add a *bit of voltage* to the signal. So, how do we do that? We can do it with a simple pot – eg a 10kΩ pot.

MANIPULATING VOLTAGE SIGNALS

Let's look at how pots can be wired into place to change the output of sensors that use analog voltages.

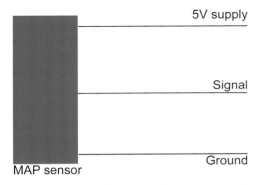

We'll start with a MAP sensor. The top connection is the 5V regulated supply from the ECU, the middle connection is the signal output, and the bottom connection is ground.

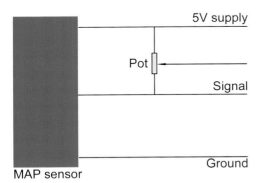

We'll now wire a 10kΩ pot between the signal wire and the 5V wire. Note how the connections have been made to each end of the pot's resistance track. (This might look like we're connecting the signal wire straight to 5V, but we're not – the resistance of the pot is so high that almost no current flows through this connection.)

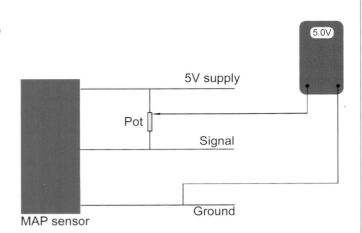

Let's put a multimeter in the circuit, measuring the voltage on the wiper of the pot. (Note how the other probe of the multimeter goes to ground – voltages are almost always measured with respect to ground.) The pot has been set so that its wiper is at the top – close to the 5V supply. This is the same as connecting the multimeter probe to the 5V supply, so the meter reads 5.0V.

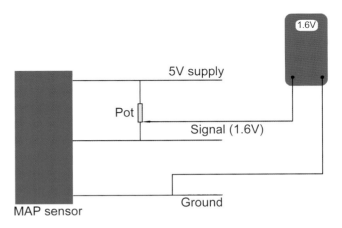

Now we've moved the pot wiper so that it's down the bottom – effectively connecting it straight to the signal wire. Since the signal is at 1.6V, that's also what the meter reads.

WORKSHOP PRO: MODIFYING THE ELECTRONICS OF MODERN CLASSIC CARS

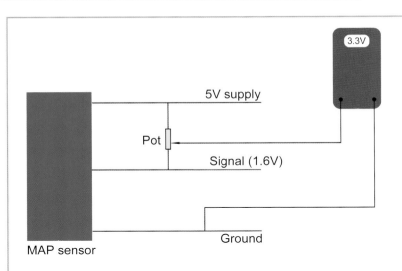

But now we've moved the wiper up just a bit. The multimeter is reading a combined signal – the signal voltage plus the little voltage we've added to it. So the meter reads 3.3V – we've added 1.7 volts to the signal.

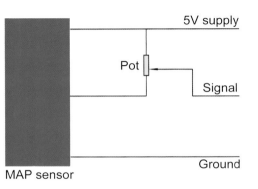

There's just one step left. If we connect the ECU signal wire to the pot wiper rather than to the MAP sensor, we can dial-in any addition to the sensor signal that we want. The signal will still rise and fall as it did before, but with an additional voltage on top. By changing the position of the pot, we can change how much voltage we add to the signal – the signal that the ECU sees.

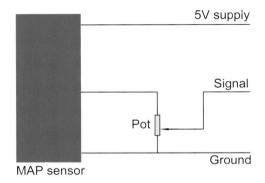

If we want to subtract a voltage, we just connect the pot between ground and the sensor signal, as shown here.

USING POTENTIOMETERS

This modification is a very powerful one that can be applied in many different circumstances. I'll show in detail a modification in a moment, but the first way I ever used this mod remains perhaps the most impressive.

I had a small 3-cylinder Daihatsu Mira turbo (660cc!) to which I'd fitted a bigger turbo and water/air intercooling. The injectors were running out of flow capability so I fitted larger injectors from a Charade GTti. But the ECU didn't know that larger injectors were being used, so over-fuelled the car.

The ECU uses as its main load input a MAP sensor, much like the one described above. By using a pot on the output of the sensor, I was able to lower the voltage (and so the load) that the ECU thought it was experiencing. This resulted in the ECU pulsing the larger injectors at a reduced rate (technically speaking, at a reduced duty cycle) and so, via the pot, the fuelling was able to be

This tiny 660cc Daihatsu Mira Turbo ran a big turbo and water/air intercooling. Larger injectors from a Charade GTtI were operated by the standard ECU, with just a simple pot on the MAP sensor used to give the correct fuelling. The car screamed out 86kW (116hp) at the wheels – more than 200hp per litre at the engine …

adjusted so that it was again correct. And, at the top end, where the engine previously ran out of injector flow, the larger injectors could keep up! With the oxygen sensor feedback loop presumably doing some trimming when in closed loop, the driveability of the car was perfect.

Let's look at another example of using the voltage adjustment, pot-based approach. This time, we're changing the amount of Exhaust Gas Recirculation (EGR) that occurs.

A 10kΩ pot can be used in nearly all cases without upsetting the original sensor behaviour. (Note: an exception is a narrow band oxygen sensor, where even a 10K pot is too great a load.). In fact, it doesn't matter much what the pot value is when being used as voltage divider in the way I am describing here – 10K, 50K or 100K. As discussed, by using a multi-turn pot, fine adjustments can be made.

As with any modifications that alter a sensor's output,

EGR MODIFICATION

It's not generally realised, but Exhaust Gas Recirculation (EGR) can be beneficial for part-throttle fuel economy (mileage). I therefore decided to increase the amount of EGR occurring on my Honda Insight.

The Insight uses an EGR valve that is electronically-controlled by the ECU. This valve, normally held shut by a spring, is opened by the action of electrical current through a coil. The amount that the valve opens is monitored by a lift sensor. The ECU monitors this lift sensor and alters the opening of the EGR valve to give the required ECU-mapped EGR flow for the driving conditions.

I used a multimeter to measure the lift sensor output when the car was being driven. This showed that the voltage from this sensor rose as the EGR valve opened by greater amounts, being around 1.2V with the valve shut and rising to about 2.5V at its peak. The values aren't very important – what is important, is that voltage rose with greater valve opening.

If this sensor signal could be altered so that the ECU was told that EGR valve opening was less than it actually was, the ECU would compensate by opening the valve more – that is, EGR would increase.

To increase the amount of EGR that occurred, the circuit shown here was used to reduce the sensor's output voltage – note how it is identical to one of the diagrams already discussed.

The car was then driven in a variety of situations, with the increase in the amount of EGR being finely adjusted by turning the pot positioned in the car. (I used a 10-turn pot, which allowed very fine adjustment to be easily carried out while on the move.)

With too much EGR flowing, in some driving situations the car could lurch and stumble, eg when passing over the crest of a slight rise and lifting-off a little. Fine-tuning of the increase in EGR consisted of adjusting the pot to the point where driving behaviour was virtually identical to standard. However, the final configuration ran a lot more EGR than standard.

In carefully tested urban conditions, the result was a small but measurable fuel economy improvement of 3 per cent. (In highway conditions there was no change.) This might not sound like much of an improvement, but the Honda Insight hybrid is among the most fuel-efficient cars in the world – so to be able to make any improvement is impressive. Second, the modification cost almost nothing and was easily installed.

The electronically-controlled EGR valve on this car uses a lift sensor that detects the opening distance of the valve. By subtracting a voltage signal from the feedback, the actual opening of the valve could be modified.

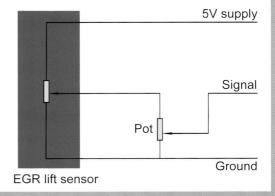

A pot was added to the EGR valve lift sensor output to allow the signal to be tweaked.

WORKSHOP PRO: MODIFYING THE ELECTRONICS OF MODERN CLASSIC CARS

you need to be aware that the ECU will change all of its outputs that rely on that sensor's input. In the case of the Honda Insight's EGR valve, the lift sensor is used only for EGR feedback. However, other sensors (eg a MAP sensor) are used by the ECU for lots of things. For example, reducing the output of the MAP sensor will not only alter the pulsing of the injectors, but will also advance the ignition timing.

Another point to keep in mind is that the technique does not add a constant voltage to the signal. Instead, it does better than this by adding (or subtracting) a constant percentage. For example, if you set the pot to add 20 per cent to the sensor voltage, this percentage stays the same across the whole sensor output range. This is one reason that the pot approach works so well in practice, and to state the point again, is something those who dismiss this technique usually fail to realise. The use of a pot to tweak the signal of a voltage outputting sensor is one that is very powerful. It is also very simple to install, extremely cheap, and allows fine-tuning of the modification.

DISABLING OVER-BOOST CUTS

Another circuit that can be added to a MAP sensor is one that prevents its output signal rising beyond a certain ceiling. This is most frequently done on turbo cars where the boost has been increased sufficiently that the system is imposing a boost-cut – that is, shutting down the engine.

The circuit consists of two 1N914 diodes and a 10kΩ pot. Its sensitivity can be adjusted by rotating the pot – this should be set so that the circuit clips (stops the signal rising any further) at as high a manifold pressure as possible. As with any of these techniques, the power over the car's electronics is very great, so extreme care should be taken that the engine doesn't suddenly run lean during the setting-up procedures.

PROJECT PRIUS

About 15 years ago, I extensively modified a first-generation NHW10 Toyota Prius. Why? I liked the idea of a challenge, and I loved the fact that Prius was the first of what would obviously be a new breed of cars. But with no less than six ECUs, a 288V battery, two electric motor/generators – and a control system that frequently switched off the engine, incorporated regenerative braking and used electronic throttle control – it was always going to be a challenging project. The story of that project also shows what can be achieved using very cheap electronic hardware on a complex car.

The NHW10 Prius. Compared to later models, it had substantially less petrol and electric power but it was still a full hybrid vehicle. Modifications I performed included altered regen braking and a turbocharged engine. (Courtesy Toyota)

PRIUS DRIVELINE

The NHW10 Prius uses a 1.5-litre four-cylinder engine closely related to the engine used in the Toyota Echo. However, it has much less power than the Echo – just 43kW (58hp). The low power output is because the engine revs to only 4000rpm and uses what is called an Atkinson Cycle. Compared with the conventional Otto Cycle, an Atkinson Cycle engine delays the intake valve opening time, resulting in a reduced intake charge. So despite the geometric compression ratio of the Prius being a sky-high 13.5:1, the cylinder pressures on the compression stroke don't really reflect this. This approach benefits efficiency because at lower loads, the throttle is open wider for a given power output, so reducing pumping losses. To allow the degree of 'Atkinsoning' to be altered on the fly, the engine ECU alters intake valve cam timing. The other vital ingredient in making this process work is the use of an ECU-controlled electronic throttle. In this car, the driver's

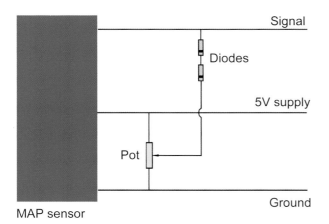

The boost-cut device uses two 1N914 diodes and a 10kΩ pot. Note that the diodes are polarised – they must be orientated as shown. The pot sets the voltage at which the circuit clips, ie prevents it rising further.

torque request often has little to do with the actual throttle angle selected by the ECU.

Atkinson Cycle (sometimes called Miller Cycle) engines have a deficit in low-rpm torque. But in the Prius, there's a 30kW AC electric motor ready to provide maximum torque at zero road speed. In fact, the gearbox (called the Power Split Device) actually contains two electric motor/generators. Along with the engine, these are connected to an epicyclic gear train. The engine output is split between the wheels and one of the generators. The generator charges the high voltage battery or alternatively, feeds the other electric motor that in turn helps drive the wheels. This electric motor can also receive power from the high voltage battery to either assist the petrol engine or propel the car on its own. The PSD's gear ratio is a result of the balance between the speeds of the engine, the electric motor/generators and the wheels that depends on how much force is applied by each. This gives the effect of a continuously variable transmission.

One of the electric motors also acts as a quiet and powerful starter for the engine, allowing it to be stopped and started smoothly as needed. The other generator is used to regeneratively recover energy from the car during braking and store it in the battery for later use. When the driver lifts the accelerator pedal, the engine's fuel supply is cut off.

MODIFYING THE REGEN
One of the first electronic modifications I performed was to increase the amount of braking regeneration. How I did it is covered in Chapter 6 – it's a technique that has lots more applications than modifying just Prius braking!

TURBO INTERCOOLED
On the road the Prius, despite its low power and 1240kg (2730lb) mass, doesn't feel particularly slow. The electric motor's low-speed torque and petrol engine integrate seamlessly, giving punchy performance. Well, in all but one driving situation, anyway. That situation is climbing a long, steep country road hill. Initial performance is fine but after a while, the high voltage battery becomes drained, decreasing the amount of electric power available. The car slows, the engine revs automatically increase, and you can find yourself with the accelerator pedal flat to the floor, just crawling along. At this point a rather cute tortoise symbol lights up on the dash – although biologically incorrect, we christened her 'Myrtle the Turtle.' (The Prius being described was sold new only in Japan. Myrtle was banished for the overseas models.)

And it just so happens that I then lived at the top of a long, country road hill that had gradients as steep as 16 per cent – and after a while, Myrtle's presence became quite annoying. By feathering the throttle and watching the movement of electrical current in and out of the high voltage (HV) battery (the colour dash LCD shows this), it was possible to negotiate this hill with a speed over the crest of 55km/h (34mph) or so. But lose concentration, and that dropped to 47km/h (29mph). That's awfully slow. (It should be stressed that the later model Prius cars don't suffer from this hill-climbing problem.)

I considered adding another HV battery pack but because of their control systems, this is an extremely complex move – nothing like as simple as wiring the two batteries in parallel. That left increasing engine power as the best option, and so I turbocharged and intercooled the engine.

And then the fun started …

Fascinatingly, the hybrid control system coped with the increased power remarkably well. Presumably because Toyota's engineers could never predict exactly how much power the petrol engine would generate (as with all engines, this varies with atmospheric conditions, individual engine build quality and so on), the hybrid system had enough flexibility in its control system to direct any excess power being developed by the engine into charging the HV battery. Note: into the HV battery, not to the wheels …

So the odd situation developed where with the HV battery level at (say) half, the performance of the car was standard – despite the turbo! But what about up the big hill? Ah, well there the car was transformed. Rather than the HV battery dropping in charge, going up the hill it actually *increased* in charge! And always having lots of battery voltage – and so electric motor power – resulted in the speed over the crest of the very steep hill increasing from a low of 47km/h (29mph) to a stunning 86km/h (53mph) – over 80 per cent faster.

I fabricated a thick-walled exhaust manifold for the turbo. The mechanical addition of the turbocharger was straightforward – it was getting the air/fuel ratio suitable for the forced aspiration that caused me to pull my hair out!

WORKSHOP PRO — MODIFYING THE ELECTRONICS OF MODERN CLASSIC CARS

But let's go back a little. When the turbo was fitted, the air/fuel ratios needed to be changed – a turbocharged engine (especially one with a 13.5:1 static compression ratio, Atkinson or no Atkinson) needs richer than standard mixtures. The Prius uses an air/fuel ratio of 14.7:1 *all the time* – the engine ECU monitors the output of the two oxygen sensors to constantly hold this air/fuel ratio. So how to change this?

The first step was to fit a one-dimensional kit interceptor (covered later in this chapter) to allow alteration of the airflow meter output. However, monitoring the mixtures with a MoTeC air/fuel ratio meter showed that the Prius ECU is extraordinarily quick at learning around any changes made in this way. Alter the mixtures to 12.5:1 and within five or so seconds, the mixtures are back at 14.7:1! The same thing occurred if fuel pressure was increased – back go the mixtures to stoichiometric.

Hmmm, so what about disconnecting the oxygen sensors, substituting an appropriate looking 0-1V square wave signal on the ECU oxygen sensor inputs, and then altering the airflow meter output voltage? A pair of 555-based circuits was constructed and the system wired-up. But the ECU immediately picked up that something was wrong with the oxygen sensors, and went into a default mode – which with the added airflow of the turbo, resulted in mixtures even leaner than 14.7:1! Aaaaagh!

Well then, what about disconnecting the oxygen

This 555-based circuit was built in an attempt to simulate the fluctuations in narrow-band oxygen sensor voltage normally seen by the ECU. If successful, this would allow the oxygen sensors to be disconnected, stopping the ECU 'learning around' mixture changes. However, the ECU immediately saw through the pretence, outputting the same mixtures as when there is no oxygen sensor input at all.

sensors (easily achieved just on high load with a voltage switch) and then using the interceptor to alter the mixtures? Again, no success – and this time, the mixtures appeared to vary randomly.

About this time, I upgraded the fuel system with a new in-tank pump, external adjustable pressure regulator and a return-line to the tank. That allowed me to run higher fuel pressure (which initially gave correct mixtures with the oxygen sensors disconnected) with the interceptor used to tweak the resulting mixtures. But, yet again, the mixtures were not consistent.

After many weeks of work, I finally devised an effective system. Two fuel pumps and two fuel pressure regulators were used to allow the running of two different fuel pressures. A solenoid allowed electronic switching between the two different pressures. The lower of the two pressures was set so that, even when running a little turbo boost, the ECU could keep the mixtures at 14.7:1, and the oxygen sensor feedback loop operated as normal. Then, when a pre-set load was reached, the voltage switch monitoring the airflow meter voltage switched out the oxygen sensors and activated the solenoid increase in fuel pressure. The resulting mixtures were then fine-tuned by the interceptor working on the airflow meter output.

With this approach, the mixtures were consistent, economical (the car was still in closed loop with 14.7:1 mixtures for the vast majority of the time), but with appropriately rich mixtures used at full turbo boost.

Incidentally, ignition timing was never a problem. I never heard the engine detonate, even when (briefly!) running 100kPa (15psi) boost. Ninety-five octane fuel was used (the car was designed to run on 91) and an efficient air/air intercooler was fitted.

Auto throttle shutdown

About this stage I started to relax. Ahhh, this was nice. The world's only turbocharged, intercooled Prius with modified regen braking … Those many hours of work were well worth it. And that was the case until I discovered that the hybrid control system's adoption of turbo power wasn't as seamless as I'd first thought.

Initially, I'd decided the abrupt engine shut-down that occasionally occurred at full power was an ignition problem – and had replaced the sparkplugs with a colder heat-range Iridium design. But then, while watching the boost gauge, I saw what was happening. At full power, the hybrid system would momentarily close the electronic throttle. Whether that's to protect one or both of the electric motor/generators – or for some other reason – I still don't know. But the result was a huge power loss perfectly timed to occur when overtaking a big truck …

If the shutdown was a result of excessive engine

3. ENGINE MANAGEMENT INTERCEPTORS

The world's first turbo Prius. The plumbing at bottom-left goes to the air/air intercooler. The turbo transformed hill-climbing performance, and, with other modifications, actually improved open-road fuel economy.

power, I could just drop turbo boost at higher revs. But the problem was that I couldn't. The wastegate spring pressure on the turbo meant that 50kPa (7psi) was as low as I could go. But there had to be another way of dropping boost – and there was. By using a solenoid to control the boost pressure feed to a recirculating blow-off valve, the blow-off valve could be made to leak, bleeding boost from the compressor outlet back to the inlet. The result was decreased boost. As finally configured, boost rose to 50kPa (7psi) and then, above the pre-set load point, smoothly dropped to 30kPa (4psi). This made no difference to on-road performance – and there were no longer any throttle shut-downs.

Auto engine off

While compared with many turbo applications the Prius turbo is not working particularly hard, there is one characteristic of its hybrid control that, if left unaddressed, could quickly kill the turbo bearing. When the throttle is lifted, fuel flow to the engine is stopped. So, approaching a red traffic light, the engine stopped running as soon as you backed-off – and stayed off until the lights went green and you applied the accelerator.

A turbocharger relies on engine oil flow for lubrication and partly for cooling, and so the engine should not be turned off until the turbo has had time to cool. If the turbo has been spinning hard, an early engine switch-off can cause oil in the turbo bearing to coke. So how could the Prius petrol engine be kept running after a boost event?

This model Prius has two air-conditioning modes.

In High mode, the engine is forced to run continuously. In Normal mode, the engine is allowed to switch off whenever the hybrid ECU decides it should be off. When High mode is selected, the air conditioner system tells the hybrid ECU that it should not switch off the engine by means of an 'engine on' request signal. This signal is very simple – above 4V means keep the engine running, below 1V means it's OK to switch it off. So, by feeding 5V to the 'engine on' input of the hybrid ECU, the engine can be kept running. This was achieved via a timer triggered by a voltage switch that monitored boost pressure. If boost pressure was high, the timer was started, so keeping the engine running if the boost event was followed by the car stopping.

Outcome

I've run out of space to cover all the mods made to the car – they included a rear sway (anti-roll) bar, electronically modified electric power steering (covered in Chapter 4), on-dash mixture indication, electronically controlled high pressure intercooler water spray (Chapter 5), underfloor aerodynamic changes, and plenty of other bits and pieces.

And the results? Well, the worst aspect of the car had previously been its country road hill-climbing ability – and that was massively improved. But what about the *raison d'etre* of the Prius: fuel economy? The modified Prius had *better than standard* fuel economy. On an open road cruise, fuel economy of the modified car was improved by about 13 per cent.

WORKSHOP PRO: MODIFYING THE ELECTRONICS OF MODERN CLASSIC CARS

HOW TO DESTROY AN EXPENSIVE PRIUS BATTERY PACK

Most of the modifications to the Prius were very successful, but there was very nearly a disaster of such proportions my house could have burned down. As it was, a valuable Prius high voltage battery pack was destroyed.

In this Prius model, the high voltage battery consists of nickel metal hydride D-sized cells. In fact, there are 240 of them wired in series, each (as normal with Ni-MH cells) having a voltage of 1.2V, and so giving a total voltage of 288V. The battery is arranged in 40 sticks of six cells, with each stick having a quoted spec of 7.2V and being of 6 amp-hours capacity.

At one stage in the modification of the car, I'd needed to quickly develop a battery charger suitable for the HV battery pack. The need occurred after the car had been off the road for many weeks – the HV battery level had dropped sufficiently that the ECUs wouldn't allow the car to start. That simple charger had comprised a bridge rectifier working directly across the 240V AC power supply, feeding the battery through a 3Ω, 500W resistor (a high-power light bulb).

That charger had worked, but I wanted the better one to keep a spare battery pack charged. The second design of charger used a transformer wired in auto-transformer configuration to boost mains voltage to 267V. Following that was a 600V 10A bridge rectifier, which fed the high voltage Prius battery through five, 15Ω, 10W ceramic dropping resistors wired in parallel. Additions included monitoring neon lights, fuses, and an internal fan. The unit was mounted in a sturdy metal box, an external power-point style earth leakage circuit breaker was used – and I was pretty proud of the result.

But I had completely miscalculated the battery capacity, and so made a huge error in calculating the required battery charging time. It was late in the afternoon when I finished building the charger. I connected it to the battery pack (using my protective high voltage rubber gloves to clip the crocodile clips directly to each end of the series string of cells), switched on the charger, and monitored current and voltages. They were fine – the battery was charging at 2A. Considering this was meant to be just a trickle charge, I had intended leaving the pack charging all night.

But thank God the fire happened before I went to bed.

After checking on the battery each hour, I left it for a

The spare Prius HV battery pack being charged. From left, the safety switch equipped power pack, custom high-voltage charger, multimeter measuring current flow, and, at the rear, the complex battery pack that includes its own ECU and huge cooling fan.

few hours. Then my wife and I started smelling something. I then lived in a multi-storey wooden house on stilts; the 'garage' was the open concrete pad directly under the house. That's where the battery was charging.

I walked outside to find the house completely enveloped in a thick, acrid, rolling fog. I rushed downstairs and saw clouds of smoke gushing from under the woollen blanket I had thrown over the pack to stop stray animals sniffing the high voltage terminals. When I pulled the blanket back, I could immediately see a brilliant red glow deep inside the battery pack; as I watched, flames 150mm (6in) long burst out. Thick white smoke was boiling from every opening.

I yanked on my high voltage rubber gloves – they were luckily still close by – and pulled the circuit breaker out of the back of the battery pack. I turned off the charger and pulled off the charging leads before man-handling the burning, corrosive, high voltage battery pack (it's very heavy but I didn't even notice the weight) 10 or 15 metres (yards) to a clear area of lawn. As I turned it end over end, I could feel that the thick plastic casing had softened – and even through the gloves, the heat of the plastic was obvious.

The flames subsided, but smoke kept pouring out. After 15 minutes or so, the smoke diminished, but hissing gas could still be heard periodically venting from the batteries.

3. ENGINE MANAGEMENT INTERCEPTORS

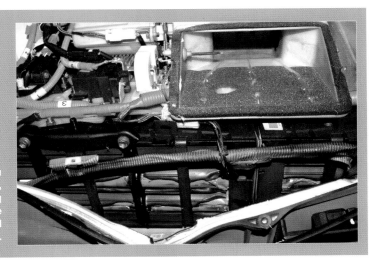

When the battery pack had cooled, an inspection showed the batteries charred and nearby wiring melted, the plastic case distorted. Salvageable were just items like the battery ECU, current monitor and circuit breakers.

The charger? Yep, she was a good one. But only to charge a Prius HV battery for half an hour or so, *not* for four solid hours …

And here's what happens when you bypass the internal safety devices, make a huge mistake in calculating battery capacity – and charge the battery. One charred and ruined high voltage Prius battery pack.

ONE DIMENSIONAL KIT INTERCEPTOR

Back in 2004, when I was a major contributor to electronics magazine *Silicon Chip*, I came up with the idea of one-dimensional interceptor that could be used on voltage signals. Why, I thought, did interceptors always have to use two axes (eg airflow meter voltage and rpm) when in many cases, the signal was sufficiently 'information rich' that modifying just the one signal would work fine?

The example I was thinking of was airflow meter voltage. A high airflow meter voltage indicates that airflow is high – and so, therefore, is load. There is no situation where airflow meter voltage is high but load is low. In effect, the airflow meter signal has *within it* both throttle opening and engine speed information.

The magazine employed a brilliant electronics engineer named John Clarke, and he designed and developed a 1-D interceptor that matched my specifications. It used a digital hand controller, had 128 adjustable load sites and could be configured to work for 0-1V, 0-5V and 0-12V signals.

At any of the load sites, the voltage that the ECU saw could be raised or lowered, and the device – that I called a Digital Fuel Adjuster (DFA) – interpolated between sites so the signal changed smoothly (rather than in steps).

The device was available both as an electronics kit and pre-built, and many were bought and used. Before we published the story, I tested the interceptor on the airflow meters of a wide range of cars – and it worked very well.

The DFA allowed both real-time and non-real-time adjustment. This meant that you could be driving the car and change the voltage outputs of the airflow meter, immediately seeing how this affected the engine's behaviour. For example, you could hold the car at one load and then move the airflow meter voltage up or down for

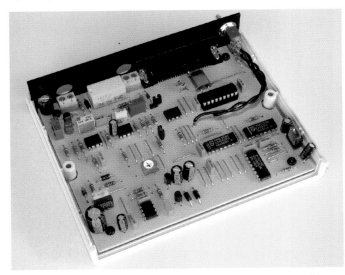

The original Digital Fuel Adjuster (DFA) 1-D programmable interceptor. It worked brilliantly on airflow meter output signals, but could also be used on any car sensor that outputs a voltage.

The original DFA hand controller, together with MoTeC air/fuel ratio meter and boost gauge. The instruments are pictured in my Maxima V6 Turbo running a huge airflow meter bypass (covered later in this chapter).

WORKSHOP PRO: MODIFYING THE ELECTRONICS OF MODERN CLASSIC CARS

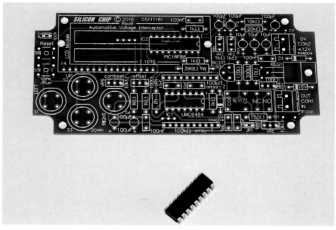

The current version of the interceptor is available only as a PCB and pre-programmed PIC – you need to get the rest of the components together and build it. The simple voltage adjustment can work extremely well.

MAP SENSORS?

In terms of intercepting signals, a MAP sensor system does not work in the same way as an airflow meter. Unlike an airflow meter, which provides all that the ECU needs to know in order to calculate how much fuel to add, the MAP sensor signal on its own is useless for this purpose.

Why? Well manifold vacuum (in a naturally-aspirated car) will drop to zero whenever the throttle is fully opened – whether the revs are at 1000 or 6000rpm. Going on just the MAP sensor information would result in the full-throttle mixtures being right at 6000rpm – and hugely over-rich at 1000rpm.

To calculate the required addition of fuel, the ECU must have internal logic that also takes into account rpm. Therefore, the ECU needs to know both the MAP value and the rpm before it can calculate the required fuel addition.

It's why simple intercepting and altering the signals coming from an airflow meter works much better than doing the same for a MAP sensor.

that load point, using an air/fuel ratio meter to show how these changes affected the mixtures. This real time mode was called RUN.

You could also use the DFA in VIEW mode, that is, without the engine having to be under load (or even running, for that matter). The VIEW mode let you scroll through the load points, change the up/down adjustments that had been made, or put in new adjustments. VIEW mode was good for smoothing the curve of changes, or quickly getting major adjustments into the ballpark before fine tuning occurred.

Both RUN and VIEW modes were selected from the hand controller. A third mode – LOCK – was selected by a switch on the main unit. It was used when you wanted to prevent inadvertent changes being made to the map.

Fast-forward to 2009 and the magazine developed a new version of the project, this time called the 'Voltage Interceptor for Cars,' then in 2017 they produced a version called the 'Automotive Sensor Modifier,' that incorporated the hand controller into the device.

I haven't been associated with the magazine for many years, but the functionality of the interceptor has changed little over that time. What has changed, rather

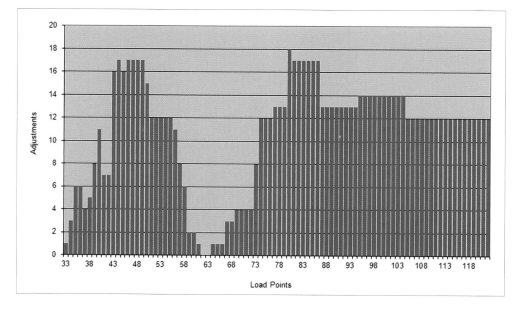

This is the map of DFA adjustments to the airflow meter voltages that were made on a 1985 BMW 735i. The vane airflow meter spring had been tightened, resulting in mixtures that were a bit lean. This explains why nearly all the corrections increased the voltage coming from the airflow meter. In fact, it was only at load sites around 63 that no changes were made – at this load, the air/fuel ratio was as desired. The BMW ran in open loop all the time – there was no oxygen sensor.

3. ENGINE MANAGEMENT INTERCEPTORS

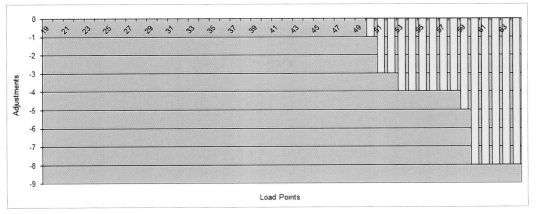

This DFA map shows the changes made to the high-load mixtures on a 1998 Lexus LS400. (This particular map was done with a prototype DFA with only 64 load sites.) As can be seen, the top-end airflow meter output voltages were reduced, so leaning-out the otherwise very rich mixtures. Performance measurably improved.

unfortunately, is the availability of kits. Rather than being able to buy the interceptor either prebuilt or as a full kit, you can buy only the printed circuit board (PCB) and the pre-programmed PIC. You then need to source all the other components yourself. They're commonly available (eg pots, resistors, diodes, etc) but the task of making one of these interceptors has become harder.

So why am I mentioning it here? For the money (it will cost about one-tenth the cost of a commercial interceptor) you can gain very good results. At the time of writing, the PCB and PIC were available from www.siliconchip.com.au.

AIRFLOW METER BYPASS

When modified engine power reaches a certain level, the airflow meter needs to be upgraded. The point at which that level is reached depends mostly on how much extra power the engine is developing, and how big the airflow meter was to start with. So what are the common upgrades? These are an aftermarket or original equipment upsized airflow meter, or a move to programmable management (which lets you ditch the airflow meter for a MAP sensor). However, these approaches are expensive.

But there is an alternative. You get to keep the original airflow meter, but the system can flow far more air. You get to keep the original airflow meter, but its output scaling is changed so that you don't hit a voltage ceiling. Want more advantages? You can map your mixtures at the same time, the approach is cheaper than anything else I have seen that can achieve this, and it's all pretty easy to implement.

Magic? Not really – it uses the 1-D interceptor described above and some simple mechanical changes. And don't think that the results are sub-par – I used the technique on a Nissan Maxima V6 Turbo and got excellent mixture accuracy (as measured by a MoTeC air/fuel ratio meter), driveability at least as good as factory – and *halved* the flow restriction of the intake.

Airflow meter upgrades

Most airflow meters are hot-wire designs called 'mass airflow' meters, often abbreviated to MAF. I described them in Chapter 2, but in brief this is how they work. The air rushing into the engine flows around a platinum wire that's been heated by electricity passing through it. The airflow cools down the wire. The ECU tries to keep the wire at the same hot temperature at all times, and can work out how much electricity it needs to pass through the wire to do this. The more electricity it needs to use, the more air that must be being breathed by the engine. Hot-wire meters automatically compensate for intake air temperature variations, and use onboard electronics to allow them to output a signal in the 0-5V range.

Some airflow meters cause quite a lot of airflow restriction. Exactly how much depends on:
- the detail of their internal design
- their diameter
- their design flow, versus the flow that's actually being drawn by the modified engine

Some airflow meters have heat-sink fins protruding into the airflow, some have very small internal diameters, and some have close-knit wire screens at each end. And all these things harm flow capability.

Measuring the amount of restriction being caused by an airflow meter is easy. It's also an important step to take before replacing or modifying it, because if the airflow meter has plenty of flow capacity for your engine, what are you going to gain by spending the money? (Except if the meter's reaching maximum voltage output, which I'll come to in a moment.)

The easiest way of measuring the actual restriction is to plumb a manometer (just a sensitive pressure measuring instrument) to the intake system after the airflow meter and take a full load measurement. Then do the same before the airflow meter. The difference in the figures is the pressure drop across the airflow meter – incontrovertible evidence of how well it flows. The lower the pressure drop, the better its flow. This can be done at near zero cost – you can make a manometer from a plastic drink bottle, a length of dowel and some plastic hose. (You'll find more on these techniques in my book *Optimising Car Performance Modifications*.)

Testing I carried out on one car showed that the hot-wire airflow meter had a restriction of 14.5 inches of water – or nearly 50 per cent of the total intake system restriction. That's pretty major – time for an airflow meter upgrade!

The other problem can come about if the airflow meter hits its output ceiling. For example, you've done some modifications to your engine and you find that the meter is hitting its maximum output (eg 4.8V) nowhere near full load. The meter is measuring as much air as it can, and after that it's all a straight line – even when the load rises still further. Mix that output with bigger injectors, and once the ceiling has been reached, the injector output is the same straight line. Not good. Again, it's time for an airflow meter upgrade.

Swaps

The most common approach to airflow meter upgrades is to swap-in a larger airflow meter. Some cars have aftermarket big-bore designs specifically made for them – but always at a high price – while original equipment makers often have a family of airflow meter sizes. For example, in the Nissan family you can look at 2-litre turbo SR20DET airflow meters being upgraded to RB25DET 2.5-litre turbo meters or even VH45DE V8 meters.

In fact, with the Nissan Maxima that was my guinea pig, I first looked at doing just this sort of swap. The trouble is, it's both expensive and not as straightforward as it first looks. First, take the cost. Really big airflow meters are rare, so immediately you're looking for something that lots of other people also want. And that means only one thing – they're expensive. Or maybe you don't worry about going for a 'name' airflow meter, but just head off to the car dismantlers to see what big meter you can nab. That sounds fine until you've made your diameter measurements, inspected the internals carefully, paid for it, and then got it home. Hmmmm, I wonder what all these pins are for?

Airflow meters use a variety of pin-outs, and without (a) knowing exactly what car model the airflow meter is from, and (b) having access to a decent workshop manual, you're not going to be able to easily find out. (And many workshop manuals don't even show the pin-outs of the airflow meter – they just tell you the colour codes and then the wires disappear into the ECU. Which wire is the signal, which is power, which is earth, which is hot-wire burn-off …? It's often quite hard to work it all out – I've been there and tried that!) Plus, you'll often also need to get part of the loom if you're to get the right matching plug – and many car dismantlers won't cut looms.

In short, this option isn't nearly as great as it's often said to be. And even when you've done all of the above, you'll still usually need to have your ECU remapped to suit the new airflow meter.

Modifications

Some people polish and modify the insides of the airflow meter, removing the protective screens and smoothing the end result. And that's fine if you want to do a lot of work, and don't really want to know what the end result voltage outputs are. Certainly, if you're going to modify an airflow meter, you'll want to check the air/fuel ratios before and after the mods – although it must be said, I've had very good flow improvements from just simple screen deletions. But if you need to take the modifications further, I don't think it's a great idea.

Bypasses

But what about a completely different approach? One that keeps the standard factory airflow meter – complete with screens if you want – and so avoids all those sourcing and wiring problems. Think about it for a moment – the reason that you would want to change an airflow meter all comes back to one thing: too much air is being forced to flow through the meter. You don't want to reduce the airflow to the engine (because then you'll also decrease power), so why not simply allow some of the engine's airflow to *come from another source*?

Does all the engine's airflow have to come through the airflow meter? The answer to that is simply 'no'. In fact, most hot-wire airflow designs *don't* measure all the air that's flowing through them. Instead they measure a *proportion of that air*, and then output a signal that takes into account the rest that's also passing. For example, some airflow meters use a tiny 'side passage' to actually sense the airflow and are then calibrated on the basis that if *this* much air is passing down this side passage, then *this* much air must be passing through the meter as a whole.

So what actually is a bypass, then? It's simply a way that some of the air entering the engine can bypass the airflow meter. How much air flows through the meter and how much flows through the bypass depends on factors including:

- How restrictive the meter is at different loads
- How big the bypass is
- The way the plumbing is arranged (eg bends)

But (and this is a really important point): in all cases, the more air flowing into the engine, the more air that will always flow through the meter! So all that we need to do is to recalibrate the meter so that the flow through the meter accurately reflects the flow through the whole system – ie the total flow through both the bypass and the meter. (As you can see, this system is quite like those airflow meters that internally measure flow through one passage and then work it out for the total flow.)

3. ENGINE MANAGEMENT INTERCEPTORS

RECALIBRATING THE AIRFLOW METER OUTPUT

Those of you who have ever left a hose-clamp loose between the airflow meter and the engine will know how badly the engine then ran – so what am I talking about when I say you can have a big bypass feeding air around the airflow meter?

The trick is in the recalibration – the airflow meter will have a lower output at all loads, one that needs to be boosted until it matches what the ECU expects to see (or at least, one that results in a suitable air/fuel ratio). However, that's not all: the airflow meter voltage needs to be increased disproportionately at higher loads. This is because as the flow through the system increases, more and more air will travel through the low restriction bypass rather than through the airflow meter. (It's like two roads, one that has lots of speed humps and the other that is smooth and flat. As the traffic increases, more and more traffic will head through the smooth road as traffic bunches up over those speed humps.)

Let's look now at doing an airflow meter bypass, step-by-step. The mixtures were tuned with the DFA kit-based 1D interceptor.

THE CAR AND INTAKE SYSTEM

This Nissan Maxima runs the VG20ET 2-litre V6 turbo engine. During its life, the car was fitted with an intercooler, intercooler water spray, a blow-off valve, a new exhaust and a new modified airfilter box. Despite its elderly-for-the-time lines, I thought it a great car.

So why still the 20 inches of pressure drop? Following the airbox was a hot-wire airflow meter that in turn connected to a large resonant chamber (top left). From the chamber, a long duct connected to the turbo intake. The pressure drop still being recorded comprised the flow losses through the airflow meter, through the resonant chamber, and along the intake pipe from the chamber to the turbo. (The pressure drop through the flat panel filter? Not measurable – the pressure drop with the filter in place or removed remained the same, as is so often the case.) Given that the resonant chamber was nicely built inside (the air has to do a U-turn but the exit duct is equipped with a bellmouth and the insides look good), and that the pipe between this chamber and the turbo has no sharp bends, it seemed likely to me that much of the restriction was in the airflow meter.

The airbox had already been modified, with the lid cut away so air could reach the filter across its full area. A thick foam rubber strip then sealed this assembly to a bonnet opening, so that only cold air could be inhaled. Making this simple modification reduced the total intake system peak load pressure drop from 30 inches of water to 20 inches of water. (With pressure drop measurements, the lower the numbers, the less restriction to flow. This figure was recorded just in front of the turbo.)

At this stage I had never actually removed the airflow meter (it was fairly hard to reach) but since the engine power in standard form is only about 110kW (148hp), I figured that it was probably a small one. I sourced a new airflow meter from a car equipped with a 3-litre Nissan six-cylinder engine developing 150kW (200hp), and then used a workshop manual to work out which pin on the new airflow meter did what. (As mentioned above, even for an airflow meter from a known car, this often isn't as simple as it first appears.) I then pulled out the Maxima's airflow meter – only to find that it was the same diameter as the one I'd bought to replace it – oops! And furthermore, the replacement airflow meter looked as if it may well flow less, as it had many alloy heatsink fins projecting into the airstream. Time for a rethink. Getting hold of a larger hot-wire meter was out of the question – the budget would have been blown out of the water, and plus, I couldn't afford the time to go from place to place trying to find a larger meter for which I could get the pin-outs. That's when the bypass approach was adopted.

WORKSHOP PRO — MODIFYING THE ELECTRONICS OF MODERN CLASSIC CARS

THE BYPASS LAYOUT

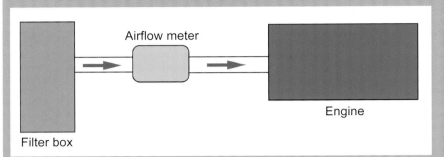

So how to organise an airflow meter bypass? This shows the standard system – the airfilter box feeds the airflow meter, then in turn connects to the engine's intake manifold.

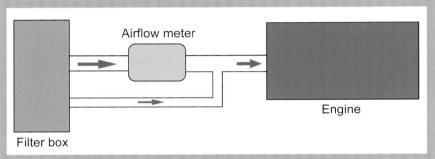

The first approach I considered was to run a second tube from the airfilter box to the intake after the airflow meter. As shown here, that tube doesn't have to be the same size as the original – it depends on how much air you want to bypass the airflow meter.

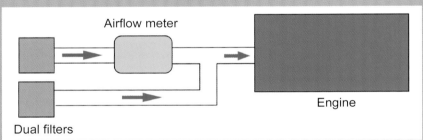

However, I finally decided to get rid of the airbox completely and instead use two new filters – one on the main intake and the other on the bypass. As can be seen here, the bypass was sized the same as the main intake duct.

I bought two oiled cotton pod filters and a ~75mm (3in) plastic flange. I then used a hole-saw to cut a hole in a relatively flat wall of the resonant intake box, and did a little panel beating to give a surface that the flange could bolt up against.

To make sure that the seal was good, I cut a gasket out of sheet rubber and sandwiched this between the flange and the box (yellow arrow). The bolts holding the flange in place were inserted from within the box and then doubled nuts (tightened against one another) were used to lower the likelihood of a bolt coming loose and then floating through the turbo compressor. These nuts are shown by the red arrow.

3. ENGINE MANAGEMENT INTERCEPTORS

With the box back in place, the approach can be more clearly seen. The top opening is the new one created for the bypass, while the lower opening is the original that connects to the airflow meter.

Some short lengths of thick-walled plastic pipe were then used to join the flanges to the two pod filters – one on the intake flowing to the airflow meter, and the other on the bypass.

VANE AIRFLOW METERS?

I've discussed hot-wire airflow meters in this section, because that's the type that I fitted the bypass to in the prototype installation. However, I can't think of any reason why the technique wouldn't also work with vane airflow meters – it's just that I haven't tried it.

However, note that when a vane airflow meter is fully open (ie at maximum load), the restriction of this type of airflow meter is actually *very low*. It is best to measure the cross-sectional area of the airflow meter and compare that with the rest of the intake plumbing. If the cross-sectional area of the vane airflow meter is much smaller than the rest of the plumbing, only then do a pressure drop test.

The idea that 'of course' a vane airflow meter is restrictive is not matched by my testing.

TUNING

As you would expect with such a large bypass, the Maxima wouldn't start or run without the DFA 1-D interceptor in place. The interceptor was wired-in at the ECU, with the wire coming from the airflow meter connecting to the 'in' terminal and the wire from the ECU connecting to the 'out' terminal. With power and earth supplied, the electrical installation of the interceptor was then finished – wiring such an interceptor is quick and easy. At this stage, the oxygen sensor was also disconnected so the tuning wasn't confused by the ECU's learning behaviour.

Cranking of the engine showed on the interceptor controller the load sites being output by the airflow meter. I increased the correction at these and all surrounding load sites – but I initially increased these outputs by far too much. I had thought that a massive correction would be needed at all loads, but, as I subsequently found out, only very small corrections were needed near idle. In fact, with the interceptor set to its coarse mode of operation, a +6 correction was all that was needed to have the car idling happily. (For those who are interested in more numbers, measurement showed that the ECU needed a 2.8V signal from the ECU to run properly at idle, and the actual bypassed airflow meter output was 2.4V. The +6 correction brought the airflow meter voltage up from 2.4 to 2.8V.)

With the car idling happily, I plugged in the same +6 correction at higher load sites – that is, I worked ahead as much as possible, putting in figures based on the corrections being used at the lower value load sites. I also attached a MoTeC air/fuel ratio meter to the exhaust so that I could see exactly what I was doing. At this stage, the interceptor map looked something like this – the car idled at load site #30 but I put in the correction for load sites 25-29 for better starting and idling behaviour.

Load site	Adjustment
25	5
26	5
27	5
28	5
29	6
30	6
31	6
32	7

With this much tuning done, I decided to hit the road. My driveway was very steep, and it took me three attempts to get up it – the air/fuel ratio varying between 10:1 and 18:1! However, some more load site adjustment of sites between 33 and 38 (shown below) gave good light load driveability. At these loads I was aiming at an air/fuel ratio of around the mid-14s (eg 14.3-14.9:1), the near-stoichiometric ratio used for best emissions performance.

Load site	Adjustment
33	7
34	7
35	7
36	8
37	9
38	9

The next loads involved going lightly into boost, where I set the interceptor to give air/fuel ratios in the mid-13s. Note how the amount of correction needed to the airflow meter output is increasing with load as more and more intake air takes the bypass route (table below).

Load site	Adjustment
39	9
40	9
41	10
42	10
43	10
44	11

Above load site 45 (see below) the engine was on substantial boost, with the peak load site being 58. (Always put in numbers above the maximum load site, in case the engine is in a situation where it develops more power than when being tuned – eg on a very cold day.) At these loads, I set the interceptor to provide an air/fuel ratio going into the mid-12s, and then progressing into the high-11s at absolute peak load.

Load site	Adjustment
45	12
46	12
47	12
48	12
49	13
50	15
51	15
52	15
53	16
54	16
55	16
56	16
57	16
58	16
59	16
60	16
61	16
62	16

FLOW TESTING

The bypass and interceptor were fitted in order to reduce intake flow restriction – so how well did it all work? Testing needed to take into account that there have been two major changes here – the replacement of the stock air-filter box with the two pod filters, and the use of the bypass. To (mostly) separate the effects of these two changes, two tests were undertaken.

Firstly, the bypass was completely blocked and the interceptor tune returned to normal. (Just removing power to the interceptor did this – it was then bypassed.) This test was designed to show the gains made by the removal of the airbox and the fitting of the pod filter to the end of the airflow meter. However, full load testing showed that these changes had *made no difference at all to the intake system flow restriction!* It still remained at a measured 20 inches of water.

The bypass was then re-opened, the interceptor switched back on and the testing done again. This time, the total intake system restriction had dropped to just 10 inches of water – the use of the airflow meter bypass and extra pod filter had halved the total intake restriction! Furthermore, the maximum output voltage of the airflow meter was then well down from its ceiling voltage – the airflow meter was now probably upsized to a 260kW (350hp) design …

Remember that the total measured restriction includes that caused by the U-turn the air needed to take through the resonant box, and the flow down the resonant box to the turbo feed pipe. It's likely that the restriction of just the combined filters, bypass and airflow meter was only 2-3 inches of water.

The use of a bypass in conjunction with a low-cost interceptor is a cheap and very effective way of upgrading the airflow meter capability, and at the same time giving control over air/fuel ratios. In this case, I chose to use a very large bypass, but a smaller one could be used instead, which would allow the airflow meter to work across a broader proportion of its original range.

COMMERCIAL INTERCEPTORS

All interceptors alter the signals that the ECU sees – and in response, the ECU changes in its output behaviour. That is the same as we saw occurring when making simple changes with resistors, pots and the 1-D interceptor, but commercial interceptors are much more sophisticated in that they can:
1. map individual load sites (that is, change outputs at specific combinations of rpm and MAP)
2. change frequency signals as well as analog voltages, giving control over ignition timing and not just fuelling
3. usually also control turbo boost and some other factors

However, *all* interceptors have major limitations, especially when the engine's mechanical modifications are large. One problem is that changing a single input variable (eg sensed engine load) may cause the ECU to alter more than one of its outputs. For example, as well as making the mixtures leaner, with a reduced load signal the ECU will probably advance the timing (lighter loads require more timing advance). That's fine if you are leaning-out mid-range mixtures (where you'd normally advance the timing anyway), but not so good if you are leaning out full load mixtures (where you would *not* normally advance timing!). Also, changing the timing of the crank-angle sensor may not only alter ignition timing, it may also (in cars with variable cam timing) change cam timing!

With an interceptor, you cannot alter rev limiters, adjust commanded idle speed, alter knock sensor sensitivity, or easily change electronic throttle control. Finally, it is not usually possible to alter air/fuel ratios when the system is operating in closed-loop – that is, the input of the oxygen sensor is being used by the ECU, usually (but not always) to keep the mixtures around 14.7:1.

MAKING A NEW AIRFLOW METER
If you don't want – or have room for – an airflow meter bypass, you may instead want to make your own upsized airflow meter. Here's how I did that. The mixtures were again successfully tuned with the DFA kit-based 1-D interceptor.

Here was the starting point – a Toyota airflow meter which is integrated into the airfilter box.

In this design, the sensing element can be removed by undoing two screws. Note the rubber O-ring (arrowed) that is used to seal the assembly into its mounting tube.

(continues overleaf)

WORKSHOP PRO MODIFYING THE ELECTRONICS OF MODERN CLASSIC CARS

An accurate measurement of the internal diameter of the airflow meter shows it's 45.2mm (1.8in), giving a cross-sectional area of 1600mm² (2.5in²).

The cut T-piece is heated with a heat gun until it becomes pliable and is then placed on top of the thick-walled joiner section ('A'). The two pieces are pushed together in a vice so that the lower, cut section of the T-piece opens-out to match the external diameter of the joiner piece. A socket placed in the vertical arm of the T-piece keeps its diameter intact while the plastic is soft.

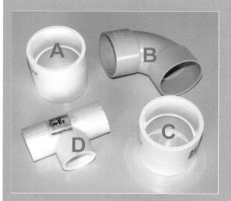

The next step is to buy some plastic plumbing. A mixture of plumbing types was used in this design: (A): a thick-walled pressure plumbing joining section to form the main body of the new airflow meter; (B): 50mm storm water bend (the flared 'socket' ends are cut off and used to sleeve down the main body); (C): a second thick-walled pressure plumbing joining section, used to form the mounting flange; (D): 26 x 25mm thick-walled T-piece, used to form the airflow meter sensing element support tube. You'll also need PVC pipe glue, a can of black spray-paint and some sandpaper of different grades. The way these bits and pieces come together will become clearer in a moment!

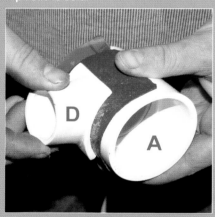

The process is difficult to see in the previous photo, but this is what you want the result to look like. Here, the T-piece ('D') is being rubbed on some fine sandpaper wrapped around the joiner piece of pipe ('A') to give the T-piece mating section its final shape, allowing it to nestle perfectly on the larger pipe.

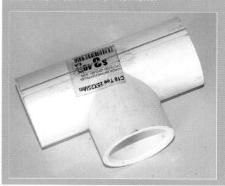

The 26 x 25mm T-piece (piece 'D' in the picture above) is cut with a hacksaw along the red line. A similar cut is made symmetrically on the other side of this section of tube to form an open U, so the lower section of the T-piece can sit on top of another tube.

To provide a mounting flange for the sensing element, a section of the wall of the spare joiner pipe ('C') is cut out, heated and then flattened by being compressed between two pieces of wood. A hole is cut in the middle of this section, and then it's glued (using the PVC pipe adhesive) to the top arm of the T-piece. I used a heavy hammer to hold the assembly in place while the glue set. When shaping this mounting plate, remember to leave clearance to the wiring plug.

3. ENGINE MANAGEMENT INTERCEPTORS

Next, prepare the main body of the airflow meter. The joiner pipe ('A') is shortened at each end with a hacksaw – if electrical tape is wrapped around the pipe to form a guide, it makes it a lot easier to cut it 'square.' (It's shortened at both ends rather than just one, because there's a small flange inside the joiner piece that pipes butt up against, and this should remain in the middle of the assembly.)

Rubbing the ends in a circular motion on smooth concrete will allow you to perfectly square them up. Once they're square, the rough edge can be smoothed with sandpaper.

Because the joiner pipe (fast becoming the main body of the airflow meter!) is larger in diameter than the ~50mm internal diameter wanted in this application, it needed to be sleeved down. When cut off, the socket end of standard 50mm stormwater fittings is a tight squeeze inside the pressure pipe fitting – these were cut off 'B', sanded down a little in diameter, then pushed home, one from each side. Once in place, they can be cut off flush.

The next step is to cut a hole in the side of the main body of the airflow meter – this is where the sensor element will protrude through into the airstream. This cutting can be done with a hole-saw and then final adjustments made with a round file.

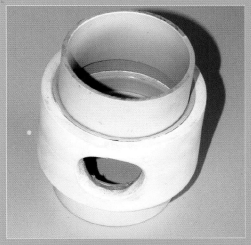

Here's the main body of the airflow meter. Two new sections of standard 50mm stormwater tube have been pushed into each end of the assembly to provide the plumbing connections to the airflow meter.

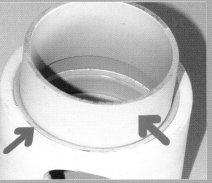

In close-up, the cut-off socket-ends from the stormwater pipe fitting can be seen sleeving down the internal diameter (red arrow) and the pushed-in sections of 50mm stormwater pipe provide the plumbing connections (green arrow). These pipe sections are a very tight fit, so pipe glue is unnecessary.

(continues overleaf)

WORKSHOP PRO: MODIFYING THE ELECTRONICS OF MODERN CLASSIC CARS

A check of the internal diameter of the main body of the airflow meter shows a diameter of 51.8mm (2.0in), giving an increase in cross-sectional area of 31 per cent over standard.

Place the T-piece over the hole cut in the main body, and then glue them together with the pipe adhesive. Place the airflow meter sensing element in the housing to help align the two sections. Hold it in a vice while the glue is setting (give it 24 hours).

The O-ring, that sealed the sensing assembly in place on the original airflow meter, can be used again to perform the same function. If it doesn't fit in the upper arm of the T-piece, thicker O-rings are available. If it's a little tight, the inside diameter of the mount (arrowed) can be filed or sanded away. The airflow meter sensing element is held in place with screws and nuts.

Once the new airflow meter is finished, it can be sanded smooth, using wet-and-dry paper used wet. An excellent finish is possible if you're patient and do lots of sanding.

The airflow meter body can then be painted using a spray-can. Plastic plumbing of the sort shown here is made from a thermoplastic called PVC (polyvinyl chloride). Its maximum continuous operating temperature is 80°C (176°F). In an underbonnet environment, temperatures can exceed this level. This is especially the case near exhaust pipes and turbos, directly behind the radiator and close to the engine block. Therefore, the approach shown here should not be used where the airflow meter is located in a hot area of the engine bay. Furthermore, heat-shielding of the airflow meter is a good idea (as it is in all applications where the airflow meter would otherwise be exposed to excessive heat). All that said, the plastic airflow meter was durable in use.

So with all these apparent disadvantages, why use an interceptor?

In short, when the factory engine management software is not 'cracked,' using an interceptor is an easy and effective approach. For example, in the Honda Insight, as described in the next section, there was no tuning software available for the factory engine management system. Replacing the original ECU with an aftermarket programmable ECU would immediately cause problems running the digital dash; easiest in this case is to use an interceptor. (Or you replace the dash as well, which I later did when I went to full programmable engine management.)

Fitting & tuning an XEDE interceptor

ChipTorque's XEDE interceptor is the example I'll be using here. It was first released way back in 2002 – it's a mature and well-developed product. (Incidentally, the name is pronounced 'exceed.')

The tuning software for the XEDE is freely available, and the interceptor is available with two high current outputs to allow it to drive an extra injector, boost control solenoid, etc. Optional extras include:

The XEDE interceptor has freely available programming software. It can alter fuel, timing, boost – and lots of other parameters.

- a bypass plug that allows the car to run should the XEDE be unplugged

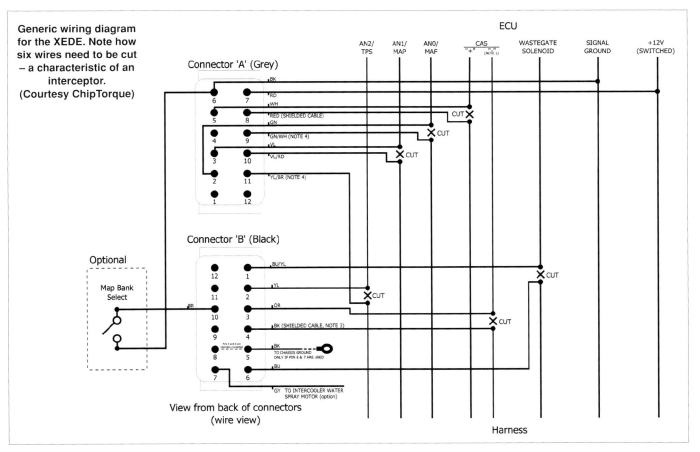

Generic wiring diagram for the XEDE. Note how six wires need to be cut – a characteristic of an interceptor. (Courtesy ChipTorque)

WORKSHOP PRO: MODIFYING THE ELECTRONICS OF MODERN CLASSIC CARS

- a map changeover switch that allows the selection of two different tuning maps
- an extra high current controllable output

The XEDE is normally pre-configured to suit your car's specific sensors, and has a base tuning map loaded. The latter is for modified cars – the map attempts to take the specific modifications into account and represents a starting point for further tuning. The XEDE is laptop programmable. Unless you are using a very old laptop that has a serial port, to connect to the XEDE you will need a serial to USB converter, and possibly an extension cable.

The Honda to which the XEDE was being fitted is the 2000-model hybrid petrol/electric car that we've met previously. In this particular car, the high voltage battery had died. It was not worthwhile replacing this battery – but, unusually for a hybrid, the Honda could still function without the hybrid system working. However, in this condition the Check Engine and IMA (Integrated Motor Assist) dash warnings were illuminated, and the rev limit dropped by about 1000rpm.

Without the electric assist, low-rpm torque is markedly reduced, and power is reduced throughout the rev range. The plan was to remove the oxygen sensor feedback loop, and then use the XEDE to tune the air/fuel ratios and ignition timing through the whole load and rev range, aiming to achieve optimal performance and fuel economy without the hybrid system working.

Fitting the XEDE
If you are fitting the XEDE in an unusual application, ChipTorque will probably need a scope trace of the crank-angle sensor output, so that the hardware and software of the XEDE can be configured to suit. (If the car model is one commonly modified, ChipTorque will already have this required information.)

In the case of the three-cylinder Honda Insight, I used a Fluke Scopemeter to capture the waveform from the crank-angle sensor and two camshaft position sensors, directly photographing the scope screens and sending the pics to ChipTorque. The company then configured the XEDE to suit.

Before you can fit the XEDE (or any other interceptor)

The modifications were made to a 2000 model Honda Insight. (Courtesy Honda)

you must have access to a good wiring diagram showing the inputs and outputs of your car's ECU. At minimum these need to include the load sensor input (airflow meter or MAP sensor), crank-angle and throttle position sensors. If you wish to control turbo boost, you'll also need access to the wastegate wiring.

I have the full factory manuals for the Honda Insight and used their very clear diagrams to locate the correct ECU wires. Connections were made to the crank-angle sensor, throttle position sensor and MAP sensor. Note that the interceptor actually *intercepts*: each of these wires is cut and the signal is routed into and out of the XEDE. In addition to these connections, you'll also need switched 12V and sensor ground – both able to picked-up directly from the ECU.

If you have purchased the optional bypass plug, after wiring the XEDE harness into place you should be able to connect the bypass plug and then start the car – it should run as it did before your wiring additions. If the car runs differently, check your wiring!

Making the PC connection
With the XEDE wired into the car and the car able to run properly (or at least run as you would expect at this stage), connect the laptop. If using a USB-to-serial adapter, load the software provided with the adapter. Plug the cable into the XEDE harness, open the XEDE tuning software (called XMap) and then turn on the ignition.

A 'connected' or 'not connected' line will show at the

base of the XEDE XMap software window. If the laptop is not connected to the XEDE, try different Comm Ports as available through the XMap Set-Up menu.

For tuning I initially used a small notebook PC that does not have a disc drive. I transferred both the USB adapter and XMap software from a PC to the notebook via a USB memory stick. While the notebook worked fine initially, I subsequently found that lots of communication errors were occurring. I tried a different USB-serial converter but the problem remained. It was only by changing to a different laptop was I able to fix the problem.

Checking

To do any tuning you will need to have a wideband air/fuel ratio meter monitoring exhaust gas. With the meter working, and the car idling, open the XMap fuel map and alter a value. You know where the engine is operating on this map by the coloured yellow boxes (these move as you blip the throttle); simply insert higher values into these boxes. You should then see on the meter the air/fuel ratio change in response. If the car is in closed loop, this change will be rapidly 'learned around' by the ECU. Alter the values up and down at the current load site and confirm that you can change mixtures (even if only for short time) in both directions.

Next, alter the current value of ignition timing – ignition timing is on another map. When changing timing you should hear the engine's idle speed alter – eg a small advance will normally increase idle speed. Again, this change may be quickly learned around by the ECU, returning engine idle speed back to standard.

Also check that the engine speed readout on the XMap software matches the actual engine speed.

The point of doing these things is to ensure that the interceptor is connected correctly and is working. (I once watched someone spend at least an hour on a dyno, tuning a car fitted with a new cam. It took that long to realise that the ignition timing wasn't changing on the basis of the tuner's laptop inputs …)

With everything wired into place and working, now you're ready to start tuning.

The plan

The plan with the Honda was to remove the oxygen sensor feedback loop, and then use the XEDE to tune the air/fuel ratios and ignition timing through the whole load and rev range, aiming to achieve optimal performance and fuel economy with the hybrid system inactive.

After confirming that the air/fuel ratios and ignition timing could be altered (that is, the XEDE was connected and working properly), the factory ECU oxygen sensor signals were disconnected. So why disconnect these sensors? The Honda runs a factory wideband sensor before the cat and a narrow band sensor after the cat. Tuning the air/fuel ratios with this feedback loop in place would have been problematic; the ECU would have learned around these changes at all loads except full throttle/high rpm (where the ECU reverts to open-loop operation).

The signal from the wideband cat could have been intercepted and altered, however my intention was to radically change the mixtures, and I thought that the ECU might have well spotted that the air/fuel ratios were no longer right, and made corrective changes, even with the oxygen sensor being intercepted (eg by the ECU comparing at certain loads the expected air/fuel ratios with the sensed air/fuel ratios).

Note that while an OBD analysis would clearly spot the disconnected oxygen signals (and also that the hybrid system wasn't working), a roadside sniffer would likely still pass the car. That has been my actual experience, where a well-tuned car with no oxygen sensor feedback passed a random roadside emissions test without problems.

Testing

The first step was to drive the car with the hybrid system not working and see what the air/fuel ratios were in this form. Using a MoTeC air/fuel ratio meter, it could be seen that in all conditions except full throttle, the car ran an air/fuel ratio of about 14.7:1, moving at full throttle to about

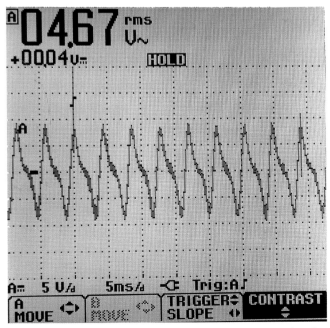

The output waveform of the crank sensor was captured and emailed to ChipTorque so that the XEDE could be configured to suit.

WORKSHOP PRO: MODIFYING THE ELECTRONICS OF MODERN CLASSIC CARS

The bypass connectors. By using these, it's easy to bypass the interceptor so the engine runs as standard.

12:1. These mixtures were much the same, as the car ran as a working hybrid; the only clear exception is the lack of factory lean cruise, where the air/fuel ratios could run as lean as 25:1!

On a fairly demanding open-road test section, the non-hybrid Honda produced a fuel economy of about 4.2 litres/100km; that compares with about 3.8 litres/100km it achieved over the same stretch as a working hybrid. (Those figures are both far worse than the 2.8 litres/100km I have previously gained with the working hybrid on a flat freeway – as I said, this stretch of country road is fairly demanding.) So away from city confines, on a hilly country road, the fuel economy without the hybrid system working was about 11 per cent poorer than with it functional.

The next step was to disconnect the oxygen sensor feedbacks. In this condition the air/fuel ratios around idle and at low loads remained at about 14.7:1, they were a bit richer at moderate loads (say 13.5:1) and at high load and full throttle reverted to mid-12s. In other words, as you'd expect with a well programmed management system, the car was still safely driveable even without the oxygen sensor feedback.

Tuning

As previously indicated, the XEDE has control in this car over the air/fuel ratios (via interception of the MAP sensor) and the ignition timing (via interception of the crank-angle position sensor).

I'll start with the fuelling. The default map for the XEDE in this configuration is to place engine rpm on the bottom axis and load (as determined by the MAP sensor) on the vertical axis. At each load 'site' (ie each combination of manifold pressure and engine speed) there is a number that can be altered up or down, so changing the MAP sensor signal seen by the ECU (and therefore the fuelling) at that spot.

The numbers used on the axes of the tables in XMap can be altered as you wish. That's very important, so let's take a look at that idea. At its simplest, being able to set the scaling of the axes lets you use the full axis lengths. For example, if your engine redlines at 7000rpm, you can set the rpm labels on this axis to 1000, 2000, 3000, 4000, 5000, 6000 and 7000. On the other hand, if it redlines at 5000rpm, you can alter the axis to finish at 5000rpm.

You can set the MAP values so that throttle lift (ie min MAP value) and full throttle (max MAP value) occur at respective ends of the axis. Since these values will vary depending on the car and the sensor, being able to set them on the axis works well.

However, even better, you can expand the axis scales at the points of greatest use. For example, if you find you need greater tuning resolution at around 2000rpm (say when a turbo engine is coming on boost) you can use finer rpm divisions of 1000, 1500, 2000, 2500 and then for the rest of the rev range revert back to coarser divisions of 3000, 4000, 5000, 6000 and 7000. The maximum number of divisions you can have on each axis is 20, giving a total potential of 400 tuning points on each map. (Note also that the system interpolates between adjacent points.)

In the case of the Honda, I watched the fuel tuning map while the car was being driven, seeing by the location of the yellow highlighted active spots where the greatest tuning resolution would be needed – and so where it would be desirable to insert more divisions.

But what was the outcome I was chasing? What I wanted was an idle air/fuel ratio of about 14.7:1, a full throttle air/fuel ratio of about 12:1, and light load cruise as lean as I could make it. By 'light load cruise' I mean the air/fuel ratio that occurs at the sort of loads found when driving at a constant speed in the 60-100km/h (about 35-60mph) range – especially on flat roads or when travelling slightly downhill.

In a conventional car, the combination of MAP and throttle values for these diverse driving conditions (ie idle, high load and cruise) would all be markedly different, so allowing the tuning of the different modes to easily occur. In other words, to set lean cruise: drive along in light-throttle cruise, look at the sites being accessed, lean out the mixtures at those sites, and there you have it. However, because the Honda has so little power (especially without its electric motor assist), cruising at 100km/h (~60mph) is uncomfortably close in load to that achieved when accelerating moderately hard through the

3. ENGINE MANAGEMENT INTERCEPTORS

A MoTeC air/fuel ratio meter was used during all tuning of the interceptor. Don't try to tune without seeing what you are doing!

gears. Therefore, lean out the mixtures that occur at the MAP/rpm points that correspond to 100km/h (~60mph) cruise, and the car then develops the staggers when it accelerates through this point when driving round town!

This is just one example – in driving reality I found lots of difficulties where the map values of different driving conditions unfortunately coincided. In fact, after a few hours of tuning based on MAP/rpm I realised that achieving my goals would not be possible with the XEDE configured in this form. So was that the end – failure? With many interceptors that would be the case, but the XEDE is much more flexible than that.

Rather than map rpm versus manifold pressure, I decided to map rpm versus throttle position. This approach is often carried out in aftermarket programmable ECUs – the outputs of the map then control injector opening times. But in my case, as with most interceptor applications, you don't have direct control over injector opening times. Instead, you have control over what load the ECU *thinks* is occurring – via the intercepted MAP sensor or airflow meter signal. So is it possible to use the XEDE to map throttle position versus rpm – and then to output values that alter the MAP sensor readings that the ECU sees? The short answer is: yes!

Throttle position vs rpm

In the 'Edit this map' facility (achieved by right-clicking on the table) is an option for selecting the Load Variable. ('Load Variable' is better thought of as 'vertical axis variable' – it doesn't actually have to be based on load.) The variables are labelled as shown on the wiring diagram that comes with the XEDE – so in my case, TPS (throttle position) is labelled as AN2. To make throttle position the vertical axis variable, just select 'AN2 In' in the Load Variable box. Leave the Input Variable as 'AN1'

(that's MAP) and the Output Variable as 'AN1 Out' (that's intercepted MAP).

In this case it was easy to check that load was now based on throttle position by turning on the ignition but not starting the engine, then moving the throttle and seeing the yellow (active) boxes move up the vertical axis.

With this change in mapping approach, it was time to again hit the road. This time the way the map operated was far more 'logical' for this car and engine: at any rpm, putting your foot right down could be set to give rich mixtures, idle could be set for stoichiometric (ie 14.7:1) and light load cruise could be leaned right out.

Mixtures

The Honda Insight uses idle and 'normal use' mixtures of 14.7:1, rich mixtures (eg 11.8:1) at high load and, most unusually, mixtures as lean as 25:1 in short-burst cruising conditions. Most engines will not even run on mixtures as lean as 20:1, so the Honda engine is rather special. In the standard car such lean mixtures are used for only short periods. This is because with these mixtures, NO_x emissions rise; the Honda uses a special cat converter to adsorb them and then, through running slightly richer, get rid of them. Presumably, the cat holding capacity dictates the length of time the car can run in lean cruise.

But with the hybrid system not working, the oxygen sensor feedback disconnected, and no requirement to pass a full emissions test, I was in a position to run whatever mixtures were desired. The best option seemed to be 14:7:1 at idle, the standard ~12:1 at high loads, and as lean as possible in very light load cruising.

As recounted earlier, with the oxygen sensors disconnected and the hybrid system switched off (and so the car registering fault codes), the Honda runs air/fuel ratios around idle and at low loads of 14.7:1, at moderate loads mixtures a bit richer (say 13.5:1), and at high load and full throttle reverts to mid-12s. Therefore, with the exception of the cruise area where I desired much leaner mixtures, the air/fuel ratios were largely correct.

So how lean could I go? Using the XEDE, the load sites corresponding to light-load cruise were leaned-out. Note that the engine in the Honda is so low in power that 'light load cruise' involved throttle positions of up to 50 per cent and revs of up to 4000rpm! The greatest leaning-out was made at load sites comprising around 15-20 per cent throttle and at less than 2000rpm. At these sites the car happily ran on air/fuel ratios of about 21:1. Mapping these mixtures was largely achieved on the road in less than 30 minutes; fine-tuning to cater for as many driving conditions as I could think of took a few more hours.

Timing

Revising the ignition timing was more difficult, primarily

LOOKING AT THE TABLES

So let's see how it was working. Shown below is the fuel map. The vertical axis shows throttle position, increasing as you go upwards. The bottom axis shows engine rpm, increasing to the right.

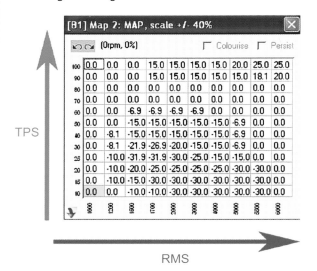

Note how the chosen axis scaling for the TPS is 10, 15, 20, 25, 30, 40, 50, 60, 70, 80, 90 and 100 per cent – more resolution has been placed at small throttle openings. On the rpm scale these divisions are used: 1000, 1200, 1500, 1700, 2000, 3000, 4000, 5000, 5500, 6000 – again more resolution at low revs.

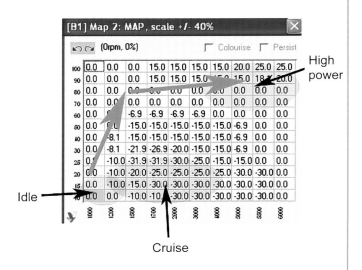

The orange arrow (above) reflects the way that the car is driven when accelerating briskly (well, what passes for brisk in this car!). You can see that the throttle is at 70 or 80 per cent before the rpm has risen above 1200, so the fuel map stays in the unmodified area (and at high revs passes into an enriched area).

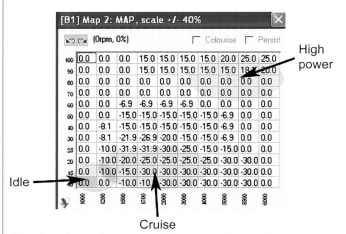

The view above shows three different driving conditions – idle, cruise and high power. Note that high power occurs at both high throttle angles and high revs; idle at low throttle angles and low rpm; and cruise at moderate rpm and low throttle angles. This differentiation of the engine operating conditions allows the mixtures to be tuned separately for the different conditions. Note the proliferation of negative numbers in and above the 'cruise' highlight – this is where fuel has been pulled out of the system.

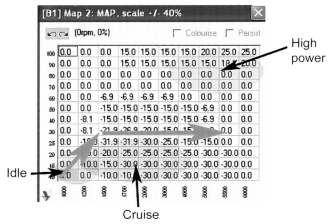

This orange arrow shows slower acceleration – the throttle has been opened a maximum of only 30 per cent, keeping the system operating in the area where a lot of fuel has been pulled out. That gives sluggish but economical performance. Therefore, with the mapping that has been undertaken, if you want to accelerate briskly, you put your foot down. It's easy to remember and feels quite natural.

3. ENGINE MANAGEMENT INTERCEPTORS

because I didn't have a real-time read-out of the timing being used.

As stated previously, intercepting and reducing the MAP sensor signal not only makes the car run leaner, it also advances the timing. In this application that is good, as leaner mixtures require more advanced timing. But was the timing being advanced enough? That can really only be assessed by further advancing the timing and seeing what happens – too much and the engine will detonate, not enough and the performance will be sluggish.

The XEDE allows only plus/minus 10 degrees of timing change to be made. That is usually enough, but in my situation, I found that adding 10 degrees improved lean load performance and didn't cause detonation. However, as that was the limit of what could be added, I can't tell if that was sufficient. Through most of the rest of the map I added 5 degrees.

In this situation it would be very good to have a readout of what the factory knock sensor was doing – for example, was timing being pulled out as the ECU detected incipient detonation? (Such a readout requires a dedicated tool that reads Honda data through the OBD port – at that time, these tools were expensive and rare, but not any more!)

However, even without being able to gain as much timing advance as may have been required, the added timing allowed the leanest of cruise mixtures to be leaned-out even further, with air/fuel ratios of 25:1 run at very light loads (eg 10 per cent throttle) at around 2000rpm.

So in conclusion, what to make of all these tuning changes? By using the interceptor, it proved quite possible to run extremely lean air/fuel ratios in some driving conditions – and to transition smoothly and effectively to much richer mixtures as demanded power and response increased. As a result, fuel economy was able to be improved – in fact, to the point of matching the open-road fuel economy achieved over a test loop when the car was working as a full hybrid (although not matching city fuel economy).

An interceptor of this sort is a good 'in-between' step for someone who wants to be able to lap-top tune their car, without having to program everything from scratch (as is often the case with a programmable management system). The majority of the original engine management system (and its tune) remains in place, and the small number of variables that can be changed – while sometimes a limiting factor in what you can achieve – also means that the learning curve is not too steep. After all, if something isn't running right, it's very easy to go back to the standard tune and start again. Wiring an interceptor into place is also not too hard, with only about dozen wires to connect (plus power and ground).

WORKSHOP PRO: MODIFYING THE ELECTRONICS OF MODERN CLASSIC CARS

Chapter 4

Programmable engine management

- Selecting a programmable management system
- Wiring programmable engine management
- A piggy-back engine management installation
- Tuning programmable engine management
- Tuning on the road
- Going beyond the normal in ECU mapping

WORKSHOP PRO: MODIFYING THE ELECTRONICS OF MODERN CLASSIC CARS

To many people, fitting programmable engine management is the holy grail – and in many ways, that's understandable. With programmable engine management, you can tune every aspect of the engine to perform as you wish. That includes not only fuelling and ignition timing, but also rev limit, idle speed, variable intake manifold changeover point, and variable valve timing (continuously adjustable, or two-step changeover). In addition, you can control turbo boost, water/air intercooler pump speed, the radiator fan, the air-conditioning compressor, and the alternator charge rate. Even if your car did not originally have electronic throttle control, you can add it.

With a good programmable management system, and with sensible mechanical modifications, you can have a car that drives as sweetly as one that is unmodified, but has tremendously more power than standard.

And if you're at the other end the performance spectrum and love hyper-miling, you can tune the engine to deliver outstanding fuel economy (mileage), including bespoke lean cruise modes and throttle-lift injector shut-off.

If you're like me, and have a boot in both camps, you can even fit a switch that allows you to change between 'economy' and 'power' engine tunes.

However, programmable engine management can also be very different to what I have described above. In fact, when I worked at an online magazine, I photographed about a hundred modified cars a year. At the photo shoot location – it might have been a boat ramp or deserted industrial estate – I'd be snapping the car in various positions. Inevitably, the car would need to be moved half a dozen times to get different angles. If, when the car needed to be moved, it had to be cranked and cranked to get it started, the journalist I was working with would say, "Ah, must have programmable management!". Why? Because getting cold- and hot-start right with a programmable management system might take you literally a whole year of tweaking – but most workshops spend a few minutes tuning the starting maps.

So while there are some brilliant aspects of programmable management, it's also the sort of approach where you (or a workshop) will need to put in a lot of effort to gain a car that drives even as well as standard.

SELECTING A PROGRAMMABLE MANAGEMENT SYSTEM

Selecting the correct programmable management system for your car is an important decision. The system that you pick will have a large bearing on how well the car runs – a bad system will remind you of its lack of ability in all driving conditions. Poor starting, poor driveability, poor economy, poor power – these are factors that you will notice every day!

An Adaptronic programmable ECU. A fully programmable ECU allows you to do almost anything you like with an engine. (Courtesy Adaptronic)

Part of a system
Good engine control units also comprise part of a broader system that might include digital dashes, smart LED indicators and CAN expansion boxes. This way, the system can grow with you. I strongly suggest that you buy a programmable engine management unit that is part of a larger 'eco-system' of compatible parts.

Legality
If legality is relevant, investigate this area very carefully. Some jurisdictions prohibit programmable ECUs in road cars, simply because once one has been fitted, there is no control over the emissions of the car. That is, it is easy to change emissions from those that may have been legal when the vehicle was tested, to those that are totally illegal! Other authorities allow the use of programmable ECUs, but only if the car undergoes a full emissions compliance test procedure. (Obviously the ECU maps are not supposed to then be changed.) Still other authorities require that the car has full original equipment OBD functionality.

In any case, if low emissions are a requirement, the programmable ECU that is selected will need to be able to work in closed loop, using the input of an oxygen sensor to keep mixtures close to stoichiometric during cruise and idle conditions. To pass emissions testing, the ECU will also need to be quite sophisticated in both its software and mapping.

Ease of installation
Is there software available that makes the standard engine management system in your car fully programmable? If so, the ease of installation and the sophistication of the finished system makes this a very attractive proposition.

If no system exists to make the factory ECU

4. PROGRAMMABLE ENGINE MANAGEMENT

When selecting a programmable ECU, it is advantageous if it is part of a complete system of parts that will easily connect to one another. (Courtesy Haltech)

programmable, is there a plug-in compatible system that uses the original loom, sensors and output actuators? Such a system has *huge* advantages in ease of installation, and the new ECU can be easily removed and replaced with the standard ECU if a return to standard is required (assuming that the engine modifications aren't too extreme).

If neither of these approaches is viable, is the selected ECU compatible with at least the standard engine position (crankshaft and camshaft) sensors? Interfacing with these sensors causes most of the sensor problems when installing a programmable ECU, so it is preferable that no modifications of the standard position sensors are required.

An electronic control unit that will plug straight into the factory wiring loom can potentially save many hours of work.

A programmable ECU that will plug straight into the standard wiring loom, and use the standard sensors, is vastly easier to install than one that needs custom wiring. (Courtesy Haltech)

Required features

What do you actually require of the programmable ECU? At minimum, full control of the fuel and ignition is needed. In a road car, good idle speed control (one that can cope with varying loads from the power steering and air-conditioning) is another requirement.

How many extra outputs are needed? For example, if you wish to switch on a gearshift light, actuate a variable length intake manifold and control turbo boost, you will require extra ECU outputs. These outputs fall into two types: those that are either on or off (eg suitable for the control of a dual-length intake manifold) and those that can be pulse width modulated (eg a boost control solenoid).

On a sophisticated car, all of the following control outputs may be required:
- turbo boost
- sequential turbo operating valves
- blow-off valve
- intake manifold
- camshaft timing
- radiator fan
- shift light
- intercooler water spray
- water injection
- active aerodynamics
- water/air intercooler pump
- carbon canister purge valve
- air-conditioning clutch
- electronic throttle control
- fuel pump speed

While this list may seem long, using the ECU to control many functions in a car has great advantages over using separate switches or controllers. The first of these advantages is that the sensors and associated wiring are

65

WORKSHOP PRO — MODIFYING THE ELECTRONICS OF MODERN CLASSIC CARS

already in place to allow the accurate determination of car operating conditions. Second, those functions that need to be controlled within strictly defined parameters can be accurately operated. For example, an intercooler fan can be switched on only when the inlet air temperature is above 45°C (113°F) and the vehicle speed is below 15km/h (9mph), or camshaft timing can be altered on the basis of engine speed and MAP pressure.

Do you require data logging? In a serious road car, this can be a useful addition. Can the ECU control the ignition system in the way that you wish to configure it – ie single coil or direct fire ignition? Will the ECU work with the original equipment coil igniters – especially important if they are built into the coils ('smart' coils)?

Does the ECU have self-diagnostics? Does it have an inbuilt scope so that you can see the pattern of crank and cam sensors? Especially helpful during setting up, self-diagnostics can save many hours when faults need to be located.

Does the ECU have CAN functionality, which will allow it to work with other devices (like digital dashes)? Are those CAN settings user-configurable?

Good-quality programmable electronic control units are available from many different manufacturers. I have had experience with MoTeC and Adaptronic, and I recommend both. Be wary of those companies that advertise their programmable management systems on the basis of enormous power outputs they achieve on drag or other race cars. Basically, it's not hard to control engines at full-load; it's *much* harder to give good driveability – the main requirement in a road car.

Service and back-up
Back-up and service are critical if you are installing and programming the ECU yourself. One way to assess the back-up is to look carefully through the software, manuals and videos before you buy the system. Manuals should be professionally produced – if the company can't even afford to put together good back-up materials, how good will their system be?

Irrespective of your experience, it's very likely that you will need to call on the ECU manufacturer or dealer for support during fitting and tuning: it's the nature of the complex beast. For this reason alone, I'd be reluctant to buy a product from a new company – unfair as that may seem, it would be a disaster to buy a programmable ECU and then find that two years down the track, there's no one to call on for support.

Good-quality programmable management systems will also be widely used across a range of different cars. If you see that the ECU manufacturer specialises in one type of car – and you have a different type – then there may be problems when you ask engine-specific questions, eg interfacing with a particular camshaft or crankshaft sensor.

WIRING PROGRAMMABLE ENGINE MANAGEMENT
Wiring engine management can be challenging and difficult. However, taken slowly, step by step, it is quite achievable – even if you do not have extensive experience in car wiring. And a reminder regarding what was said earlier – if you can buy a plug-in programmable engine management system (or one with an adaptor loom that connects straight to the original ECU plug), that will be a major time saver.

Approaches
A number of different wiring approaches can be taken. You can:
1. Use the new ECU loom to complete the entire wiring job. That is, you will need to fit plugs to the new loom that are compatible with the existing injectors, sensors, and so on. Electrically, this approach is easiest, but sourcing all the correct plugs is likely to be quite difficult – although typically not impossible.
2. Use the new loom and splice to the old loom close to the original plugs in the engine bay. This avoids having to source new plugs or terminals that match the existing hardware.
3. Use the new loom and splice to the old loom close to the original ECU. If the ECU is mounted in the cabin (the case in most cars of the era covered by this book), this approach takes the wiring joins away from the high-vibration environment of the engine bay and places them in the weather-proof cabin. However, it also makes 'sorting' the wiring more difficult.

My preference is to splice the new loom to the old, making these connections to the loom near the original ECU. It gives the neatest job (bar sourcing all new plugs and

Wiring programmable engine management is often a challenge – there are just so many wires! This part of the loom comes prefitted with injector plugs. (Courtesy Holley)

4. PROGRAMMABLE ENGINE MANAGEMENT

pins), but can get tricky if you don't have very good wiring diagrams.

Information
Before you can start the wiring job, you will need information about the:

- programmable ECU plug pin-outs
- programmable ECU wiring loom colours/labels
- original car's ECU plug pin-outs
- original car's engine management wiring diagram.

It is vital that you keep detailed and accurate records of the wiring – *as you do the wiring*.

EXAMPLE – WIRING THE THROTTLE POSITION SENSOR

It is difficult in the space available to describe the wiring of every sensor and output actuator. So, I'll look at a single example – the standard throttle position sensor being wired from a Honda to a MoTeC ECU. The process is much the same with all programmable ECUs and cars.

Step 1
Examine the standard car's wiring diagram for the throttle position sensor (TPS). It is shown below.

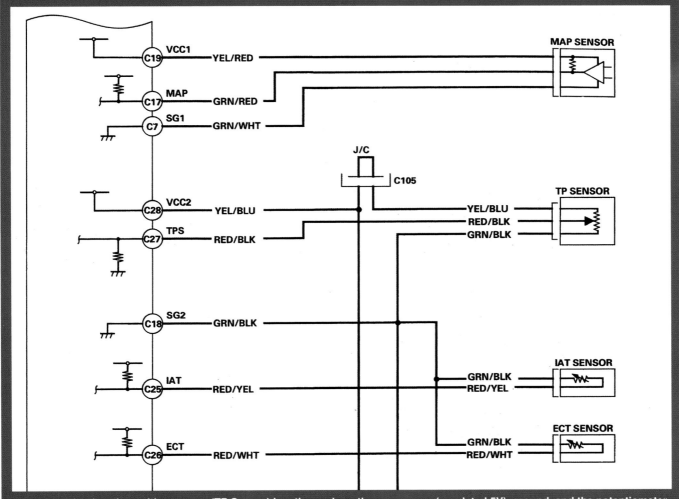

Note how the throttle position sensor (TP Sensor) has three wires: they are power (regulated 5V), ground, and the potentiometer wiper that moves when the TPS moves. The last is called the signal wire and is shown here as being red/black and connecting to ECU pin C27. Tracing the other wires from the TPS (and so moving off this extract) shows that yellow/blue is the 5V feed and green/black is ground. (Courtesy Honda)

WORKSHOP PRO — MODIFYING THE ELECTRONICS OF MODERN CLASSIC CARS

Step 2

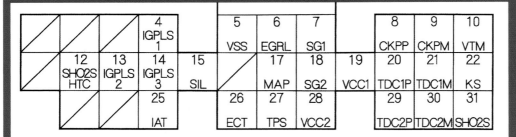

Wire side of female terminals

Check that the correct wiring colour codes are present on the correct ECU pins. Here we're looking for the TPS signal wire (red/black) on Pin 27 – TPS. At the same time, we'd also check the wiring colour codes and ECU pin-outs for the TPS 5V feed and TPS ground.
(Courtesy Honda)

Step 3

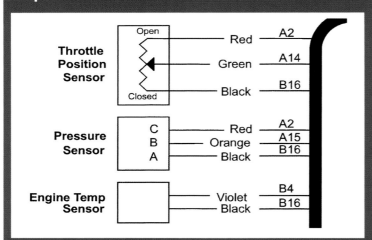

Now let's look at the diagram for the programmable ECU. From this we can see that the following loom colours apply:
Power – red
Ground – black
Signal – green

Therefore, summarising the wiring connections that need to be made:

TPS	Honda	MoTeC
Power	Yellow/blue – C28	Red – A2
Ground	Green/black – C18	Black – B16
Signal	Red/black – C27	Green – A14

Step 4

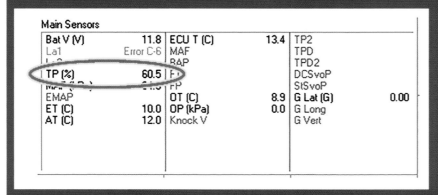

Make these connections, and then check that the new ECU can see the signal. That is, that when the throttle is depressed, the TPS signal changes appropriately. By far the best approach is to do one sensor or output actuator at a time, and then immediately test that it is working correctly. It is of course much quicker to make all the connections at once, but if something then doesn't work, fault-finding can be a nightmare.

4. PROGRAMMABLE ENGINE MANAGEMENT

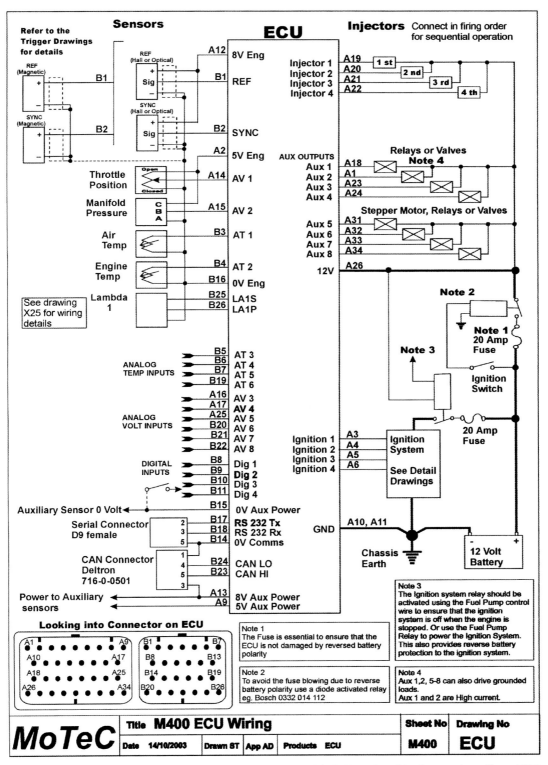

The wiring diagram for MoTeC M400 electronic control unit (ECU). Note that this diagram is really an ECU 'pin-outs' diagram – no wiring colour coding is shown. (Courtesy MoTeC)

WORKSHOP PRO: MODIFYING THE ELECTRONICS OF MODERN CLASSIC CARS

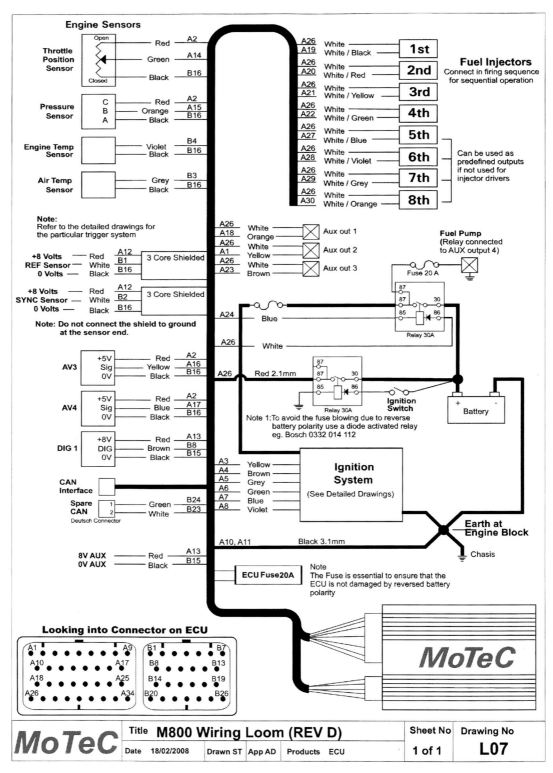

The wiring diagram for MoTeC 'Hundred series' wiring loom. To install the electronic control unit, this and the diagram on the previous page need to be used in conjunction with one another. (Courtesy MoTeC)

4. PROGRAMMABLE ENGINE MANAGEMENT

EXAMPLE RUNNING CHART OF WIRING CONNECTIONS AND NOTES

The following table shows an example of a 'running chart': in this case, for wiring a MoTeC M400 programmable electronic control unit (ECU) to a 2000 model Honda Insight. A similar chart should be constructed and maintained whenever an ECU is being fitted to a car. Without it, fault-finding and wiring changes become nightmares!

	MoTeC ECU channel	MoTeC M400	Honda Insight	Function	Notes
Main power feed ECU		A26 Red A10 A11 Black	B1 Yellow/black B2 Black	Ignition switched 12V + Main ECU ground	
Analog voltage inputs					
Throttle position sensor (TPS)	AV1	B16 Black A2 Red A14 Green	C18 Green/black C28 Yellow/blue C27 Red/black	Ground TPS TPS 5V TPS signal	
Manifold absolute pressure (MAP) sensor	AV2	A2 Red (C) A15 Orange (B) B16 Black (A)	C19 Yellow/red C17 Red/green C7 Green/white	MAP 5V MAP signal MAP ground	Modified Commodore plug used at sensor C = White B = Green A = Black
Boost control dash pot	AV3	A16	Green		Left pot
Ignition trim dash pot	AV4	A17	Green		Right pot
Analog temperature inputs (may also be switch inputs)					
Intake air temp (IAT)	AT1	B3 Grey B16 Black	C25 Red/yellow	IAT signal	Black of MoTeC loom to B16
Engine coolant temp (ECT)	AT2	B4 Violet B16 Black	C26 Red/white	ECT signal	Black of MoTeC loom to B16
Engine oil temp	AT3	B5 Black Black to sensor ground	Blue/yellow Blue	Oil temperature sump	
Digital inputs					
Speed sensor	Dig input 3	B10 Yellow	C5 Blue/white to White/blue		Pulls to ground to turn on
A/C request	Dig input 4	B11 Brown	A27 Blue/black to Blue		Provisional
	MoTeC ECU channel	MoTeC M400	Honda Insight	Function	Notes
Crank/cam inputs					
Crank position (ref)	Ref	B1 White signal Black	C8 Blue C9 White	Signal Ground	
Camshaft position sensor (sync)	Sync	B2 white signal Black	C20 Green C21 Red C29 Yellow C30 Black	TDC#1 signal TDC#1 ground TDC#2 signal TDC#2 ground	See Miro Module attachment

(Chart continues overleaf)

WORKSHOP PRO MODIFYING THE ELECTRONICS OF MODERN CLASSIC CARS

Outputs					
Fuel injectors	First injector fires Second injector fires Third injector fires	A19 white/black A20 white/red A21 white/yellow	B11 Brown – Inj 1 B4 Blue – Inj 3 B3 Red – Inj 2	Firing order: 1-3-2	First injector output drives injector 1 Second injector output drives injector 3 Third injector output drives injector 2
Ignition	First plug fires Second plug fires Third plug fires	A3 Yellow A4 Brown A5 Grey	C4 White C14 white/black C13 white/green	Firing order: 1-3-2	First injector output drives coil 1 Second injector output drives coil 3 Third injector output drives coil 2
Auxiliary outputs					
Oxy sensor heater	Aux 1 Lambda heater	A18			ECU grounds to turn on heater. Heater fed via relay. Relay power straight from battery. Relay on/off from heater relay connection. LH kick panel fuse. Relay behind glove-box.
Water/air intercooler pump	Aux 2 output Aux Table	A1 Yellow	White/green (12V power from Yellow/black) Black/green (Aux 2)		ECU grounds to turn on pump. Pump fed via relay. Relay power straight from battery. Relay on/off from heater relay connection. LH kick panel fuse. Relay behind glove-box.
Boost control valve	Aux 3 Boost control	A23 Brown White (12V supply)	Red/yellow (ECU power) White/red (Aux 3)		ECU grounds to turn on valve
Fuel pump relay	Aux 4 Fuel pump	A24 Blue	A15 Green/yellow		ECU grounds to turn on pump relay
Idle speed control	Aux 5 Idle control	A31 Black	B23 Black/blue		Ground to turn on. Wired to Bosch valve from original throttle body location.
VTEC valve	Aux 6 Aux table	A32 Brown	B12 Green/yellow to Yellow/white		12V to turn on
Exhaust gas recirculation	Aux 7 Aux table	A33 Unknown colour*	B7 Pink to orange/green		12V to turn on
Engine speed for electric power steering	Aux 8 Tacho	A34 Grey	A19 Blue	Gives power steering from stationary	1 pulse per rev

* This chart is a real working example. It was only when preparing it to use in this book that I realised that I hadn't recorded one of the wiring colour codes – hence the 'unknown colour' mentioned here.

4. PROGRAMMABLE ENGINE MANAGEMENT

A PIGGY-BACK INSTALLATION

In the above section I covered the wiring approach when installing a programmable management system that completely replaces the standard ECU. But what about a 'piggy-back' configuration, where the new ECU controls only fuel and ignition?

Let's take a look at the wiring involved:

John's first step was to very carefully sort out an appropriate wiring diagram for the car and another for the factory ECU pin-outs. In this case, he didn't have one specifically for the 180SX, so used an N14 Pulsar diagram. Getting the best possible diagram for the car being worked on is critical to success.

The car was a Nissan 180SX, equipped with the SR20DET engine. It arrived from Japan modified with a ball-bearing turbo (thought to be a GT-25), 540cc injectors, large intercooler, larger Z32 airflow meter, pod filter and exhaust system. The installation was being performed by John Nash, then of ChipTorque.

The ECU being installed was the MicroTech LT-8, a simple programmable ECU that uses a hand controller.

As is common on many modified 'grey market' imports, the factory ECU wiring loom had already been altered. In this case it appeared as if an interceptor had been fitted and then later removed. The significance of this when fitting a new ECU piggy-back style is that the factory ECU will still be used to control idle speed and some other parameters – the new ECU controls fuel and ignition. If the standard ECU isn't working properly, a new aftermarket ECU won't make these aspects any better! So before starting the new job, check that there aren't any wires left cut and dangling from previous work.

WORKSHOP PRO — MODIFYING THE ELECTRONICS OF MODERN CLASSIC CARS

John photocopied the wiring diagrams so that he had the three sheets easily to hand: the wiring diagram for the MicroTech, the wiring diagram for the 180SX (in this case the nearest diagram possible, anyway), and the pin-outs of the factory ECU.

The injector feeds were connected to the 180SX wiring harness by cutting the original harness near to the factory ECU and then connecting (by means of stainless steel crimps covered in heatshrink) the injector feeds from the MicroTech. The MicroTech harness (which is long enough to extend to the injectors in the engine bay) was shortened so that the new ECU could be mounted close to the factory ECU without needing to have a long loom bundled away.

The first step was to wire in the four injectors. The MicroTech ECU provides a power and an ECU-switched ground for each injector. However, because the 180SX already supplies 12V to each injector (and also uses a switched ground to turn on the injector) only one wire from the MicroTech needed to be hooked up to each injector. Here, John is checking the colour-coding of the injector wiring, confirming that the diagram he is working with is correct in this regard.

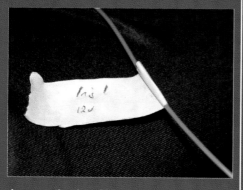

The leftover 12V MicroTech injector feed was carefully labelled. Any wires that will be left unused should be labelled as you go along. This avoids the possibility of later inadvertently selecting one and connecting it up! Note that when doing the wiring for the injectors, the firing order of the engine needs to be kept in mind. The MT-8 uses sequential injection, so the injectors need to be connected in the order in which they will fire.

The ignition connections were next. The 180SX uses direct fire ignition (ie one coil per plug) controlled by an ignition module. The not-quite-close-enough wiring diagram that was available for the car wasn't really clear in this area, so to make absolutely sure that the wire John was tapping into at the ECU-end of the loom was one of the wires reaching the ignition module, he did a continuity test using a multimeter. This is also an excellent approach to confirm that you are not making a simple mistake – eg working one pin across on the connector of the factory ECU.

4. PROGRAMMABLE ENGINE MANAGEMENT

However, with the colour codes changing across the ignition module, and with the factory wiring diagram a little unclear, John took a common-sense approach and made temporary connections for the ignition module feeds, using bullet connectors. The worst that could happen is that the engine would run very poorly (eg on only a few cylinders), in which case swapping the wires would then quickly give the right results. Once the correct wiring was ascertained, the connectors could be replaced with proper connections.

Rather than take main power and ground connections from the factory ECU, John ran new cables right to the battery. This avoids any voltage drop problems which may result in the car being hard to start. A blade-type fuse was mounted close to the positive terminal. The ignition-switched power feed (which basically just wakes up the ECU) was taken from the original ECU loom.

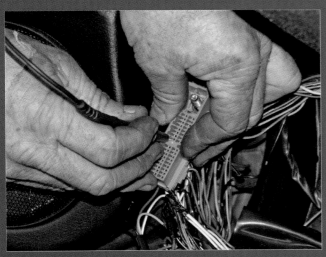

Other connections that needed to be made included crank-angle sensor, engine coolant temperature, and throttle position. In general, for these connections John wired the MicroTech in parallel with the factory ECU. In other words, the coolant temperature sensor was connected to both the MicroTech and factory ECU. This allows the tachometer to work, the car to have correct idle speed control, etc. However, in operation the systems aren't always happy when worked in this way, so in some cases the factory ECU may have to be disconnected. More on this in a moment.

As with many aftermarket ECUs, the MicroTech uses an onboard MAP sensor. This is fed by a rubber hose running from the ECU to the plenum chamber. Here, John is working the hose around the engine bay.

The MicroTech has the facility to take an intake air temperature gauge, however the 180SX doesn't have such a separate sensor as standard (it's probably integrated into the airflow meter). Interestingly, John used an adjustable resistance wheel to feed an input into this facility, so that the MicroTech was actually seeing a value. The installation (or otherwise) of a new intake air temperature sensor was left to the tuner.

WORKSHOP PRO: MODIFYING THE ELECTRONICS OF MODERN CLASSIC CARS

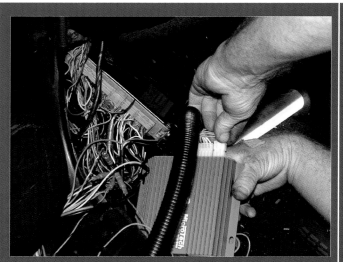

The new MicroTech ECU could then be plugged into its loom. Wiring the aftermarket ECU piggyback style in this way removes the need to run the new loom right into the engine bay (except for the power and ground). The existing loom, including the often hard to find plugs, are all used. The underbonnet view also looks near-standard.

The MicroTech comes pre-programmed with base maps. That means than when the wiring is correct, the key can be turned and the car started. In this case, the engine ran very rich in warm-up – so rich that it was staggering. John initially suspected that his ignition module wires (the ones fitted with temporary bullet terminals to allow easy swapping) were incorrect, but in fact this was not the case. When leaned-out, the car warmed up fine.

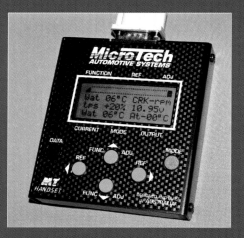

The next step is a vital one. Using the hand controller, and with the ignition switched on but the engine not cranked, John checked that the sensor data being displayed on the controller made sense. For example, here the water temperature is showing just 6°C (43°F) – and it wasn't that cold in the workshop! The throttle position sensor (TPS) is at 20 per cent – but the throttle was closed. Each of these problems needed to be sorted before the car was started. In this case, the coolant temperature connection had to be configured so that it fed only the MicroTech, not both ECUs. The TPS had to have earth and +5V feeds from the MicroTech, rather than the minimalist signal wire connection initially made. John takes the view that it is better to start off with the minimum of wiring connections (consistent with the system operating well), and then make further changes as necessary.

The 180SX could then go to the dyno room, ready for tuning.

4. PROGRAMMABLE ENGINE MANAGEMENT

TUNING PROGRAMMABLE MANAGEMENT

While wiring-up a programmable ECU takes time and effort, the really big commitment from a person working at home is tuning the ECU. So what expectations regarding air/fuel ratios and ignition timing does an engine have?

Air/fuel ratio requirements

A well-tuned engine used in normal road conditions has an air/fuel ratio (AFR) that is constantly varying. At light loads, lean AFRs are used, while when the engine is required to develop substantial power, richer (lower number) AFRs are used.

A rule of thumb is that maximum power is developed at AFRs of 12.5:1-14:1, maximum fuel economy at 16.2:1-17.6:1, and good load transitions from about 11:1-12.5:1. However, engine AFRs at maximum power are often richer than the quoted 12.5:1, especially in forced-induction engines, where the excess fuel is used to cool combustion and so prevent detonation. In addition, in high-swirl engines, the AFR may be able to be run leaner than 18:1 at light loads, with resulting improvements in fuel economy.

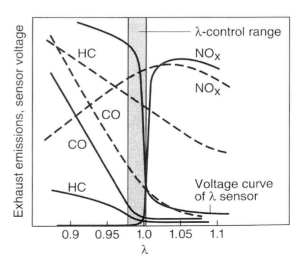

Another classic Bosch diagram, showing the effect on emissions of changed air/fuel ratios. Again, the air/fuel ratio is shown in Lambda ratios. There is a lot to see on this diagram! Firstly, note how NOx (oxides of nitrogen) rises rapidly with lean air/fuel ratios (high Lambda numbers), both with and without catalytic treatment. Conversely, HC (hydrocarbons – unburnt fuel) rises at rich air/fuel ratios (low Lambda). CO (carbon monoxide) falls linearly with leaner air/fuel ratios on the rich side of stoichiometric (Lambda = 1), which is why this gas was measured to indicate air/fuel ratio richness before oxygen sensors came into widespread use. The switch in output level of a narrow band oxygen sensor either side of stoichiometric can also be seen. (Courtesy Bosch)

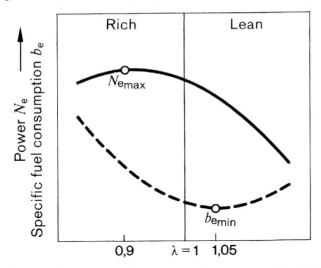

This classic graph from Bosch shows the effect of air/fuel ratio on power (Ne) and specific fuel consumption (be). The air/fuel ratio is shown as Lambda (multiply Lambda by 14.7 to get the air/fuel ratio for normal petrol (gas)). It can be seen that maximum power occurs at 0.9 Lambda (13.2 AFR) and maximum economy at 1.05 Lambda (15.4 AFR). However, you cannot blindly apply these figures to all engines. A high-boost turbo engine, for example, would be dangerously lean at an AFR of 13.2:1 at full load, and in some engines, running much leaner than 15.4:1 gives better fuel economy. (Courtesy Bosch)

Cranking and idle

The amount of fuel that needs to be added during cranking can best be determined by experimentation. This enrichment may be configured by just a one-dimensional variable based on engine coolant temperature, or it may be able to be controlled in a more sophisticated manner. Examples of the latter include post-start enrichment and enrichment decay time.

Cold start is one of the dirtiest times in regard to emissions, and so, if emissions requirements are to be met, a sophisticated ECU with multiple starting enrichment and decay maps should be used. Reducing the cold start enrichment while increasing the cold acceleration enrichment will reduce the total amount of emissions. Some factory systems open the idle air bypass during cold deceleration, presumably to act as a form of exhaust air injection.

The AFR required for a smooth idle will depend on the engine's combustion efficiency and the camshaft(s) used. Some heavily modified engines with very 'hot' cams will require an AFR as rich as 12-12.5:1 for a smooth idle, while others will run happily at 13-13.5:1. Engines with modified cams that are fitted with sequential injection management systems can run leaner idle mixtures than systems using bank or group fire.

WORKSHOP PRO: MODIFYING THE ELECTRONICS OF MODERN CLASSIC CARS

Fuel ET Comp (% Trim)													
🔧	ET °C	-10.0	0.0	10.0	20.0	30.0	40.0	50.0	60.0	70.0	80.0	90.0	100.0	110.0
		25	20	14	10	10	5	5	2	0	0	0	0	5

Intake air temperature fuel correction chart, expressed as a percentage trim change. As you'd expect, a lot more fuel is needed for the engine to run at very low intake air temperatures. The extra fuelling at very high intake air temperatures is for safety.

Those engines that can be configured to run in closed loop at idle will use an AFR of about 14.7:1 when fully warmed. Keeping the engine AFR as close to stoichiometric as possible will benefit emissions because the catalytic converter works most efficiently at this ratio.

Cruise

Light-load cruise conditions permit the use of lean AFRs. Ratios of 15-16:1 can be used in engines with standard cams, while engines with modified cams will require a richer 14:1 air/fuel. If a specific lean cruise function is available, AFRs of 17 or 17.5:1 can be used, normally at the standard light-load ignition advance. Remember, an engine in a road car will spend more time at light-load cruise than in any other operating condition. The AFR used in these conditions will therefore determine to a significant degree the average fuel economy gained, especially on the open road.

High load

A naturally-aspirated engine should run an AFR of around 12-13:1 at peak torque. Rich AFRs can be used to control detonation, and this is a strategy normally employed in forced-induction engines. Thus, on a forced-induction engine, the mixture should be substantially richer – 11.6-12.3:1 on a boosted turbo car, and as rich as 11:1 on an engine converted to forced aspiration without being decompressed.

As is also the case for ignition timing, the AFR should vary with torque, rather than with power.

In the engine operating range from peak torque to peak power, a naturally-aspirated engine should be slightly leaner at about 13:1, with the forced-induction factory engine about 12:1 and an aftermarket supercharged engine staying at about 11:1.

Acceleration

During acceleration, the engine requires a richer mixture than during steady-state running, with the extra fuel provided by acceleration enrichment. Under strong acceleration, the AFR will typically drop 1-1.5 ratios from its static level.

The amount of acceleration enrichment that is required is normally found by trial and error, and this is best done on the road. The acceleration enrichment should be leaned-out until a flat spot occurs, then just enough fuel to get rid

Fuel Accel Sens			
🔧	Lean	0.0	0.1
MAP kPa	180.0	5	15
	160.0	5	15
	140.0	5	15
	120.0	5	15
	100.0	4	15
	80.0	3	20
	60.0	3	25
	40.0	3	25
	20.0	3	25
	0.0	3	25

Acceleration enrichment map based just on manifold pressure. The '0.0' column is for normal running, and the '0.1' for lean cruise, where a lot more acceleration enrichment is needed.

of the flat spot should be added. This approach usually gives the sharpest response. Note that both over-rich and over-lean acceleration enrichment will result in flat spots.

Over-run

In road-going vehicles, deceleration enleanment is used to reduce emissions and improve fuel economy. This normally takes the form of injector shut-off, with the shut-off often occurring at mid-rpm (such as 3000-4000 rpm), and the injector operation restarting at 1200-1800 rpm.

Ignition timing requirements

Firing the spark at the right moment is critical to gaining good power, emissions and economy. The period between the spark firing and the complete combustion of the fuel/air mix is very short – on average, only about 2 milliseconds. Ignition of the fuel/air mix must take place sufficiently early for the peak pressure caused by the combustion to occur just as the piston has passed Top Dead Centre (TDC), and so is on its way down the cylinder bore.

If the ignition occurs a little too early, the piston will be slowed in its upward movement. If it occurs too late, then the piston will already be moving downwards, so reducing the work done on it. If the spark occurs *much* too early, the ignition pressure wave can ignite the mixture in various parts of the combustion chamber, causing detonation.

If the composition of the mixture were constant (and it isn't!), the elapsed time between ignition and full combustion would remain about the same at all rpm. If the ignition advance angle was set at a fixed angle before

4. PROGRAMMABLE ENGINE MANAGEMENT

TDC, then combustion would be shifted further and further into the stroke as the engine speed increased. This is because the faster moving piston would be further down the bore by the time combustion actually occurred. To prevent this, the ignition advance must increase as engine speed rises.

The other major factor affecting the amount of advance required is the engine load. At light loads when lean mixtures are used, the speed of combustion is slowed, and so more ignition advance is needed.

Although engine speed and load determine the best timing for the combustion of the mixture, the following factors are also relevant:

- design and size of the combustion chamber
- variable valve operation
- position of the ignition spark(s) in the chamber
- fuel characteristics
- emissions levels required
- engine coolant and intake air temperature
- the safety margin required before detonation occurs

These factors mean that the ignition advance has to vary with load and rpm. If optimal advance is to be used, the ECU must also be able to sense detonation that might occur on lower-octane fuel or in very hot conditions. Ignition timing maps on engines that use knock sensing are not fixed in value – the timing can alter depending on the input of knock sensor(s). The input of the knock sensor is an ignition timing correction; additional corrections of the main ignition timing map should be available from an intake air temperature sensor and other maps.

Ascertaining the best ignition timing by juggling these interrelating factors on paper is nearly impossible. Instead, making real-time changes to the ignition timing while using an effective detonation detection system is the only practical way of seeing the most appropriate ignition timing at all loads and revs.

Cranking and idle

Some programmable engine management systems have a default cranking advance of 15°, a value about midway through the range of appropriate cranking advances. Smaller engines with faster cranking speeds need a greater ignition advance (up to 20°), while slower cranking speeds of a high-compression engine will require less advance (down to 10°).

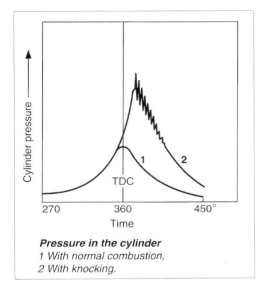

The in-cylinder pressure trace, with and without knock (detonation). Note that the pressure rises much higher when knocking is occurring, and that the pressure also violently fluctuates. You do not want this occurring! (Courtesy Bosch)

Pressure in the cylinder
1 With normal combustion,
2 With knocking.

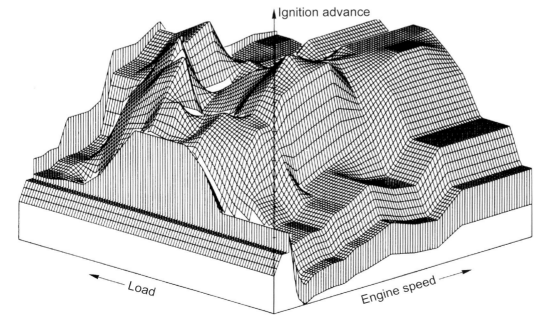

The required ignition timing varies with engine load and speed. Mapping to this level of detail requires both a sophisticated engine management system and a lot of time. (Courtesy Bosch)

The compression ratio of the engine will also determine the likelihood of kickback on starting. Engines with a low static compression ratio of 8:1 will accept an ignition advance of anything from 0-20° without kickback. A 10:1 compression ratio will reduce this to 15°, 11:1 to around 10-12°, and race engines using very high compression ratios of 12-13:1 can sometimes tolerate no cranking ignition advance at all.

Most engines will idle happily with an ignition advance of 15-32°. This is a very wide range – some engines will certainly not be happy at 32°, and others won't be at 15°! An overly high amount of ignition advance for a given engine will result in lumpiness at idle, excessive hydrocarbon emissions, and sometimes exhaust popping, while too little advance will also cause lumpiness.

Timing that is more advanced at slightly lower engine speeds than idle is sometimes used to help stabilise idle. This is effective because when the engine starts to slow down, the greater ignition advance causes the engine to produce more torque, so increasing engine speed. Many standard management systems use ignition timing as a major element in controlling idle smoothness, with an increase or decrease in rpm at idle responded to by a change in timing advance.

Cruise

At light loads – as in normal everyday cruise conditions – an ignition advance of 40° or more will improve responsiveness and economy. This advance can be used successfully on many engines: even those with high compression ratios, if they are being run on high-octane fuel.

One factor limiting the cruise ignition advance is the maximum ignition timing attack rate provided by the ECU – that is, how fast the timing can change. If very advanced timing is being used with light loads, and the attack rate is not high, there may be slight detonation when the engine load suddenly increases.

The more modified the camshaft(s), the less advance that is able to be used in light-load conditions (the limiting factor in this case being driveability, rather than detonation). However, timing in the range of 35-40° is still usually used. Engines with good combustion chamber design will be able to run up to 45° in these conditions.

Fuel economy and engine responsiveness are both very much affected by light-load ignition timing.

High load

The torque output of a given engine is proportional to average cylinder pressures. So, the full-throttle ignition timing advance that is used should relate to the torque curve, rather than the power curve.

The maximum ignition timing that can be used at peak torque is usually limited by the occurrence of detonation. A detonation limit is always the case in forced-aspirated engines, but not always the case in naturally-aspirated engines. As an example of the latter, one Porsche flat-six developed best power with a maximum advance of 8°, but the engine did not detonate at even 27° of advance! An older Mercedes V8 engine was able to run 38° at high rpm, peak load without audible detonation. However, best results came from a full-load advance of 28°.

On a modified engine having increased compression and modified cams, a peak torque advance of 28-36° can often be used. In a factory-forced induction engine using a little more boost than standard, the peak torque timing will be around 18-22°, while in a naturally-aspirated engine converted to forced induction without internal modifications, timing should be well back at about 10°.

As already indicated, most forced aspiration engines – and many naturally-aspirated engines – develop best performance when the ignition timing is advanced close to the point of detonation. Therefore, great care should be taken when setting the full-load ignition timing. Optimal ignition timing is that which gives a lack of detonation, the lowest exhaust gas temperatures, and maximum torque. If the engine uses reliable knock sensing and the ignition timing can be retarded quickly at the onset of detonation

Ignition timing retard chart showing timing advance removed as intake air temperature increases. For safety on this small turbo engine, I pull off a lot of timing at high intake air temperatures and high loads. In normal use, these temperatures should never be experienced.

Ign AT Comp (Trim)

Effcy kPa \ AT °C	-20.0	-10.0	0.0	10.0	30.0	40.0	45.0	50.0	60.0	70.0	80.0
200.0	0.0	0.0	0.0	0.0	0.0	0.0	-2.0	-3.5	-5.0	-10.0	-15.0
180.0	0.0	0.0	0.0	0.0	0.0	0.0	-2.0	-3.5	-5.0	-10.0	-15.0
160.0	0.0	0.0	0.0	0.0	0.0	0.0	-2.0	-3.5	-5.0	-10.0	-15.0
140.0	0.0	0.0	0.0	0.0	0.0	0.0	-2.0	-3.5	-5.0	-10.0	-15.0
120.0	0.0	0.0	0.0	0.0	0.0	0.0	-1.0	-3.5	-5.0	-10.0	-15.0
100.0	0.0	0.0	0.0	0.0	0.0	0.0	-1.0	-3.5	-5.0	-7.0	-7.0
80.0	0.0	0.0	0.0	0.0	0.0	0.0	-1.0	-2.0	-2.0	-5.0	-5.0
60.0	0.0	0.0	0.0	0.0	0.0	0.0	0.0	0.0	-2.0	-5.0	-5.0
40.0	0.0	0.0	0.0	0.0	0.0	0.0	0.0	0.0	-2.0	-5.0	-5.0
20.0	0.0	0.0	0.0	0.0	0.0	0.0	0.0	0.0	-2.0	-5.0	-5.0
10.0	0.0	0.0	0.0	0.0	0.0	0.0	0.0	0.0	-2.0	-5.0	-5.0

4. PROGRAMMABLE ENGINE MANAGEMENT

(and then re-introduced only slowly), more advanced timing can be used at high rpm.

An intake air temperature correction chart that quickly pulls off timing advance with increased air intake temperatures can also allow the main table's ignition timing to be fairly advanced. For example, with an intake air temperature of 120°C (248°F), the timing can be retarded by 12-15°, so providing an acceptable level of safety while still allowing good cool weather and short-burst performance. The importance of using a programmable ECU that has tables for the intake air temperature correction of ignition timing can be seen from this example.

Over-run
On deceleration (with injector cut-off working) most factory cars run retarded timing, such as 10-12°. However, in modified cars, this has been found at times to cause an exhaust burble, and, if this is unwanted, more advanced timing (20-26°) can be used.

TUNING ON THE ROAD
If you have a programmable management system fitted to your car, can you tune it on the road? The answer is that yes, you can – with some caveats. First, the car cannot be immensely powerful. It's just too dangerous if full throttle results in mass wheel-spin, or tuning at higher rpm in

EXAMPLE IGNITION TIMING MAPS

Once you have decided what air/fuel ratios to run at what loads, getting the right outcome is just a case of plugging in the numbers to achieve that outcome, as measured on a wideband air/fuel ratio meter. However, mapping the ignition timing is different. People become frustrated when there is no 'pat' answer to the required ignition timing. However, as can be seen from the following maps, the requirements really do vary substantially from engine to engine.

Ign Main (°BTDC)

Load kPa \ RPM	0	500	750	1000	1350	1500	1750	2000	2500	3000	3500	3750	4000	4500	5000	5500	6000
110.0	3.0	1.0	1.0	-5.0	-5.0	-5.0	-2.0	-5.0	-6.0	-3.0	2.0	5.0	5.0	4.5	1.0	0.0	0.0
100.0	3.0	-4.0	-5.0	-10.0	-10.5	-9.0	-5.0	-5.0	-3.0	1.0	3.0	7.0	5.0	4.0	4.0	4.0	4.0
90.0	3.0	-7.0	-7.0	-10.0	-10.5	-9.0	-4.0	1.0	3.0	8.0	8.0	8.0	7.0	5.0	4.0	4.0	4.0
80.0	3.0	-7.0	-7.0	-7.0	-10.0	-3.0	3.0	5.0	8.0	10.0	10.0	8.0	8.0	5.0	4.0	4.0	4.0
70.0	3.0	-7.0	-7.0	-3.0	-6.0	0.0	6.0	6.0	8.0	12.0	12.0	12.0	12.0	11.0	9.0	9.0	9.0
60.0	3.0	-7.0	-7.0	4.0	4.0	8.0	8.0	8.0	8.0	14.0	20.0	20.0	17.0	15.0	13.0	13.0	13.0
50.0	3.0	-7.0	-5.0	7.0	7.0	8.0	8.0	8.0	8.0	10.0	10.0	10.0	10.0	10.0	10.0	10.0	10.0
45.0	3.0	-7.0	-5.0	10.0	10.0	10.0	10.0	10.0	10.0	10.0	10.0	10.0	10.0	10.0	10.0	10.0	10.0
40.0	3.0	3.0	3.0	14.0	14.0	14.0	14.0	14.0	14.0	14.0	14.0	14.0	14.0	14.0	14.0	14.0	14.0
0.0	3.0	3.0	3.0	12.0	14.0	14.0	14.0	14.0	14.0	14.0	14.0	14.0	14.0	14.0	14.0	14.0	14.0

The ignition timing map for a three-cylinder, V-TEC, 1-litre engine (Honda Insight with added turbo). Very conservative ignition timing was required to avoid detonation, especially at low loads and revs.

Ign Main (°BTDC)

Load % \ RPM	0	500	750	870	1000	1500	2000	2500	3000	3500	3750	4000	4250	4500	5000	5250	5500	6000	6500	7000	7500
100.0	5.0	5.0	6.0	6.5	7.0	8.0	10.0	12.0	13.0	13.5	14.0	14.0	15.0	15.5	16.0	16.0	15.5	14.5	13.0	13.0	12.5
90.0	5.0	5.0	6.0	6.5	7.0	8.0	10.0	12.0	15.0	16.0	17.0	18.0	19.0	19.0	19.0	19.0	18.5	17.5	17.0	17.0	16.0
80.0	5.0	5.0	6.0	6.5	7.0	8.0	10.0	12.0	13.0	18.0	20.0	21.0	21.5	21.5	21.0	21.0	20.5	20.0	20.0	20.0	19.5
70.0	5.0	5.0	6.0	6.5	7.0	8.0	10.0	12.0	13.0	18.5	20.0	21.5	22.0	22.0	22.0	22.0	22.0	22.0	22.0	22.0	22.0
60.0	5.0	5.0	6.0	6.5	7.0	8.0	10.0	12.0	13.0	17.5	20.0	22.0	22.0	22.0	22.0	22.0	22.0	22.0	22.0	22.0	22.0
50.0	5.0	5.0	6.0	6.5	7.0	8.0	10.0	12.0	13.0	18.5	20.5	22.0	22.5	23.0	23.0	23.0	23.0	23.0	23.0	23.0	23.0
40.0	5.0	5.0	5.5	5.5	6.0	8.0	10.0	12.0	13.0	20.0	21.5	23.0	23.0	23.0	23.0	23.0	23.0	23.0	23.0	23.0	23.0
30.0	5.0	5.0	5.0	5.0	5.0	6.5	9.0	12.0	13.0	20.0	21.5	23.0	23.0	23.0	23.0	23.0	23.0	23.0	23.0	23.0	23.0
20.0	5.0	5.0	5.5	5.5	5.5	6.5	9.0	12.0	15.0	20.5	22.0	23.5	24.5	25.0	25.0	25.0	25.0	25.0	25.0	25.0	25.0
10.0	5.0	5.0	0.0	-2.5	-5.0	-6.0	-4.0	5.5	16.5	22.5	23.0	25.0	25.5	25.5	25.0	25.0	25.0	25.0	25.0	25.0	25.0
5.0	5.0	-0.5	-9.0	-23.0	-22.5	-22.5	-18.0	-1.0	8.5	20.0	23.0	24.5	25.5	25.5	25.5	25.5	25.5	25.5	25.5	25.5	25.5
0.0	5.0	-7.0	-12.5	-24.0	-25.0	-23.5	-21.0	-8.0	0.5	17.0	22.5	24.0	26.0	26.0	26.0	26.0	26.0	26.0	26.0	26.0	26.0

The ignition timing map for a standard 3.2-litre, naturally-aspirated V6 engine (Volkswagen Beetle RS VR6). Note the conservative timing, including some negative numbers. (Courtesy MoTeC)

WORKSHOP PRO — MODIFYING THE ELECTRONICS OF MODERN CLASSIC CARS

Ign Main (°BTDC)

Load kPa \ RPM	0	500	750	1000	1250	1500	1750	2000	2250	2500	2750	3000	3250	3500	3750	4000	4250	4500	4750	5000	5250	5500	5750	6000	6250	6500	6750	7000	7250
180.0	18.0	18.0	18.0	18.0	14.0	14.0	14.0	14.0	14.0	14.0	14.0	14.0	14.0	14.0	14.0	11.5	10.5	10.5	10.0	10.0	11.5	12.5	13.0	14.0	15.0	15.5	16.5	17.0	18.0	19.0
160.0	18.0	18.0	18.0	18.0	15.0	15.0	15.0	15.0	15.0	15.0	15.0	15.0	14.0	13.5	13.5	11.0	10.0	10.0	10.5	12.5	13.5	14.0	15.0	16.0	17.0	18.0	18.5	19.5	20.0
140.0	18.0	18.0	18.0	18.0	16.0	16.0	16.0	16.0	16.0	16.0	15.0	15.0	15.0	13.0	11.5	11.0	11.5	14.0	14.5	15.5	16.5	17.5	18.0	19.0	20.0	21.0	21.5
120.0	18.0	18.0	18.0	17.0	17.0	17.0	19.0	20.0	21.5	23.0	23.5	22.0	19.5	18.0	16.5	14.5	14.0	14.0	15.5	16.0	16.5	18.0	19.5	19.5	20.0	21.0	22.0	22.5
110.0	22.0	22.5	24.0	26.0	28.0	28.5	29.5	30.0	31.5	31.0	30.5	28.0	25.0	23.0	20.0	18.5	16.5	16.0	16.5	17.5	18.5	19.0	20.0	21.0	22.0	22.0	22.5	23.0
100.0	22.0	23.0	25.0	28.5	30.0	31.0	31.5	32.5	33.0	34.0	34.0	32.5	29.0	26.0	21.5	20.0	19.0	18.0	19.0	22.0	22.0	22.0	22.5	24.5	25.0	26.0	26.5	29.0	29.0
90.0	22.5	23.5	26.5	30.0	31.5	32.0	33.0	34.0	34.0	35.0	35.5	34.0	31.0	29.5	25.5	23.0	22.0	20.5	21.5	25.0	25.5	26.0	27.0	28.5	30.0	30.0	30.0	30.0
80.0	22.5	23.0	26.0	30.0	32.0	33.0	34.0	36.0	36.5	37.0	37.5	36.0	34.0	32.5	30.0	27.5	25.5	24.0	25.0	27.0	27.5	28.5	29.5	31.0	32.0	32.0	32.0	32.0
70.0	23.0	24.0	26.0	30.0	32.0	33.5	35.5	37.0	38.0	39.0	39.5	39.0	37.0	34.0	32.0	31.0	29.0	28.0	28.5	30.5	30.5	31.0	32.0	32.5	33.0	33.0	33.0	33.0
60.0	23.5	24.5	25.5	29.5	32.0	35.5	37.0	39.0	39.0	39.0	40.0	40.0	38.5	37.0	36.0	35.5	34.5	33.5	33.0	33.5	34.0	34.0	34.0	34.0	34.0	34.0	34.0	34.0
50.0	23.0	23.5	24.5	28.0	32.0	37.5	40.0	41.0	41.0	41.0	41.0	40.5	39.0	39.0	39.0	38.0	37.5	37.0	36.0	36.0	35.0	36.0	36.0	36.0	36.0	36.0	36.0	36.0
40.0	22.0	22.0	20.0	23.5	32.0	39.0	41.5	42.5	44.0	44.0	44.0	44.0	44.0	44.0	44.0	44.0	44.0	44.0	44.0	44.0	44.0	44.0	44.0	44.0	44.0	44.0	44.0	44.0
30.0	22.0	20.0	18.0	21.0	32.0	40.0	42.5	44.0	44.0	44.0	44.0	44.0	44.0	44.0	44.0	44.0	44.0	44.0	44.0	44.0	44.0	44.0	44.0	44.0	44.0	44.0	44.0	44.0
20.0	22.0	22.0	19.0	21.0	32.0	40.5	43.0	44.0	44.0	44.0	44.0	44.0	44.0	44.0	44.0	44.0	44.0	44.0	44.0	44.0	44.0	44.0	44.0	44.0	44.0	44.0	44.0	44.0
0.0	22.0	22.0	20.5	21.5	32.0	40.5	44.0	44.0	44.0	44.0	44.0	44.0	44.0	44.0	44.0	44.0	44.0	44.0	44.0	44.0	44.0	44.0	44.0	44.0	44.0	44.0	44.0	44.0

The ignition timing map for a standard turbo six-cylinder engine (BA Ford Falcon XR6 Turbo). Note how ignition timing on boost is at its lowest at peak torque. (Courtesy MoTeC)

Ign Main (°BTDC)

Load kPa \ RPM	0	500	1000	1350	1500	1750	2000	2500	3000	3500	4000	4500	5000	5500	6000	6500	7000	7500	8000	8500	9000	9500
160.0	10.0	10.0	5.0	5.0	5.0	5.0	14.0	23.0	29.0	30.0	30.0	30.0	31.0	31.0	31.0	31.0	31.0	31.0	31.0	31.0	31.0	31.0
150.0	10.0	10.0	5.0	5.0	5.0	5.0	14.0	23.0	29.0	30.0	30.0	30.0	31.0	31.0	31.0	31.0	31.0	31.0	31.0	31.0	31.0	31.0
140.0	10.0	10.0	5.0	5.0	5.0	5.0	14.0	23.0	29.0	30.0	30.0	30.0	31.0	31.0	31.0	31.0	31.0	31.0	31.0	31.0	31.0	31.0
130.0	10.0	10.0	5.0	5.0	5.0	5.0	14.0	23.0	29.0	30.0	30.0	30.0	31.0	31.0	31.0	31.0	31.0	31.0	31.0	31.0	31.0	31.0
120.0	10.0	10.0	5.0	5.0	5.0	7.0	14.0	23.0	29.0	30.0	30.0	30.0	31.0	31.0	31.0	31.0	31.0	31.0	31.0	31.0	31.0	31.0
110.0	10.0	10.0	5.0	5.0	5.0	8.0	14.0	23.0	29.0	30.0	30.0	30.0	31.0	31.0	31.0	31.0	31.0	31.0	31.0	31.0	31.0	31.0
100.0	10.0	10.0	5.0	5.0	5.0	7.0	16.0	23.0	29.0	30.0	30.0	30.0	31.0	31.0	31.0	31.0	31.0	31.0	31.0	31.0	31.0	31.0
90.0	10.0	10.0	5.0	5.0	5.0	8.0	17.0	23.0	29.0	30.0	30.0	30.0	31.0	31.0	31.0	31.0	31.0	31.0	31.0	31.0	31.0	31.0
80.0	10.0	10.0	5.0	5.0	5.0	8.0	16.0	23.0	29.0	30.0	30.0	30.0	31.0	31.0	31.0	31.0	31.0	31.0	31.0	31.0	31.0	31.0
70.0	10.0	10.0	5.0	5.0	5.0	8.0	16.0	21.0	26.0	30.0	30.0	30.0	32.0	32.0	32.0	32.0	32.0	32.0	32.0	32.0	32.0	32.0
60.0	10.0	10.0	5.0	5.0	5.0	8.0	15.0	20.5	24.5	30.0	30.0	31.5	33.0	33.0	33.0	33.0	33.0	33.0	33.0	33.0	33.0	33.0
50.0	10.0	10.0	5.0	5.0	5.0	7.5	12.0	20.0	24.5	29.0	30.0	32.0	34.0	34.0	34.0	34.0	34.0	34.0	34.0	34.0	34.0	34.0
45.0	10.0	10.0	5.0	5.0	5.0	7.5	12.0	17.0	23.5	28.5	30.0	31.5	34.0	34.0	34.0	34.0	34.0	34.0	34.0	34.0	34.0	34.0
40.0	10.0	10.0	5.0	5.0	5.0	7.5	12.0	17.0	23.5	27.5	30.0	31.5	35.0	35.0	35.0	35.0	35.0	35.0	35.0	35.0	35.0	35.0
0.0	9.5	10.0	10.0	5.0	5.0	8.5	12.0	20.0	25.5	30.5	32.5	35.0	36.0	36.0	36.0	36.0	36.0	36.0	36.0	36.0	36.0	36.0

The ignition timing map for a supercharged 1.5-litre engine (Kawasaki Ultra 250X personal watercraft). This engine runs a lot of advance at high boosted loads. (Courtesy MoTeC)

Ign Main (°BTDC)

Load g/s \ RPM	0	500	1000	1500	2000	2500	3000	3500	4000	4500	5000	5500	6000	6500	7000
280.0	1.5	1.0	1.0	1.0	1.0	1.0	0.0	1.0	1.0	1.5	1.5	1.0	1.0	0.5	0.5
260.0	3.0	2.0	2.0	3.0	2.5	2.5	3.5	4.0	4.0	4.0	4.5	4.0	3.0	2.5	0.5
240.0	3.5	4.0	4.0	4.0	4.0	4.5	6.0	6.5	7.5	8.0	7.5	7.5	7.0	5.5	2.5
220.0	4.0	4.5	4.5	5.0	5.5	6.5	8.5	9.5	10.5	11.0	11.0	11.0	10.5	8.5	6.5
200.0	4.0	5.0	6.0	6.5	7.5	8.5	11.0	12.0	13.0	13.5	13.5	13.5	13.0	12.0	11.0
180.0	5.0	6.0	8.0	9.5	11.0	11.5	14.0	16.0	16.5	16.0	16.0	16.5	16.5	15.5	15.0
160.0	8.0	8.0	9.0	10.5	13.0	15.0	17.0	19.5	19.5	19.0	19.0	20.0	19.5	18.5	18.5
140.0	8.0	9.5	9.0	10.0	15.0	18.5	20.0	22.0	23.0	23.5	24.0	24.0	23.0	22.5	21.5
120.0	8.0	11.0	9.0	10.0	16.0	21.0	24.0	25.0	25.0	25.5	25.5	25.5	26.0	26.0	25.0
100.0	8.0	11.0	10.0	10.0	17.0	22.0	26.5	28.0	28.0	28.0	28.0	28.0	28.0	28.0	28.0
80.0	8.0	11.0	10.0	10.0	16.5	24.0	28.0	29.5	29.5	29.5	29.5	29.5	29.5	29.5	29.5
60.0	8.0	11.0	10.0	10.0	17.0	23.5	28.5	30.0	30.0	30.0	30.0	30.0	30.0	30.0	30.0
40.0	8.0	11.0	10.0	10.0	16.5	24.0	28.5	30.0	30.0	30.0	30.0	30.0	30.0	30.0	30.0
20.0	8.0	11.5	10.0	10.0	16.0	24.0	28.5	30.0	30.0	30.0	30.0	30.0	30.0	30.0	30.0
0.0	8.0	11.5	10.0	10.0	15.5	23.5	28.5	30.0	30.0	30.0	30.0	30.0	30.0	30.0	30.0

The ignition timing map for a turbo 2-litre engine (Mitsubishi Evo 7). Note how small the ignition advance is at high loads. (Courtesy MoTeC)

Ign Main (°BTDC)

Load % \ RPM	0	400	800	1200	1600	2000	2400	2800	3200	3600	4000	4400	4800	5200	5600	6000	6400	6800	7200	7600	8000
100.0	20.0	20.0	20.0	20.0	20.0	21.0	23.0	25.0	26.0	30.0	32.0	32.0	32.0	32.0	32.0	32.0	32.0	32.0	32.0	32.0	32.0
90.0	20.0	20.0	20.0	20.0	20.0	21.0	23.0	25.0	26.0	30.0	32.0	32.0	32.0	32.0	32.0	32.0	32.0	32.0	32.0	32.0	32.0
80.0	20.0	20.0	20.0	20.0	20.0	21.5	23.0	25.0	26.0	30.0	32.0	32.0	32.0	32.0	32.0	32.0	32.0	32.0	32.0	32.0	32.0
70.0	20.0	20.0	20.0	20.0	20.0	22.0	23.0	25.0	26.0	30.0	32.0	32.0	32.0	32.0	32.0	32.0	32.0	32.0	32.0	32.0	32.0
60.0	20.0	20.0	20.0	20.0	20.0	22.0	23.0	25.5	26.0	30.0	32.0	32.0	32.0	32.0	32.0	32.0	32.0	32.0	32.0	32.0	32.0
50.0	20.0	20.0	20.0	20.0	20.0	23.0	24.0	26.0	26.0	30.0	32.0	32.0	32.0	32.0	32.0	32.0	32.0	32.0	32.0	32.0	32.0
40.0	18.5	22.0	19.0	20.5	20.0	23.0	24.0	26.0	27.0	30.0	32.0	32.0	32.0	32.0	32.0	32.0	32.0	32.0	32.0	32.0	32.0
30.0	16.5	22.0	19.0	18.0	20.0	23.0	24.0	26.0	28.0	33.0	34.0	34.0	34.0	34.0	34.0	34.0	34.0	34.0	34.0	34.0	34.0
20.0	15.0	22.0	20.0	18.0	22.0	23.0	26.0	31.0	35.0	36.5	37.5	38.0	38.0	38.0	38.0	38.0	38.0	38.0	38.0	38.0	38.0
10.0	15.0	22.0	20.0	18.0	24.0	28.0	31.0	36.3	38.5	42.5	44.0	44.0	44.0	44.0	44.0	44.0	44.0	44.0	44.0	44.0	44.0
0.0	22.0	22.0	20.0	18.0	24.0	28.0	31.0	36.8	40.0	42.0	44.0	44.0	44.0	44.0	44.0	44.0	44.0	44.0	44.0	44.0	44.0

The ignition timing map for a Gen 5 Toyota 3S-GE 2-litre engine fitted with individual throttle bodies and using the factory variable valve timing. Note the large amount of ignition advance that can be used, even at peak torque. (Courtesy Bill Sherwood)

4. PROGRAMMABLE ENGINE MANAGEMENT

fourth, fifth or sixth gears results in warp speed. Second, you need a mix of roads – from heavy urban traffic all the way through to near-empty blacktop.

You'll also need some things organised before you begin, so let's start there. In addition to the wide-band air/fuel ratio meter and detonation listening system described in Chapter 3, you'll need a sparkplug spanner, and an assistant.

1. **Sparkplug spanner** – Periodically when tuning, you should pull a sparkplug and inspect it closely. Plenty of charts exist on the web that will show you what sparkplugs look like when the engine is detonating (for example, tiny globules of aluminium deposited on the porcelain insulator), or when the plugs are overheating (erosion of electrodes), or when the mixtures are rich or lean – and so on. 'Reading' sparkplugs is probably among the oldest of engine tuning techniques – and is still quite valid. Over the years I've been in some very high-level engine dyno workshops – and those mechanics still pull out sparkplugs for inspection.
2. **An assistant** – You also need an assistant – either to drive the car, or listen to the engine and operate the laptop. Note that the assistant doesn't actually need to know a lot about tuning!

My approach is to drive the car, feeling its behaviour and watching the AFR display, while someone else operates the laptop. Based on what I can see, hear and feel, I ask for specific tuning changes. Taking this approach, the assistant obviously needs to be able to find their way through the maps and make the required changes, but a detailed knowledge of tuning isn't needed.

Tuning sequence
Having a plan of attack is vital. Whether you're starting with a base map tune or tuning from scratch, the tune should pass through a series of stages.

The hardest areas to tune really well in a road car are idle and light loads. That surprises many people, but in comparison, high loads are basically just a case of running reasonably conservative ignition timing and lots of fuel! The differences in tuning between (say) 100 per cent throttle at 6000rpm, and 80 per cent throttle at 5000rpm, are relatively slight.

If you have access to an empty road or racetrack (or a dyno) for a limited time, schedule that for later in the tuning process and then spend that time on high loads – the lower loads may well take ten times as long, but they can be more easily done on populated roads.

So, what sort of tuning steps should you go through?

Idle
1. Start with the engine warm. (Before tuning the idle, you may have to apply throttle to keep it revving so that it actually gets warm.)
2. Ensure that the system is set up as simply as possible – at this stage, no exhaust gas recirculation (EGR), no closed loop feedback, no warm-up enrichment running.
3. Ensure that the sites at and around idle are all at the same values, both for fuel and ignition – then when you do make changes, change all these sites together. (If the active site is swapping back and forth across different values, oscillations in idle behaviour can occur as the engine is altering in fuel or timing.)
4. Without having any active idle speed control activated (ie turn off proportional integral derivative (PID) loops, etc), see how well you can get the idle speed to work, tuning just the opening of the idle speed control valve.
5. Activate control loops only one parameter at a time, assessing the action of altering just that one parameter. (Many people try to change all three PID parameters at once, and can lose themselves in their own control loop!)
6. Switch loads (eg air con or power steering) on and off, and then use the various ECU corrections to keep idle stable when these loads are occurring.
7. Optimise fuel and ignition at the stable idle setting. Idle AFR on petrol will typically be at 14.7:1 AFR, and ignition timing as specified for idle in the workshop manual for the engine. (Both the case if the engine is relatively unmodified in terms of compression, camshaft and cylinder head.)
8. Wait until the next day and then start the engine from cold, ensuring that fast idle settings and fuel enrichment are correct for idle warm-up. Don't change the settings that you optimised the day before for normal, warmed-up idle!
9. Expect to make minor ongoing tweaks to idle speed control – it's hard to get it perfect straight out of the box.

Light loads
1. Again the engine should be fully warmed. At this stage, don't use closed loop feedback, acceleration enrichment or any other corrections of that sort. In a turbo car, you are at this stage tuning in the area of the map that is off boost.
2. Drive the car at the smallest possible throttle angles consistent with normal driving, starting at low revs first.
3. Advance the timing until detonation is just heard when the throttle is opened further, then retard the timing so that it doesn't occur, even with further throttle openings. As the load is reduced, timing is increased, and as the revs are increased, timing is advanced. Therefore, the most advanced parts of the timing map will be at high revs, very low loads. You are almost

certain to do no engine damage if the engine can be heard lightly detonating (through the proper listening system, not just by bare ears) at light loads. Taking this approach allows you to (i) hear what detonation of your engine actually sounds like, and (ii) get some initial timing figures to then work outwards from.
4. Tune the AFR. At light loads it will be stoichiometric (as at idle) or even leaner (higher AFR numbers).
5. When tuning light loads for good driveability and economy, the juggling act is between going too lean for good driveability (flat spots, doughy response) and using too much fuel or having poor combustion behaviour.

Medium loads
1. The process is very much the same as for light loads, except this time you should be able to recognise the sound of detonation, and also be able to hear through the listening system when the engine is changing a fraction in combustion note, *indicating it is getting close to detonation*.
2. Based on the pattern of fuel and ignition numbers developed in light loads, you should be able to pre-fill the medium load areas with 'best guess' numbers – road testing then becomes a case of seeing whether those numbers give the AFR and ignition timing behaviour that you had expected.
3. In a turbo car, you'll now be working in the area of low boost, so you should also watch intake air temperature – a higher intake air temperature is more likely to cause detonation.
4. Ignition timing is going to be more retarded (lower numbers) than at lighter loads, while the AFR should still be stoichiometric or just going into slightly richer (lower) numbers on the meter.

High loads
1. The process is an extension of the tuning you did at light and medium loads. The complete high-load tables for fuel and ignition should be pre-filled with numbers based on the changing pattern you have developed so far. For example, ignition timing will continue to be more retarded (numbers getting smaller) and AFR continue to get richer (again numbers getting smaller on the meter). If in doubt, make both sets of numbers more conservative than you actually expect will be needed.
2. Do quick bursts of power into the high-load area, ensuring that there is no detonation and that the AFR is suitably rich. In most cars, mid-12s in AFR at high loads are safe. Make only small adjustments if you are going leaner in AFR or more advanced in ignition timing.
3. Any stutters – no matter how minor – indicate a problem; don't try to 'drive through' these. Instead, back off and look carefully at the tuning maps to ensure that you don't have a rogue number that is causing the problem.
4. In a turbo car, carefully watch boost and intake air temperature numbers. Start with the lowest full-throttle boost that you can run, and build up slowly from there. For example, start by using the default low wastegate value (eg 4 psi) rather than using electronic control of boost to give higher values.
5. Pull out sparkplugs and assess whether their appearance is indicative of appropriate combustion behaviour.
6. Place the car in a condition where you gradually spend a longer period in high-load areas (eg a long country road hill, retarding the car using brakes, or a long full-throttle burst in third or fourth gears). You are especially looking for detonation – the AFR is unlikely to change, but holding the car at a high load point can indicate the need for timing that is more retarded than you found when the engine was only briefly at that load site.
7. Go back over the loads between peak torque and peak power, seeing if you can add more timing in this area without getting the harsh engine 'edge' sound characteristic of an engine nearing detonation. If you can add more timing, you'll probably also need to then add more fuel.
8. Drive at high revs, low load – and then go straight to full throttle from this point. This driving behaviour requires probably the fastest change in ignition timing, so listen carefully for detonation.
9. If you have a turbo car where you can easily change peak boost, set the peak boost to varying levels (eg changing in 2psi increments) and then check that the fuel and ignition are correct right through the rev range for these different boost levels.
10. The use of a video camera (eg a smartphone) to record AFR numbers, rpm and manifold pressure during acceleration runs allows you to watch the numbers later at your leisure. You can then make appropriate map tuning changes.

Acceleration enrichment
Acceleration enrichment is more important than many people believe. Set up correctly, it makes the engine more responsive to throttle, can bring up boost faster in a turbo car, makes gear changes smoother, and avoids jerks or flat spots when throttle is reapplied at highway speeds.
1. Assuming that the acceleration enrichment is a 3D table (eg MAP pressure versus throttle), aim to set the acceleration enrichment for the mid-throttle movements first. Note that acceleration enrichment in

4. PROGRAMMABLE ENGINE MANAGEMENT

most ECUs occurs only *during* the throttle movement, not when revs are rising at a constant throttle. Check to see if this is the case with the ECU you're working with – it changes the way you think about the numbers.

2. Drive in second gear at (say) 25 per cent throttle, then quickly move the throttle to 75 per cent. Add acceleration enrichment at this point of the acceleration enrichment map until the car responds quickly without measured AFRs becoming overly rich on the transition. Depending on the speed of response of the AFR meter, best response may come when the meter shows a little leaner in transition than you'd expect (eg 15.5:1).
3. Now do the same 25 to 75 per cent throttle movement in the other gears. These sites will typically be in other parts of the map.
4. With the mid-throttle movements sorted, pre-fill the table for the higher throttle angles. Now move the throttle from 50 to 100 per cent. You may find that these figures require tweaking – it depends in part if the car runs full-throttle enrichment as part of another tuning strategy.
5. Tune the acceleration enrichment when moving the throttle from 0 per cent open to about 50 per cent. Again pre-fill these figures, but note that they may end up needing to be different from other parts of the map. Again, do this tuning in all gears.
6. Do a final test of the acceleration enrichment by:
   * rapid movement of the throttle in all gears
   * slow movement of the throttle in all gears
   * fast gear changes (eg, 1-2)
   * slow gear changes
   * driving at 60, 80 and 100km/h in high gears and gently getting on and off the throttle.
7. When tuning acceleration enrichment, do not change the figures in the main fuel table, or they will then be incorrect for steady-state conditions. When tuning acceleration enrichment, wear the listening headphones (connected via an amplifier to a microphone in the engine bay) so you can hear if there is transient detonation occurring on fast throttle movements.

Over-run injector cut-off

Over-run fuel injector cut-off saves fuel and reduces emissions, because the injectors are switched off (or run at minimum duty cycle) when the throttle is fully lifted at higher rpm.

1. Setting this is relatively straightforward, but ensure that the revs at which fuelling resumes are high enough above idle speed that the idle speed control and fuel injector resume are not fighting each other.
2. You may need to change acceleration enrichment values at around 0 per cent throttle so that reapplying throttle doesn't give a lean spot. (It is now more likely to be lean, because no fuel was being injected with the throttle lifted, whereas previously it was at these points.)

Exhaust gas recirculation

If your car uses electronically-controlled EGR, major driveability benefits can be achieved by careful tuning of the valve's operation.

The benefits of EGR are most apparent in low-power engines (or laggy turbo engines) at part throttle, where the use of EGR means that pumping losses are reduced. This is because the engine doesn't have to try so hard to suck air past the nearly-closed throttle. Instead, EGR provides that gas 'free of charge.' In these engines, much more part-throttle torque is achieved by the use of EGR.

1. If the table axes can be configured as you wish, start with throttle position versus engine rpm. You want to be able to drive the EGR valve at varying duty cycles. You may have to configure an ECU output appropriately first (eg valve pulsing frequency, and ensuring the output has enough current capability to drive the valve. Also make sure this output is protected against back-EMF spikes, or add an appropriate free-wheeling diode).
2. Ensure that the system is working by setting the EGR valve duty cycle to 100 at idle speed. The engine should stall, or at least stagger very badly. If there is no change in engine behaviour, the system is not working.
3. Add EGR at low and medium throttle angles and revs. Increase values until the car starts to get staggers and other bad driving characteristics, then pull the numbers back until the car drives well. You will probably not be able to use EGR at idle or very low loads/low rpm (eg where the engine is on the map when driven away from a standstill).
4. High rpm, zero throttle can have lots of EGR added. This will decrease engine braking – adjust values to give you the throttle-closed coasting behaviour that you want.
5. After EGR addition, other maps may need to be

Exhaust Gas Recirculation (EGR) tuning table. EGR rates have been mapped against manifold pressure. The '0.1' line indicates lean cruise, and as can be seen, no EGR is used in that mode.

Expander Output Table 2										
MAP kPa	20.0	30.0	40.0	50.0	60.0	70.0	80.0	90.0	100.0	
Lean	0.1	0	0	0	0	0	0	0	0	
	0.0	0	0	15	25	25	25	20	10	0

revised. For example, acceleration enrichment may need to be increased if you have a large amount of EGR set for zero throttle at higher rpm. Decreased pumping losses may need small revisions to main fuel and ignition tables.
6. Check that EGR doesn't intrude negatively during a cold start, where cold idle revs may be higher than for normal idle. If EGR can be easily configured to start operating only when the engine is up to temperature, that may be a solution to EGR problems at cold start.

Turbo boost

Turbo boost in most programmable ECUs can be set up in either open loop (the wastegate valve pulsed with a predetermined map of values) or closed loop (the wastegate pulsing values modified by the actual measured boost value). Closed loop boost control is similar to closed loop idle, in that PID values may need to be set. Most people use open loop boost control, at least initially. That's what will be covered here.

1. If the ECU has a fuel or ignition cut based on peak boost, set this boost cut before doing any boost tuning. Set the cut value a fraction above the maximum boost value you intend using.
2. Drive the car with the electronic boost control valve not operating. That is, boost will be set by the spring pressure in the wastegate vacuum canister. This is normally a low value, like 4psi.
3. Set up a map with throttle position and rpm on the axes, with the output being valve duty cycle. Taking this approach rather than using just a map of duty cycle output versus rpm will give you the potential for much better part-throttle control.
4. Put some valve duty cycle numbers in the tables (starting off at low values) and see what effect they have on boost. You will then quickly learn what sort of duty cycle numbers correspond to what sort of boost levels.
5. Progressively set the full throttle numbers, creeping up slowly to the boost level you want. Test in the tallest gear you can safely drive on the street, and yet still go through most of the rev range, eg third gear. Most cars will develop higher boost levels in higher gears when controlled in open loop. Therefore, setting the boost level in first gear may give much higher boost levels in third gear.
6. At full throttle, you will typically set the valve duty cycles at 100 per cent (full open) until boost starts to rise, and then taper the duty cycle back to give the required boost. This will bring up boost the quickest.
7. At throttle positions of less than full throttle, aim to set the values so that when a constant part-throttle is held (eg 50 per cent), boost is developed progressively and is then held at a lower maximum level through the rest of the rev range. Doing this means you will have much better throttle control of engine torque output – you can still get full boost by putting your foot down, but driveability will be much better. Test by selecting (say) second gear at low revs, and holding your foot at a constant 50 per cent throttle. As revs rise, boost should come up progressively and then hold at a lower level than at full throttle. Then do the same at 25 per cent throttle and 75 per cent throttle. Adjust the map accordingly.
8. Note that there is no requirement to hold full-throttle boost at a flat level through the engine rev range. You may choose to run higher boost in the midrange and lower at the redline, or lower boost in the midrange and higher at the redline. For example, if a turbo has been added to an engine that develops peak torque at high revs, the increase in max conrod loading will be reduced if boost is tapered back at high revs.

Going beyond the normal in ECU mapping

Aftermarket programmable engine management has been around now for many years. Fuel, ignition timing, turbo boost control, idle speed control – these are all very common areas that are mapped using programmable management. But what about other, rarer functions? Functions that you make up for yourself and then build within the logic and capability of the existing system? These are seldom – if ever – described, either in the media or in the manuals that come with the gear.

Here I want to explore in detail the development of a new function that uses the programmable logic blocks that are available. This particular example uses a MoTeC M400 ECU and a MoTeC ADL3 dash. While still available new, both devices have actually been around for years – they are no longer cutting-edge technology, and so are both well-proven (and also available secondhand!).

The specialised function that was developed is lean

Aim Boost

RPM	1000	2000	2500	3000	3500	3630	3750	4000	5000	6000	7000
TP % 100.0	0	100	100	92	68	74	66	66	50	50	50
75.0	0	100	100	40	40	38	36	36	36	36	36
50.0	0	100	100	20	10	10	10	10	10	10	10
25.0	0	0	0	0	0	0	0	0	0	0	0
0.0	0	0	0	0	0	0	0	0	0	0	0

Open loop turbo boost control chart. Boost control valve duty cycle is mapped against both throttle position and rpm. Doing this results in a more linear boost delivery.

4. PROGRAMMABLE ENGINE MANAGEMENT

THE PROPORTIONAL INTEGRAL DERIVATIVE (PID) CONTROL APPROACH

PID control is used in many advanced engine management systems – to control idle speed, turbo boost and others. So what's it all about?

Interestingly, PID control systems have been around for about 100 years – they were in existence well before the advent of digital electronic control. In fact, some of the earliest aspects of PID theory occurred in the development of automatic steering systems for ships.

In the 1920s, Russian/American engineer Nicolas Minorsky studied how a human helmsman controlled a ship. He realised that the helmsman corrected the ship's heading based on three different ideas:
- the current error
- the past error
- the rate of change of error

Let's look at each of these – note that I'll do so in the order 'PDI,' which makes it easier to understand.

In a PID system, the size of the current error is responded to by the 'P' term – the proportional control. If the error is great, a larger correction is made than when the error is small. So, taking the example of the helmsman steering the ship, if the ship is a long way off course, the helmsman swings on a lot of steering correction. To put it simply, the correction by the system is achieved by P multiplied by the error. It takes into account how great the error is, and applies a *proportional* correction.

The 'P' component, however, doesn't take into account how fast the error is developing. What if a sudden gust of wind blows the ship off course? In that case, the helmsman would react by rapidly inputting a steering correction – they wouldn't just apply the same type of correction as when the ship was slowly drifting off course.

The 'D' component relates the correction to the *rate of change* of the error. So in the case of the ship, it's not the amount of error that has occurred in the ship's heading, but how fast that error is developing.

So, if we have a large error, proportional control applies correction. If we have a suddenly developing error, the derivative control reacts. But there's another kind of error that neither 'P' nor 'D' will handle – a small but fixed error.

So, what could cause a 'small but fixed' error? Imagine the helmsman steering the ship across a slight current. The error in course is only slight – so the 'P' function will do little. That's because the P function applies correction in proportion to the error, and if the error is small, so will be the 'P' correction.

The 'small but fixed' error also won't be corrected by the 'D' function, because its correction relates to rate of change – and the error isn't changing ... it's fixed!

This type of error is corrected by the integral function.

The 'I' function looks at the error over time. It makes lots of measurements of the error, adds up these up, and then provides correction that is proportional to the total of these additions, rather than the size (or rate of change) of the error.

Idle	
Parameter	Value
Proportional Gain	0.12
Integral Gain	0.250
Derivative Gain	0.10
Dead-band	75
Activate TP	3.0
Activate Ground Speed	4.0
Activate RPM	1250
RPM Filter	4
Air Con Increase	5.0
Power Steering Increase	5.0

PID (and other) tuning parameters of idle speed control.

cruise – *but the same ideas can be applied to any bespoke outcome that you want to achieve.* Innovative control approaches to cam control, EGR, electronic throttle, intercooler water pump speed, radiator fan control – the list is endless.

That people so seldom develop custom functions in this way is probably indicative of the fact that people don't realise such approaches are even possible. And to be honest, before I started exploring in detail programming of the MoTeC ECU and dash, I had no idea that you could do so much.

Unique requirement

The car in question is the turbocharged 2001 Honda Insight that you've seen a lot of in this book, and the function that was desired was a lean cruise mode.

In standard naturally-aspirated form, the little Honda uses a lean cruise function that comes into operation when the engine is under low loads and is using a fairly constant throttle. In this mode, the standard car can run as lean as an air/fuel ratio of 24:1 (and sometimes, as the photo on the next page shows, even leaner!). But with the standard ECU completely gone, how to create a lean cruise mode?

The engine is being run by the MoTeC M400 ECU. The M400 is a conventional programmable ECU – so as standard, it doesn't have a lean cruise function. But is it possible to create such a mode from scratch? The answer is 'yes,' but to make it really work well, maths functions in the ADL3 dash are also needed.

So how was it all done?

87

WORKSHOP PRO: MODIFYING THE ELECTRONICS OF MODERN CLASSIC CARS

The standard Insight runs air/fuel ratios that are at times extraordinarily lean for a port fuel injected petrol (gas) engine. But how to replicate a similar lean cruise mode with programmable management?

Fuel and ignition changes

Let's do the easy stuff first – lean cruise, as a specific mode, reduces the amount of fuel that is being injected and at the same time, advances ignition timing. (The latter is needed because leaner mixtures burn more slowly.) Both steps are relatively easily achieved in the M400 by using two specific functions.

To change the *fuelling* during lean cruise, a Lambda Compensation table is used. This causes the ECU to aim for leaner mixtures when this mode is activated. To change the *timing* during lean cruise, an ignition compensation table is used. This causes the ECU to add timing when this mode is activated. Both use two-dimensional tables, with the changes in fuel and ignition varying with MAP pressure.

So straightforward, huh? No – because the big issue is: *when should the car be in lean cruise and when should it not be?*

This is hugely important because when in lean cruise, the engine develops less power and is less responsive to throttle. Set it up wrongly, and the car drives like a dog.

Furthermore, at high loads, the use of lean mixtures and advanced timing could easily destroy the engine. So to maintain good driveability and safe engine operation, the lean cruise mode needs to occur only in certain, very specific driving situations.

Starting points

So what aspects need to be analysed to work out when lean cruise is appropriate?

Load and throttle angle

Load is measured by the ECU via a MAP sensor. In a turbocharged car like the Honda, vacuum is indicated by pressures below 100kPa and boost is indicated by pressures above 100kPa. (That is, atmospheric pressure is around 100kPa.) Therefore, the first logical step in determining when lean cruise should occur is to specify that *lean cruise can occur only at MAP of less than 115kPa*. That is, any boost greater than 15kPa prevents lean cruise occurring.

However, MAP alone is insufficient. Throttle position is also needed – after all, the driver flooring the throttle is immediately indicative of high load … desired, if not actual! Therefore, *lean cruise can occur only at throttle positions less than 60 per cent*. (This is a higher value than you'd normally use, because the Honda uses a small engine and very tall gearing, so throttle angles are often larger than in other cars.)

But specifying throttle openings of less than 60 per cent is insufficient. What happens when the throttle is fully lifted? In that case, the ECU is programmed to reduce fuelling to zero – that is, complete injector shut off then occurs. If throttle is then reapplied, and lean cruise mode can then immediately occur, the engine would transition from no fuel (on no load) to very lean mixtures (on light load). The result? Doughy response and potentially a 'lean jerk' as there is insufficient fuel for the transition.

Therefore, *lean cruise can occur only when throttle position is over 4 per cent for 2 seconds*. (Note in this

Lambda Table (Lambda)

Load kPa	RPM 0	1000	1500	2000	2500	3000	4000	5000	6000	7000
180.0	1.00	1.00	0.85	0.85	0.85	0.85	0.82	0.80	0.80	0.80
160.0	1.00	1.00	0.85	0.85	0.85	0.85	0.85	0.80	0.80	0.80
140.0	1.00	1.00	0.95	0.95	0.88	0.85	0.85	0.85	0.80	0.80
120.0	1.00	1.00	1.00	1.00	1.00	0.92	0.92	0.92	0.92	0.92
100.0	1.00	1.00	1.00	1.00	1.00	1.00	1.00	1.00	1.00	1.00
80.0	1.00	1.00	1.00	1.00	1.00	1.00	1.00	1.00	1.00	1.00
60.0	1.00	1.00	1.00	1.20	1.20	1.20	1.20	1.20	1.20	1.20
40.0	1.00	1.00	1.00	1.20	1.20	1.20	1.20	1.20	1.20	1.20
30.0	1.00	1.00	1.00	1.20	1.20	1.20	1.20	1.20	1.20	1.20
20.0	1.00	1.00	1.00	1.20	1.20	1.20	1.20	1.20	1.20	1.20

The Lambda 'aim' chart for normal (ie non lean cruise) conditions. Multiply Lambda by 14.7 to get an equivalent air/fuel ratio for petrol (gas).

4. PROGRAMMABLE ENGINE MANAGEMENT

Ign Main (°BTDC)

Load kPa \ RPM	0	500	750	1000	1350	1500	1750	2000	2500	3000	3500	3750	4000	4500	5000	5500	6000
110.0	3.0	1.0	1.0	-5.0	-5.0	-5.0	-2.0	-5.0	-6.0	-3.0	2.0	5.0	5.0	4.5	1.0	0.0	0.0
100.0	3.0	-4.0	-5.0	-10.0	-10.5	-9.0	-5.0	-5.0	-3.0	1.0	3.0	7.0	5.0	4.0	4.0	4.0	4.0
90.0	3.0	-7.0	-7.0	-10.0	-10.5	-9.0	-4.0	1.0	3.0	8.0	8.0	8.0	7.0	5.0	4.0	4.0	4.0
80.0	3.0	-7.0	-7.0	-7.0	-10.0	-3.0	3.0	5.0	8.0	10.0	10.0	8.0	8.0	5.0	4.0	4.0	4.0
70.0	3.0	-7.0	-7.0	-3.0	-6.0	0.0	6.0	6.0	8.0	12.0	12.0	12.0	12.0	11.0	9.0	9.0	9.0
60.0	3.0	-7.0	-7.0	4.0	4.0	8.0	8.0	8.0	8.0	14.0	20.0	20.0	17.0	15.0	13.0	13.0	13.0
50.0	3.0	-7.0	-5.0	7.0	7.0	8.0	8.0	8.0	8.0	10.0	10.0	10.0	10.0	10.0	10.0	10.0	10.0
45.0	3.0	-7.0	-5.0	10.0	10.0	10.0	10.0	10.0	10.0	10.0	10.0	10.0	10.0	10.0	10.0	10.0	10.0
40.0	3.0	3.0	3.0	14.0	14.0	14.0	14.0	14.0	14.0	14.0	14.0	14.0	14.0	14.0	14.0	14.0	14.0
0.0	3.0	3.0	3.0	12.0	14.0	14.0	14.0	14.0	14.0	14.0	14.0	14.0	14.0	14.0	14.0	14.0	14.0

The normal (ie non lean cruise) ignition timing chart.

Lambda Comp (Lambda)

MAP kPa \ Lean	0.0	0.1
100.0	0.00	0.00
90.0	0.00	0.00
80.0	0.00	0.00
70.0	0.00	0.35
60.0	0.00	0.50
50.0	0.00	0.60
40.0	0.20	0.60
30.0	0.20	0.70
20.0	0.20	0.60
10.0	0.00	0.20

The additional correction to 'aim' Lambda values made when in lean cruise (right-hand column headed '0.1'). These Lambda figures are added to the ones in the table (page 88) to gain the final lean cruise Lambda aim figures.

Ign Comp 1 (Trim)

MAP kPa \ Lean	0.0	0.1
100.0	0	0
90.0	0	-3
85.0	0	3
80.0	0	21
70.0	0	34
60.0	0	44
50.0	0	44
40.0	0	44
30.0	0	19
20.0	0	10

The ignition timing correction figures when in lean cruise mode (right-hand column headed '0.1'). Again, these figures are added to the existing ignition timing (table above) to give the final lean cruise advance. And yes, that does give a variation in ignition timing of over 50 degrees …

parameter the use of the time aspect (2 seconds) as well as throttle position.)

Gear

The Honda uses a five-speed manual transmission and, as already mentioned, very tall gearing. Let's consider a real world driving situation. If you're trickling along in a long line of slow-moving traffic at 30km/h, the car will be in second or third gear. In that situation, lean cruise would not be desired. Throttle response would be poor, and so keeping up with other traffic that is varying in speed would be clumsy.

But here's the issue: with only the above two criteria of load (less than 100kPa) and throttle position (4-60 per cent), in this 'traffic jam' situation, the car would go into lean cruise. So how can that be avoided? By monitoring the gear and allowing lean cruise to occur only in fourth and fifth gears.

The calculation of gear is made by the ADL3 dash. The ADL3 is linked to the ECU via a CAN bus that gives it access to all the ECU inputs and outputs. In this case, using the engine rpm and road speed inputs and an internal look-up table that specifies gear ratios, the dash can calculate the selected gear. This was initially done so that 'selected gear' could be displayed on the dash, but the calculated gear can also be used as a lean cruise input.

Engine warm-up

Running lean cruise when the engine is still cold will result in poor driveability. That's because the engine needs richer mixtures during warm-up. Even engines that will tolerate 14.7:1 (stoichiometric) during warm-up will not be happy with leaner mixtures in that time. Therefore, lean cruise can occur only at coolant temperatures over 80°C for 5 seconds.

The logic, so far

Let's take stock. So far, lean cruise mode will be entered only if the following criteria are met:
- MAP less than 115kPa
- Throttle between 4 and 60 per cent
- Coolant temperature over 80°C for more than 5 seconds

> **USER FUNCTION**
>
> These parameters (fourth or fifth gears, engine over 80°C, etc) are used by the ADL3 dash to activate Lean Cruise Mode via what is called a 'User Function'. User Function allows a string of variables to be examined, and an output energised only when all the variables meet the required conditions. This function is usually for simple tasks (like activating the radiator fan when the coolant temperature is over 80°C), but, as shown here, it can be used to set up much more complex functions than that. Up to eight variables can be used in the one logic string, and multiple logic strings can be activated.

This list might seem sufficient – but it isn't. Remember, doughy response occurs when the engine is running very lean mixtures. That is, a quick movement of the throttle results in not much happening in terms of increased torque output from the engine. That, in turn, gives poor driveability. So what can we do about this?

Rate of throttle change

Here's where it starts getting very interesting – and where with the MoTeC hardware, again the ADL3 dash needs to become involved. The dash, by using its internal 'differentiate seconds' function, can assess how fast the throttle is being moved. This value (let's call it 'Accelerator Rate of Change') can then be used as an input into the lean cruise decision making.

Therefore, *lean cruise can occur only if the Accelerator Rate of Change is low.*

In fact, this factor is most important in transitioning *out* of lean cruise – there is good throttle response at all times, because the car cannot be in lean cruise if the throttle is being moved rapidly.

So it's starting to sound good. But not quite yet.

Imagine another driving situation – this time the car is being driven for fun. And when being driven for fun, lean cruise should not occur! But how do you determine when a car is being driven for fun?

Driving style

By monitoring throttle rate of change, we can get a feel for how the car is being driven. But not just instantaneous rate of change, as described above, but *rate of change over a period.* If the throttle is often being moved fast, the car is being driven for fun – and so lean cruise isn't wanted.

We can assess this again using the ADL3 dash, again using the 'Accelerator Rate of Change' value. However, this time we mathematically square the Accelerator Rate of Change value (so that it's always positive), and apply a large filter to the value, so that it changes only slowly. As the speed of movement of the throttle (either opening or closing) slows, so this value gradually drops towards zero. We can call this value 'Driving Style.'

(To make it easy for me to remember, I refer in my notes to Driving Style as the '*emptying bucket* rate of change of throttle.' That is, the analogy is of a bucket with a small hole in it – water is placed in the bucket each time the throttle is moved quickly, but when no water is being put in, the bucket slowly empties.)

So the next parameter is added – *lean cruise can occur only if the Driving Style value is below a certain value.*

Transition to lean cruise

So let's summarise the factors that we're using to determine whether the car is in lean cruise or not:

- MAP less than 115kPa
- Throttle between 4 and 60 per cent
- Coolant temperature over 80°C for more than 5 seconds
- Accelerator Rate of Change low
- Driving Style value low

But there's still one more factor to consider. As so far described, the transition into lean cruise occurs as a distinct step. That is, when the criteria are all met, and the car goes into lean cruise. That transition immediately reduces engine torque output for a given throttle position: the power at those revs therefore can be felt to drop. So, from the driver's point of view, when the car goes into lean cruise, something has gone wrong! That is, the driver thinks: why has power just dropped?

The way to avoid this 'step' transition is to *slide* into the mode – to implement it over a few seconds rather than abruptly. This is achieved by filtering the value that indicates to the ECU that it should go into lean cruise – thus the value is not on/off but slides in over a short period.

Outcome

And the outcome? A lean cruise function that appears to think for itself. Accelerate hard through the gears and the car's air/fuel ratios are as you'd expect. But back off and drive gently, and within a few seconds of driving gently, the car slides into lean cruise mode. Accelerate when in lean cruise with a quick foot movement and the air/fuel ratios are immediately returned to standard.

Drive hard, and even when you drive more gently for 20 or 30 seconds before again driving hard, the management system has not gone into lean cruise during the gentle driving … it 'knew' you were likely to again drive hard.

In lean cruise and you back right off? When you reapply throttle the response is as you'd expect … air/fuel ratios normal before lean cruise is again slid back into.

If you are using a programmable ECU (or ECU and dash combo, as here), there may be possibilities that you've never envisaged. With programmable management systems having the capability to develop user-definable variables (as was done here with the 'Driving Style' factor), and have user-definable decision making based on multiple variables, there is a flexibility that is truly limited only by your imagination.

And this is one area where the DIY tuner has a lot more opportunity to develop these areas than a workshop. Developing bespoke functions like the lean cruise mode covered in this story takes a lot of time – but if you're working from home, you're not paying anyone an hourly rate! If you currently run a sophisticated programmable ECU in your car, think of what you can potentially do …

Footnote: the fully programmable logic of the Adaptronic e1280s ECU described in Chapter 6 can take this customised approach even further.

4. PROGRAMMABLE ENGINE MANAGEMENT

FINDING THE CRANK REFERENCE INDICATOR POSITION

Once an ECU is configured so that it can sense both camshaft and crankshaft positions, the relationship between these two signals needs to be set in the ECU.

In MoTeC software, this critical number is specified in degrees of crank rotation. It is called the crankshaft index position – abbreviated to CRIP. (Other ECUs require the same information, but may use different nomenclature.)

Without an accurate CRIP, the ignition spark will be incorrectly timed, and the sequential injection will also be incorrect. In fact, if the CRIP is greatly in error, the engine won't even start. So how do you ascertain what the CRIP is for a particular engine? Here I'll use the example of a Honda three-cylinder engine.

You will need to have a degree wheel and a timing light.

Measuring the CRIP

To calculate the CRIP, three factors needed to be ascertained:
1. the crank rotational position at which the camshaft signal is output;
2. the crank rotational position at TDC on the firing stroke of Cylinder #1;
3. the number of crankshaft degrees between steps 1 and 2 above.

Let's now take each step one at a time.

1. Crank position at camshaft signal output

Unfortunately, on the Honda, the crank rotational position at which the camshaft signal is output could not be found electronically. Why not? Because until the engine is spinning fast enough, no signal is output! Instead, this position needed to be directly observed.

Making this tricky is the fact that the Honda actually has two teeth on the cam, each sensed by two sensors. An electronic conversion box was built that output a signal only when both teeth were opposite their respective sensors – giving one output per cam rotation. In this situation, it would have been good if both camshaft position sensors could be removed – then a visual check could have been made when the two cam teeth were opposite their respective sensors. However, on this engine, this would have involved a great deal of work to gain access, especially to the rear sensor. Instead, just the front camshaft position sensor was removed, and the cam teeth spinning past it were observed down the hole.

With an assistant watching carefully, I turned over the crank. (Ensure the crank is turned in the correct direction!) By observation, the assistant could then ascertain where in their rotation the adjoining teeth were on the cam. By getting the pattern mentally worked out in this way, the assistant could then state when each tooth must be near its respective sensor – despite being able to see only one tooth at one sensor.

2. Crank rotational position at TDC on the firing stroke of Cylinder #1

The crank rotates twice per firing stroke on Cylinder #1, so how to find out when the cylinder was at TDC on its firing stroke, not exhaust stroke? I could have lifted the rocker cover and observed the valves directly, but a local mechanic suggested an easier way.

The sparkplug of Cylinder #1 was removed, and a tight-fitting rubber hose pushed into the sparkplug recess. By placing a finger over the end of the hose, the pressure rise could be easily felt as the piston rose on the firing stroke. (On the exhaust stroke, there was no pressure increase – the exhaust valves were open.)

3. Number of crankshaft degrees

So the process was this.
1. Crank rotated until the assistant could surmise that the two cam teeth were each opposite their respective sensors;
2. Degree wheel on crank marked at this point;
3. Crank rotated until pressure felt rising in rubber hose coming from Cylinder #1 (indicative of piston rising on firing stroke);
4. Crank then stopped at TDC mark on crank pulley;
5. Number of crank degrees that the crankshaft had rotated from Step 1 to 4 read off degree wheel.

In the case of the Honda, the CRIP was 694 degrees. In other words, TDC of Cylinder #1 occurs 694 crankshaft degrees after the cam position sensor output first occurs. (Or, if you want to look at it another way, the camshaft sensor output occurs 26 degrees before TDC on Cylinder #1's firing stroke.)

Checking and fine-tuning the CRIP

The next step is to set the ECU to give a constant, fixed ignition timing output. This number can be specified in the MoTeC software as 'Test Advance.' In the Honda's case, I set it at 12 degrees before TDC.

With a timing light working with Cylinder #1, the engine is then cranked. If the timing light doesn't show 12 degrees of ignition timing (or whatever number you have set – 0 degrees would be another easy one to use), then the CRIP needs to be changed in the software.

The aim is to use a CRIP that gives a measured timing corresponding to the ECU-dictated timing (in this case, 12 degrees). In other words, an ECU-set timing of (say) 12 degrees actually is 12 degrees when measured with the timing light on the crank pulley. Fine-tuning the CRIP during cranking in this way will normally give accurate enough timing for the engine to run. When the engine is running, and so the timing light is firing more frequently, the CRIP can then be fine-tuned in 0.5-degree increments – this way, it can be set dead-on.

WORKSHOP PRO MODIFYING THE ELECTRONICS OF MODERN CLASSIC CARS

Chapter 5
Other engine bay modifications

- Moving the battery to the boot (trunk)
- Installing a diagnostics socket on non-OBD cars
- Auto air conditioner controller
- Five channel data-logger
- High pressure intercooler water spray
- Adding a speed-controlled fan to your intercooler
- Electronic adjustable radiator fan switch
- Voltage boosters

WORKSHOP PRO: MODIFYING THE ELECTRONICS OF MODERN CLASSIC CARS

This chapter covers modifications that involve the engine or engine bay – but not engine management systems, that have been covered in previous chapters.

MOVING THE BATTERY TO THE BOOT (TRUNK)

Often the battery needs to be moved to create space in the engine bay for performance modifications, such as the installation of an intercooler – either an air/air core or a water/air heat exchanger. So what is the best way of moving the battery to another location? Let's take a look.

In this car the battery was moved to the boot to make room for an underbonnet intercooler (right).

UNDER THE BONNET (HOOD)

Here's what the typical battery mounting arrangement of the standard car looks like. The battery is retained by two vertical rods and a crossover bracket. The rods are anchored to a battery tray, which in turn is bolted to the bodywork. Small diameter electrical clamps are used around the battery posts to connect the battery to the car's wiring. Next to the positive clamp are fusible links.

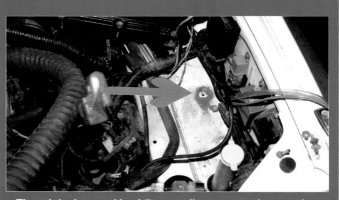

The original ground lead (it normally connects the negative terminal of the battery straight to the engine) needs to be terminated on the bodywork. Find a convenient bolt (here one of the threaded holes from a bolt that had held the battery tray in place has been used) and remove the paint from around the hole. You want to see bare, shiny metal all around this spot.

The first steps were to remove the battery and then unbolt its tray. In some cars, the tray will be spot-welded – rather than bolted – into place. To remove this type of tray, the welds will need to be drilled out – or if it's not in the way, it can be left in place and a new battery tray bought for the relocation project.

The earth battery post clamp looked like this. It was easy enough to use a hacksaw to cut off the loop (ie cut along the red line), leaving a well-terminated lug that could then be bolted to the bodywork.

5. OTHER ENGINE BAY MODIFICATIONS

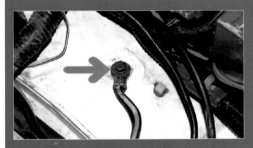

The ground strap bolted in place. Only the one positive cable runs to the new battery location, so the ground connections here and at the battery need to be very good if there's not to be a voltage drop between the battery and the electrical loads.

The positive battery clamp was also cut off in the same way, leaving a pair of lugs. (One connected to a cable heading straight to the starter motor, the other to a cable going to the main loom.) Using a nut and bolt, these lugs were connected to a new terminal that had been soldered to the new heavy-duty cable that will run to the relocated battery. If you don't have a big soldering iron, pick terminals that can be attached to the new cable by crimping or with set screws. Again, these connections need to be electrically good.

You must make absolutely certain that this connection can't come into electrical contact with the body or engine. I slipped heat-shrink over the new join and then, as seen in the lower picture, made a clamp from a piece of aluminium sheet to hold the cable in place. Some rubber fuel hose was slipped over the connection to further protect it from the clamp. So, original battery cable goes in at one end ... and new battery cable comes out at the other end. At the top right of the photo you can also see the revised ground cable connection.

RUNNING THE CABLE

There are two paths that the cable can take – through the cabin or under the floor. I chose the latter, for two reasons. Firstly, it's usually a lot easier drilling a hole in the boot floor than trying to drill through the firewall (often firewalls are double layer and/or insulated). Secondly, the cable (complete with insulation) is very thick, and clearances along a sill panel inside the cabin are often tight. This is one cable that you don't want crushed, eg under a seatbelt retaining bolt! I placed one side of the car on ramps and chocked the other wheels, allowing easy access under that side.

The easiest route from the engine bay to the floor of the car was to go around the engine bay until the cable reached the centreline of the car, then head southwards. Here the cable has been clamped into place, using the clamps available for retaining copper plumbing. From the point where the cable might be subjected to exhaust heat and/or abrasion, the battery cable has been run inside clear plastic hose. This acts as an additional protectant and adds little to the cost.

WORKSHOP PRO: MODIFYING THE ELECTRONICS OF MODERN CLASSIC CARS

Make sure that you buy at least 2 metres (yards) more cable than you think that you will need – it is very difficult to join on a little bit extra if it turns out to be too short.

The cable – complete with its clear plastic tube cover – was cable-tied into place alongside the plumbing that already runs from the back to the front of the car. Heavy-duty cable ties were used, being tightened up only when everything was positioned correctly.

At the rear of the car, the cable goes its own way – it separates from the plumbing – and from here on it was clamped into place. Note how it has been kept higher than the chassis rail, helping to protect it from impacts and flying rocks.

The cable passes through a rubber grommet in the floor of the boot (trunk). This will probably need a new hole to be drilled for the cable, but first have a good look around – often there are holes through the boot floor which are filled by rubber plugs. Don't be tempted to leave out the grommet – it's a must-have to protect the cable against the sharp metal edge.

THE BATTERY BOX

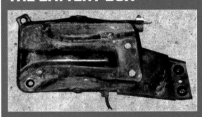

The first steps in installing the battery box was to retrieve the original battery tray, cut off the protruding brackets and give the modified tray a quick splash of paint. If the original battery tray isn't suitable for use in the box, buy another – avoid just placing the battery in the (relatively flimsy) plastic box.

Talking of battery boxes, here it is bolted to the floor of the boot. Note the large washers that have been used on the high tensile bolts – the washers help spread the load, and so resist tearing through the tray. Only three bolts have been used here, because of the shape of the tray – normally you'd use four.

Here the underside of two of the bolts holding down the battery tray (and so box) can be seen. Again large washers have been used to spread the load – one of the washers needed a flat cut on it to clear the chassis rail. Placed either side of the chassis rail in this way makes the assembly strong and rigid – important as the battery is heavy and you don't want it coming loose in any circumstances … including an accident. The third bolt holding the box in place has been attached to an aluminium plate that is supported by the rear bumper brackets, located on top of the chassis rail.

5. OTHER ENGINE BAY MODIFICATIONS

The battery terminals are of the screw-on type: the cable slips through an opening and then a bolt tightens on the cable from the side. This shows the ground cable. It's red cable like that already used but it has been wrapped in black tape to prevent colour confusion. It's grounded to the body in the same way as the cable in the engine bay.

A 75A circuit breaker is mounted on the side of the box. These breakers are available online. Without a circuit breaker or fuse, any short-circuit that develops between the battery and the original fusible link in the engine bay has the potential to cause a fire. The breaker can tolerate twice its rated current for 10 seconds, allowing its rating to be lower than the starter motor current draw.

To prevent anything coming into contact with the live terminals, a cover over the circuit breaker was made from acrylic sheet. The press-button (power off) and the reset button (power on) can still be easily reached as required. Note that unless a completely sealed battery is used, ventilation should be provided to allow the escape of gases emitted when the battery is being charged. Most boots in cars are ventilated but if there could be a problem, the battery box should be sealed (eg with foam rubber weather-sealing strips), and a hose placed from the box to outside air for ventilation. Alternatively, batteries designed for use inside the cabin of cars often have a vent tube provided – this can easily be run to the outside.

ANOTHER APPROACH

I took a different approach in another car. In this car – a small hatchback – the battery was moved to a position inside the spare wheel well. The spare wheel was then inverted and supported on a frame that allows it to be rigidly placed over the top of the battery.

This folded sheet metal tray bolts in place in the spare wheel well. The frame above it supports the inverted spare wheel.

The battery in place. The battery is a sealed design with a vent tube that was run through the bottom of the spare wheel well to the outside air. Note the box at lower left that contains the circuit breaker, push-button and digital meter display. Push the button to check battery voltage.

The wheel is held firmly in place by the frame and two horizontal bars that screw into place. You do not want the wheel sliding around over the top of the battery! Taking this approach allowed the battery to be moved to the rear of the car without it intruding into the load space. The approach also keeps the battery low in the car. A carpeted trim panel normally covers this compartment.

WORKSHOP PRO: MODIFYING THE ELECTRONICS OF MODERN CLASSIC CARS

INSTALLING A DIAGNOSTICS SOCKET ON NON-OBD CARS

In cars with on-board diagnostics (OBD) support, there's a wealth of high-quality information that can be accessed from the socket. In addition to fault codes, aspects like airflow, intake air temperature, long- and short-term fuel learning trims, and other aspects can be measured. But what if your car doesn't have an OBD socket? In that case, you might like to do what I've done here and fit a custom diagnostics socket. It won't give you read-outs in engineering units (like airflow in grams per second) but it will allow you to easily access a lot of interesting information.

So what is this diagnostic socket, then? Well, put simply, it's a multi-pin socket that's wired to certain pins of the engine ECU. By connecting a matching plug to a multimeter or oscilloscope, you can then quickly and easily access ECU sensor information – just by plugging into the socket. That's so much easier than having to pull off trim panels, locate the ECU, find the right pins, connect the wires ... and then do the measurements, only to then have to undo the whole process when you have finished!

While installation of the DIY diagnostics socket is straightforward, it's the sort of job which is best done when you are already accessing the ECU and making some connections to it. So for example, if you are fitting an interceptor, you're connecting an aftermarket tachometer, or you're connecting a voltage switch circuit that clicks over when certain sensor conditions occur, now's also the time to fit the diagnostic socket. The extra work isn't all that great – and the convenience and ease of use sure make it worthwhile.

You'll need to source a multi-pin, polarised plug and socket combination. (Polarised simply means that the plug fits into the socket with only the one orientation.) In my case I used a 6-pin DIN socket (available from electronics suppliers), because I figured that would give about the right number of outputs, and I had a matching plug and chassis mount socket already on hand. These days I'd probably use my preferred plug and socket type – Deutsch DT.

If you want to be able to easily measure nearly every ECU input, you'll need to have a correspondingly greater number of pins. I suggest that the following are good possibilities:

All applications:
- 12V
- Ground

Inputs:
- Oxygen sensor (if multiple units, any located before the cats)
- Airflow meter (or MAP sensor)
- Intake air temperature
- Coolant temperature
- Crank angle sensor
- Cam angle sensor

Outputs:
- Injector pulse
- Variable intake changeover

Now if you decide to do all of these, you'll need to have a plug/socket combination with at least 10 pins. In my application I decided to initially do only +12V, earth, front oxygen sensor and airflow meter. This approach used four of the six pins – at a later date, another pin will probably be wired for injector output. Note that pretty well all sensors are referenced to chassis ground – so only the one ground connection is needed.

Nearly all ECUs use regulated 5-volt outputs to at least some sensors. It's possible to draw a small amount of extra current from this supply without causing dramas, so if you're connecting a custom measuring instrument that requires a regulated 5V supply, you can save on some components by using the 5-volt supply that's already available at the ECU.

Incidentally, the reason that you would normally connect to the oxygen sensor that's located in front of the cat is this sensor has a more meaningful output than the post-cat one.

INFORMATION USES

OK, so what use can be made of the information from your new diagnostics socket? Obviously, what you can do with the information depends on what ECU pins you've accessed and what measuring instruments you have. However, here's a smorgasbord:

- **Oxygen sensor output:** measure with multimeter (millivolts) to assess oxygen sensor health, when the ECU is in closed and open loop, and approximate mixture strength.
- **Airflow meter output:** measure with a multimeter (volts or Hertz) to make sure that meter hasn't reached its ceiling output (eg through increased airflow caused by modifications); assess tuning changes to see if peak airflow reading alters (more air = more power!).
- **MAP sensor:** on turbo or supercharged cars, a logging multimeter can be used to very accurately track boost levels, allowing (for example) the picking of boost overshoots.
- **Injector duty cycle ECU output:** measure with a multimeter (duty cycle) to make sure that injectors aren't maxing out at 100 per cent duty cycle, or see how much more fuel flow capability is left after modifications.
- **Intake air/coolant temperature sensor outputs:** Use a multimeter (volts) to watch trends in these temps

5. OTHER ENGINE BAY MODIFICATIONS

INSTALLING THE SOCKET

A chassis mount socket is likely to give the neatest job. This DIN socket was mounted on a bracket made from 3mm (⅛in) thick aluminium bar. Why so thick? Well, you don't want the bracket flexing when you're pushing in and pulling out the plug. For the same reason, the bracket should be held in place with a strong bolt.

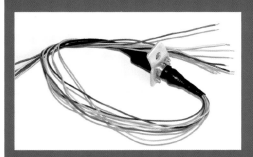

Depending on the design of socket, the wires will be either soldered or crimped into place. It's very easy to make mistakes when wiring sockets and plugs, so do as I've done here and use the same colour wires to the same pins of both the plug and the socket. That way, it's both easy to make sure that you have continuity across the plug/socket (just use your multimeter to make sure that each colour wire is continuous across the plug/socket) and it also makes it much easier to wire the system into place in the car.

Place the new socket where it can be easily accessed. Also note that the exposed pins will be carrying voltages – so either locate it where that's not a problem, or put a second, unwired plug into it whenever it's not being used for diagnostics. In this car, the glovebox lid can be pivoted down further than normal in order that the cabin air filter can be changed. That movement also reveals a good location for the socket.

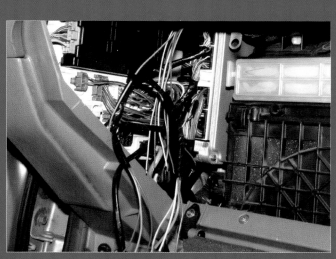

Most ECU wiring – especially when something else is being installed – looks like spaghetti junction. In addition to the DIY diagnostics socket, here a voltage switch is being installed on the airflow meter output, and an interceptor kit is also being installed – again using the airflow meter output and corresponding ECU input. For the DIY diagnostic socket, use the workshop manual and a multimeter to find the signals that you'd like to be able to monitor. Then connect one of the colour-coded wires from the diagnostic socket to each of these ECU pins, connecting to the ECU wire back a little from its plug. Use solder or high-quality crimps to make these connections, and don't forget to insulate them well.

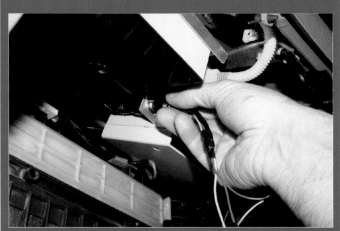

Accessing the DIY diagnostics socket is as easy as pivoting down the glovebox lid and inserting the plug.

99

WORKSHOP PRO: MODIFYING THE ELECTRONICS OF MODERN CLASSIC CARS

(volts will go down as temperature goes up).
- **Crank or cam angle sensors:** Use a dual trace oscilloscope to watch real time cam timing variations
- **Variable intake system changeover:** Use a multimeter (volts) to watch when a dual-length resonant intake system changes from one mode to another.

AUTO AIR CONDITIONER CONTROLLER

If you have a really economical, low-powered car, you probably know the feeling. Climbing a hill, holding as high a gear (or in an automatic transmission car, as small a throttle angle) as possible – and then the air conditioner compressor cuts in.

Suddenly, you have to drop back a gear or apply more throttle. And there goes your good fuel economy. Or you're performing a passing move, and you want absolutely every bit of horsepower the engine can produce. The air conditioner? For the next 30 seconds, you want it off, off, off!

Well, both scenarios are now easily catered for. By using an electronic module that constantly monitors your throttle position, the air conditioner compressor can be automatically switched off (or, sometimes even better, stopped from coming on!) when you have your foot down past a certain point. During the initial set-up, you can set the throttle angle at which the compressor becomes disabled, and you can also set how much you then need to lift the throttle before the compressor switches back on.

More power when you put your foot down, and – because when accelerating and climbing hills, you can hold taller gears, and so use less engine revs – better fuel economy as well. (And there's another reason that fuel economy is improved – I'll come back to that later.)

Here I've used the voltage switch available from eLabtronics (www.elabtronics.com), but any of the many voltage switches available cheaply online can also be used. In addition to a voltage switch, you'll also need a box to mount it in, and a 20A 'changeover' automotive relay that you can buy from your local auto parts shop.

How it works

The heart of the system is the eLabtronics voltage switch. The module is internally highly sophisticated, but has only four external wiring connections. In addition to power and ground connections, there are 'in' and 'out' connections. In this application, the 'in' terminal is connected to the throttle position sensor that is standard fitment on all cars. These sensors typically output a voltage that rises from about 0.5V to about 4.5V as the throttle is opened. The switch is able to monitor this voltage, without affecting the standard operation of the car's engine management or electronic throttle systems.

By setting an adjustable pot on the voltage switch, it can be made to trip at any throttle position you choose. In this application, you'd probably have it tripping at about one-third throttle – more on setup in a moment.

Now, how does the voltage switch turn off the air conditioner compressor?

All air conditioner compressors use an electro-magnetic clutch. This means that when power is fed to the clutch, the compressor can be driven by the engine. When power is cut off, the compressor drive wheel just spins freely. This compressor clutch power feed is the single wire that you can see going to the front of the compressor. To disable the compressor, all that we need to do is to put a normally-closed relay in this circuit, so that when the voltage switch trips, the power feed to the compressor clutch is turned off.

Wiring

Firstly, make sure that you can locate:
- the air conditioner compressor and the single wire going to its clutch – it will probably be a fairly thick wire
- the throttle position sensor, mounted on the throttle – normally, this sensor will use a connector with three wires

Once you have identified the throttle position sensor, use a multimeter to back-probe the wires (with the connector still in plugged in). Connect the black lead of the multimeter to the negative terminal of the battery (or the chassis) and use the other lead to back-probe the connector. With the ignition switched on, one wire should have a voltage signal that rises with increasing throttle movement. This is the sensor output that will be connected to the 'in' terminal of the voltage switch.

The eLabtronics voltage switch was used here, but any voltage switch that works off 12V and can sense voltages around 1-5V will work fine.

5. OTHER ENGINE BAY MODIFICATIONS

As can be seen, I used the air con compressor power feed to supply power to the voltage switch. Taking this approach makes installation easier – for example, it's simple to mount the module under the bonnet (hood) as no wires then need to be fed through the firewall.

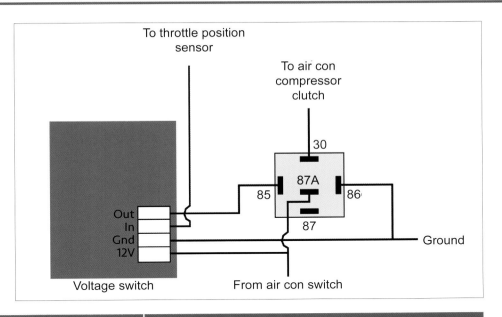

The next step is to do the actual wiring. Automotive changeover relays (properly called single pole, double throw) have standardised numbers on their terminals, making the wiring straightforward.

FITTING THE AIR CONDITIONER CONTROLLER
I fitted the Auto Air Conditioner Controller to my Honda Insight.

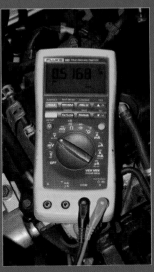

A multimeter was used to back-probe the throttle position sensor until a wire (red/black in the case of the Honda) was found that had the voltage on it that varied with throttle position. This photo shows the 'throttle released' voltage.

Next the wire to the compressor clutch was located – in the case of the Honda, it is blue/red.

The red/black wire was bared and a new wire (green) was soldered to it. The original wire should not be cut – the new wire just taps into the signal.

The compressor clutch wire was cut and extension wires (grey) were soldered to them before later being insulated with tape. Note that it is important that you mark the wire that comes from the air con on/off switch (or climate control ECU) as this is the one used to feed power to the voltage switch. The extension wires from the air con clutch need to be at least as thick in copper diameter as the original wires.

WORKSHOP PRO: MODIFYING THE ELECTRONICS OF MODERN CLASSIC CARS

The wiring between the relay and the module, and to the compressor clutch and throttle position sensor, were done next.

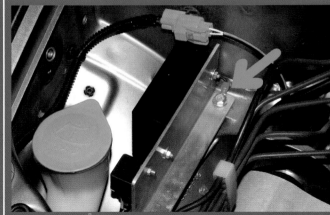

Note the ground lug (arrowed) that was bolted to the chassis when the module was installed. Both the relay and the box housing the module were mounted on folded aluminium plate.

Setup
If you are using the eLabtronics voltage switch, there are three steps in setting it up.

DIP switch
On the board there's a DIP switch to configure the module for different functions. Orientate the board so that the terminal strip is on the right and then use a ballpoint pen or a small screwdriver to set the switches so that they look like this:

			X
X	X	X	

This configures the voltage switch to turn on its output when the voltage rises above the set-point. If you have a weird throttle position sensor that works backwards (so that voltage falls with increasing load) configure the DIP switch like this:

		X	X
X	X		

Pots
The next step is to set the two pots. Orientate the board so that the terminal strip is on the right, then look at the bottom pot first. This pot adjusts the set-point. Use a small flat-bladed screwdriver to rotate this pot anti-clockwise at least 15 turns, or until it can be heard clicking. (Why so many turns? Multi-turn pots like the ones fitted don't have clear 'end stops,' so you need to ensure you've turned it anti-clockwise as far as possible.) Then rotate this pot clockwise about four full turns. By doing this, you have set the turn-on voltage to a low value.

OK, so what about the other pot? This one sets the difference between the turn-on and turn-off values. As with the set-point pot, use a small flat-bladed screwdriver to rotate the bottom pot anti-clockwise at least 15 turns, or until it can be heard clicking. Then also rotate this pot clockwise about 4 full turns. This adjusts the difference between turn on and turn off values (what's called 'hysteresis') to a small value.

Both the set-point and hysteresis are *increased* by rotating their respective pots *clockwise*.

Now you need to mount the module (housed in the box) and the relay – and do some test driving. (Remember that if any of the PCB tracks of the module – or its terminal connections – touch the car's body, nasty things will happen. The module must be mounted in a box before you can do any testing!)

If it is not clear when the voltage switch is operating (eg if it's mounted under the bonnet), temporarily wire a 12V pilot light across the relay's coil and place the light in the cabin. If the module is inside the cabin, the onboard LED will show when the module has tripped.

How you set the pots is up to you – but here are some hints.

For *maximum improvement in fuel economy*, adjust the set-point pot to the lowest value that does not trip the switch when accelerating normally (eg away from traffic lights) and when travelling at the open road speed limit. Any lower value than this will cause you to lose the air con when travelling at higher speeds, and turn the compressor on and off a lot in urban conditions. For *maximum retention of the air con*, set the pot so that the voltage switch trips only at full throttle – or near full throttle.

But what about the other pot – hysteresis? I found that for the best outcome, this should be set to a small value

5. OTHER ENGINE BAY MODIFICATIONS

– but not so small that the voltage switch chatters on and off.

Results

The results of fitting the Auto Air Conditioner Controller to the Honda Insight exceeded my expectations. On the small engine Honda, the ability to hold higher gears up hills and accelerate more easily at small throttle angles was excellent. An unexpected benefit – and probably the thing I noticed the most – was the absence of the air con cutting-in when climbing a long hill. Previously, you'd have the car set to work with a fairly large throttle angle at low-ish revs in a high gear (the best approach for fuel economy) and then part way up the hill, the air con would cut-in and all the balance would be lost.

With the Auto Air Conditioner Controller in place, time after time it was like you had a really attentive passenger with their finger hovering over the air con button, ready to turn the air con off at the first sign you could benefit from their doing so.

FIVE CHANNEL DATA LOGGER

The value of data logging when modifying cars is common knowledge. Programmable engine management systems have in-built logging facilities, and many standard management systems with aftermarket programming software can log engine signals. Even cars with only an OBD port can have simple, low speed logging performed for certain engine management parameters.

But what if your car fits none of these categories? Or you have an OBD port, but want higher speed logging?

In the past it's been hard to come up with a simple and cheap way of logging signals – but this data logger achieves just that. It's not perfect – it won't log frequency signals like engine revs for example – but the price and ease of use still makes it a very useful product. Data is logged to a laptop via a USB connection. The logger and its software are available from eLabtronics.

The logger comprises a small PCB, 75 x 50mm. It is supplied without a box, so you should either wrap it in insulating tape or find a box to place it in. Also supplied is a short wiring loom that has at one end a multi-pin connector (it plugs into Port A on the PCB) and, at the other end, seven clip-on connectors.

These connectors comprise:
- Black – ground connection
- Red – 5V supply for powering sensors (20mA max)
- Five white or yellow connectors – the signal inputs

The input signals must be in the 0-5V range. Electrically connecting the logger is very simple. Let's say that you want to monitor the voltage output of the throttle position sensor.

Here's what you do:

FUEL ECONOMY WITH THE AIR CONDITIONER CONTROLLER

Both power and fuel economy can benefit from fitment of an Auto Air Conditioner Controller. The increased power is easy to understand – the power drain of the compressor is switched off when you need maximum power at the wheels. (Note some standard ECUs already do this, but most only at 100 per cent throttle.)

But what about fuel economy? Firstly, as already indicated, the ability to hold a higher gear at lower revs improves fuel economy. But the other effect is probably even more significant. So how does this one work?

Air conditioners operate on a thermostat basis – the temperature of the evaporator (or air in the cabin) is monitored, and the air conditioner control system switches off the compressor when the correct temperature has been reached. The Auto Air Conditioner Controller puts another switch in after that, one that turns off the compressor at high throttle angles.

OK, so the air con system says: 'The cabin temperature is getting too high: run the compressor.'

But if the voltage switch says: 'Nope, the throttle angle is still too high,' the air con compressor will stay off.

When the high throttle event is over, the air temperature in the cabin will be even higher – so as soon as you lift off the throttle, the compressor will start.

Over a drive cycle of hills and valleys, or urban start/stop traffic lights, the air con compressor will therefore run far more often in times of throttle lifts – rolling up to traffic lights, or rolling down hills – than it would without the controller.

The result is that you make much more use of the 'free' energy that is otherwise expended in braking and coasting. And the result of *that* is improved fuel economy, achieved with the trade-off of a greater fluctuation in air-conditioned cabin temp.

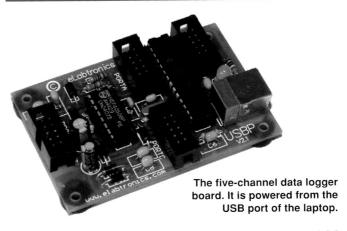

The five-channel data logger board. It is powered from the USB port of the laptop.

WORKSHOP PRO: MODIFYING THE ELECTRONICS OF MODERN CLASSIC CARS

The logger is supplied with five probes. Out of the box, the logger can monitor voltages in the 0-5V range. If you want to monitor higher voltages, use a voltage divider.

- Connect the black probe to the chassis (ground)
- Connect probe #1 to the signal output of the throttle position sensor
- Plug one end of the USB cable into the logger and the other end of the cable into the laptop's USB port. The yellow LED on the logger will light up.

This is a very simple logger – and that makes it very quick and easy to use. It can monitor up to five channels, allows you to set the sampling rate to the required application, has effectively no limit to the data it can collect, and the log file can be exported for graphing or other analysis in a spreadsheet program. The breakout box on the right shows how the software can be used.

So what uses can be made of the logger? Let's take a look at some simple modified car uses.

1. Airflow meter output

Nearly all airflow meters output a signal that is in the 0-5V range. (The exceptions are those that output a variable frequency – the logger cannot be used with these meters.)

So what use is there in displaying and logging the airflow meter output? At least two spring to mind. The first is that in a modified car the airflow meter might be 'maxing out' – the airflow demand is so high that the signal reaches a ceiling and stays there, despite engine revs (and full throttle airflow) continuing to rise. If this is the case, the graphed log trace will show this very clearly – on full throttle acceleration, the signal trace will rise until, all of a sudden, there's a level above which it never goes – it just flat-lines.

Another airflow meter logging use is to see where your performance modifications are going. If the engine is breathing more air, its volumetric efficiency has been improved. And if the airflow meter hasn't maxed out, more

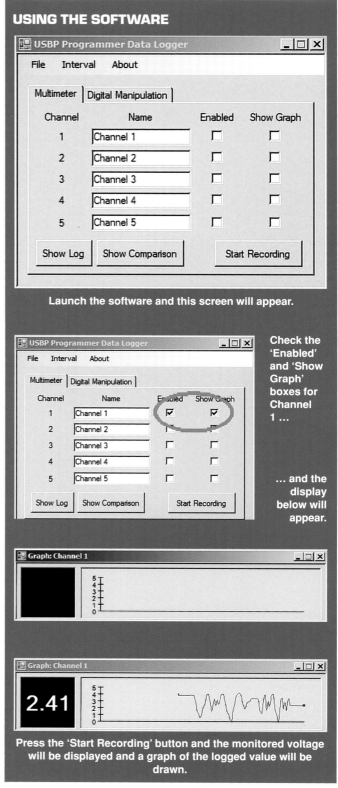

USING THE SOFTWARE

Launch the software and this screen will appear.

Check the 'Enabled' and 'Show Graph' boxes for Channel 1 …

… and the display below will appear.

Press the 'Start Recording' button and the monitored voltage will be displayed and a graph of the logged value will be drawn.

5. OTHER ENGINE BAY MODIFICATIONS

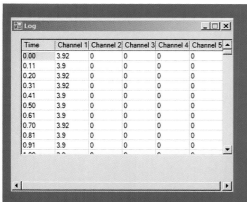

Press the 'Show Log' button and the data readings (and the time at which they're taken) will be shown in another box.

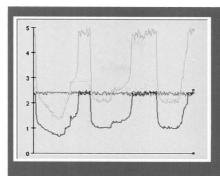

The channels can also be graphed on the one display – press the 'Show Comparison' button to do this. Note that in this logger application we don't use the 'Digital Manipulation' function of the software.

After the 'Stop Recording' button has been pressed, you have the option of exporting the log file. This is done by File > Export Log > Analog Export. The file is exported as a .csv file that can be read in Excel. Note that this is the only way of saving your logged file.

By clicking on the Interval button, the sampling speed can be set at any of eight values – from 10Hz (ie 10 times a second) to once every 10 minutes. The default is 10Hz. Note there is also a faster setting than 10Hz – called 'Max.' Maximum is the fastest that the particular PC and USB interface can run at, and also depends on how many channels are being monitored. On my laptop, monitoring one channel, 'Max' was 16Hz. In all cases, the logged record shows the actual timing values that were used when the data was collected. If more than one channel is to be monitored, additional input probes are connected and the respective channels activated in the software. Up to five channels can be monitored real time and logged.

air will be recorded as a higher airflow meter signal output. Of course, if the fuel or ignition settings are all wrong, more air won't mean more power, but if the engine breathing is improving, the potential for more power is there. After all, if the airflow reading is going backwards, you certainly won't be getting more power!

Note that this applies to both naturally-aspirated and forced inducted cars – so you can, for example, see if an intercooler (or intercooler spray, etc) is actually increasing the mass of air getting into the engine.

2. Oxygen sensor output

(Note: the USB logger has an input impedance of 50 megohms. This means that it draws very little current from the circuit it is measuring. However, some oxygen sensors may still be affected by this tiny current draw, so keep this in mind.)

When measured at the ECU, all exhaust gas oxygen sensors output a voltage signal that's in the data logger's 0-5V range. Logging of the oxygen sensor can show three things.

First, it will indicate the health of the sensor. A narrow band exhaust gas oxygen sensor should switch rapidly in voltage between a high level and a low level, the relative heights and depths being specific to different types of sensors. For example, a narrow band sensor will switch at least a few times a second between about 0.2V and 0.8V when in closed loop cruise. A very slow switching rate is indicative of a sick sensor, and a sensor that stays permanently high or low is near-dead.

The logger graph will show this switching in voltage value far better than a multimeter reading, where the numbers will just appear to be jumping all over the place!

Second, the behaviour of the sensor will show when (or if) the car's management system goes into open loop, characterised by a change in the up/down voltage movements to a steady voltage output. In open loop it's much easier to alter the mixtures, eg by an airflow meter or MAP sensor interceptor (or even by altering fuel pressure).

Finally, the value output by the sensor (yes, even a

narrow band sensor) will give you an indication of the air/fuel ratio that's being run. Sure, a narrow band sensor (as fitted to most older cars) is not ideal for tuning mixtures, but it is simply far better than trying to do it by feel alone.

(Note: If you're using a professional air/fuel ratio meter, you'll also almost certainly be able to log its output. Normally, these meters have a 0-5V output for just this purpose.)

3. Turbo boost response

To measure turbo boost you'll need to have either a car that already runs a boost pressure sensor that has sufficient range (eg the MAP sensor), or you'll need to fit an additional sensor (eg a 2 or 3 bar generic GM sensor). The logger can provide 5V power for the sensor (up to 20mA max), so if you're working on different turbo cars, you can easily move it from car to car, keeping it with the logger.

Measuring and logging boost on the road is a very useful tool. Turbo response, and the response of the boost control, can vary a lot between a dyno and the road. For example, the way boost ramps up in first gear will invariably be different to that which occurs when flooring it in fourth gear. Being able to clearly see both the ramping behaviour (ie the shape of the curve) and the maximum pressure is a lot more useful than just glancing at a boost gauge and seeing only a few values during a full-throttle run.

In addition, you can very accurately measure boost response by simultaneously measuring the throttle position sensor output. The time taken between the accelerator being fully opened and the boost level reaching its peak can be measured with the logger down to one tenth of a second.

HIGH-PRESSURE INTERCOOLER WATER SPRAY

Air/air intercoolers in hot climates benefit from the cooling action of a water spray. However, to achieve best cooling, the spray needs to comprise very small droplets that evaporate rapidly. Many intercooler sprays use coarse sprays, and so are limited in their effectiveness. However, this spray uses an ultra-high pressure pump and specific atomising nozzle and achieves excellent results.

Pump

The heart of this system is the pump – and what a pump it is! For years I'd been looking for a really good ultra-high pressure pump that could cope with water. Fuel pumps won't – they corrode internally. Multi-diaphragm pumps (as used in boat and recreational vehicle potable water supply systems) can generate good pressures (eg 7 bar – 60psi) but they're expensive and noisy. (Plus they usually have a built-in pressure switch that requires the use of a bypass if the pump isn't to continually stop and

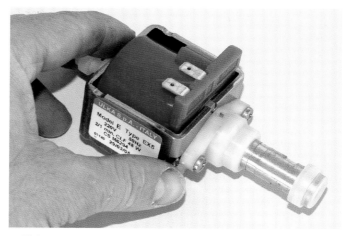

This pump is normally found inside a mains-powered coffee machine. However, when run off an inverter, it is an excellent high-pressure pump for an intercooler water spray.

start.) I even looked at the old-design Aquamist water injection pump, but the high cost always put me off. But it's the Aquamist pump that sent me off in the right direction. Rather than using rotating rollers (like a high-pressure fuel pump) or diaphragms compressed one after the other (like a diaphragm pump), the Aquamist pump used a pulsating piston. The piston, powered by an electro-magnet, slid back and forth, pushing ahead of it little bursts of water that soon added up to a very high pressure.

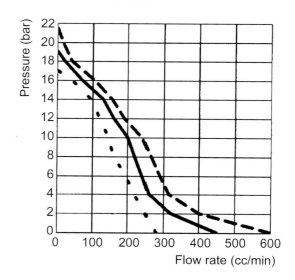

The manufacturer's performance specs for the Ulka E5EX pump, a commonly available coffee machine pump. As can be seen, flow drops off with increased pressure – but this means that you can control the flow of the pump just by changing nozzle size. The curve also shows it's possible to flow over 200cc a minute at a pressure of 10 bar (145psi) and 100cc a minute at 15 bar (218psi).

5. OTHER ENGINE BAY MODIFICATIONS

Power supply

As mentioned above, the Ulka pump needs an AC high-voltage power supply to operate. However, it draws only 50 watts and so pretty well any 12 → 240V inverter will work the pump. The one I selected from the local auto parts store was branded 'Pro User' and provides up to 150W continuous and 300W peak – far more than required in this application. Further, the inverter includes low voltage, over-temperature, and short-circuit protection.

Nozzle

The Spraying Systems Company of the US makes among the world's best water spray nozzles. It has nozzles sized to flow from hundreds of litres a minute to a few millilitres a minute – and everything in between.

The nozzles used in this application are in the Unijet small capacity range. These assemblies consist of a ¼in TT male body spray nozzle holder, a screen strainer incorporating a check valve, a spray tip and a tip retainer. All the components are top quality brass. The check valve stops the valve from dribbling when the pump is off (even if the nozzle is located below the level of the pump), and the strainer stops the tip being blocked by foreign material that might be contained in the water (however, you should still always strain the water you're adding to the reservoir).

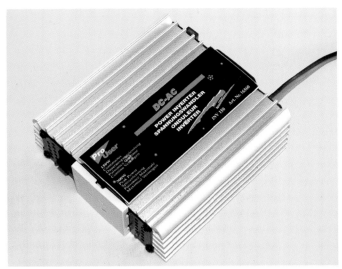

A small inverter is used to power the pump from the car 12V supply.

And guess what? Just the same design of pump is used in espresso and cappuccino coffee machines! Except instead of costing a lot, these coffee machine pumps are quite cheap. I used an Ulka branded pump. So that it can be used in coffee machines around the world, the Ulka pump is made in 110V AC and 220V AC models. In a car, simply power it from a small mains inverter that develops the appropriate AC voltage for the pump you are using.

The Ulka pump has a pressure of over 15 bar (218psi), and is actually designed to pump water – so you can see why I got excited. (In fact, when measured, the pump did even better than this, with a peak recorded pressure of 25 bar, or 360psi.)

In addition to its very high pressure, the Ulka pump has another attribute – it can provide enough suction to draw up water from a reservoir mounted below it. This feature gives greater flexibility in pump mounting position.

A high-quality atomising nozzle is needed to produce the very fine required droplets. This assembly incorporates a check valve and fine mesh filter.

HOW COOL?

Evaporating a kilogram (ie a litre) of water requires 2257 kilo-joules of energy – and that's a lot! If the nozzle flows 400 ml/minute, and if all the water evaporates, each minute 903 kilo-joules of energy are extracted. One joule per second is the equivalent of 1 watt, so fully evaporating 400ml/min of water provides a cooling power of 15 kilowatts! Even a 130ml/min spray provides a potential cooling power of just under 5kW.

The key point is that the water *must evaporate* – it is this change of state from water to a gas which absorbs the energy. If the water droplets do not evaporate, they provide almost no cooling performance. And the key to getting water to evaporate is to use very small drops – an atomised mist – which dramatically increases the surface-area-to-volume-ratio of each drop, promoting evaporation.

In addition to drop size, the rate of evaporation will also depend on the relative humidity of the air (if you like, an indicator of how much 'room' there is left in the air for evaporated water) and the temperature of the heat exchanger.

With a good enough spray, there is no technical reason why the temperature of the intercooler cannot be brought lower than ambient. After all, that's how evaporative air conditioners work ...

Used with the Ulka pump, the following spray volumes are achieved:

¼-inch tip capacity size	Orifice diameter (mm)	Core number	Approx flow with Ulka pump	
			(ml/min)	(litres/h)
1	0.51	210	130	7.8
4	1.1	220	400	24

The hose that connects the pump and nozzle has to withstand up to 25 bar, or 360psi.

Hose

It's possible to get barbed hose fittings for both the nozzle and the pump and connect them together with high pressure hose held on with hose-clamps. However, I found in bench testing that, time after time, the hose would blow-off either the pump or nozzle – we're talking *high* pressure here! You may be able to get away with running two hose-clamps at each end but to ensure reliability, I chose instead to get an industrial hose with high pressure fittings made to suit.

Note: Another approach, that I have subsequently used with these pumps, is to use 'push-fit' nylon hose with appropriate brass or stainless-steel fittings. These fittings and hose are often used in heavy vehicle air suspension systems. Push-fit fittings, as their name suggests, connect to the hose by simply sliding straight in, being then held captive by a tricky fitting.

The feed hose to the pump handles effectively no pressure, and so any hose that is durable enough to withstand the heat of the engine bay can be used.

Reservoir

Most engine bays are usually too cramped to fit in a dedicated water spray container, and so I made use of the existing windscreen washer spray reservoir. In the guinea pig vehicle, this is a substantial 6 litres (1.6 US gal) – enough to run the smaller of the two spray nozzles continuously for over 45 minutes! Even with the windscreen washer drawing from the same container, with sophisticated triggering of the spray, refills should be needed only every two or three tanks of fuel.

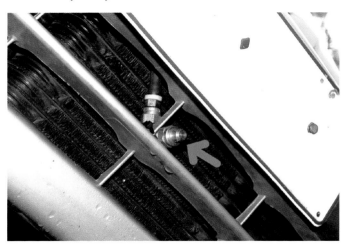

Unlike conventional intercooler water sprays, the nozzle (arrowed) is positioned to face forwards, directing its droplets into the oncoming airstream. The cooled air then passes through the intercooler located immediately behind the nozzle.

Installing the spray

The nozzle should *not* be placed as is normally done with an intercooler spray. Conventionally, a spray nozzle is aimed at the intercooler, or in some cases, across the intercooler. However, the droplets produced by this high-pressure system are so small that it's best to simply add them to the airflow in front of the intercooler. The air that passes through the intercooler will then carry the droplets with it. To best achieve this outcome, there should be about 150mm (6in) of clear air in front of the nozzle – if there is a surface closer than that, the drops will coalesce (ie join together) and the very small droplet size will be lost.

Of course, when adding tiny droplets to the airflow in this way, there is a potential problem if the frontal

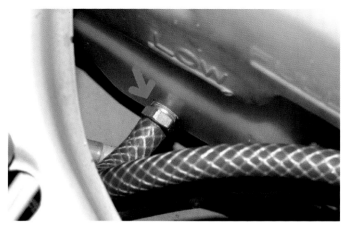

The intercooler water spray was fed from the windscreen washer reservoir via this added brass fitting.

5. OTHER ENGINE BAY MODIFICATIONS

aerodynamics of the car is poor and not much air passes through the intercooler. In that case, that's just what many of the tiny water droplets will also do! Another aspect to keep in mind is that the nozzle should be accessible so that it can periodically be unscrewed to clean the internal filter.

On the guinea pig car, the nozzle was placed in front of the centre of the intercooler and aimed forwards. When stationary, this created a cloud of drops in front of the car, but testing showed that when moving, the distribution of the droplets over the intercooler was good. To see the effect of different nozzle positions, you can temporarily tie the nozzle in place with cable-ties before making a more permanent mount after the location has been shown to work well.

When tested on the bench and supported by the hoses, the pump is very quiet. However, when hard-mounted to the car's bodywork, it can be quite noisy! Because of its rapid oscillating piston movement, the pump generates a low frequency vibration which can be picked up and transmitted. However, this is easily overcome by the use of a couple of rubber isolation mounts.

The pump doesn't readily lend itself to mounting – there are no mounting holes, for example. In this installation, the pump was held in place by the use of two sections of aluminium angle that tightly sandwiched the pump metal parts. A sleeve of high density rubber (like thin wetsuit material) was first placed around the pump. A second bracket – also made from scrap aluminium – was used to support the two rubber isolation bushes which bolted to the pump bracket. This made for a cheap, reliable and easy to fabricate mounting system.

As mentioned above, the reservoir used to supply water

The ultra-fine mist of droplets produced by the spray.

for the intercooler spray is the standard windscreen washer reservoir. But how do you tap into it for the intercooler spray supply? There are a few ways – the best depends on your application.

Easiest is to simply put the new feed tube to the pump through a hole made in the reservoir cap. The Ulka pump will draw up the water, and so the system will work fine – although it's a bit ugly. Another way is to carefully drill a hole in the base of the reservoir (best to first drill a small hole and enlarge it gradually with a tapered reamer), and then fit a tight rubber grommet. If the plastic water supply tube is an equally tight fit through the grommet, there won't be any leaks. Finally, if the wall thickness of the container is sufficient, you can do as I did and drill a hole, and then screw a brass fitting into place. If you carefully size the hole, the fitting will self-tap its thread, and the result will be leak-proof.

The mains power inverter should be securely mounted and carefully wired into place. Remember that the connection to the pump is carrying high voltage AC, so this cable must be insulated and routed carefully, using grommets where it passes through holes, and keeping it away from hot exhausts, etc. Use a 12V relay to turn on and off the DC supply to the inverter – this relay is triggered to operate the water spray. That is, when the spray is off, so is the inverter. Note: the inverter must not be mounted under the bonnet or anywhere else where it is exposed to the elements. The spray can be triggered by any method traditionally used to switch on an intercooler spray.

Testing

The first step is to make sure the distribution of the small droplets is good. This can be tested in a number of ways. One method is to have the car driven past you while you observe where the mist of droplets is going. However, in some cases, for example a top-mount intercooler, the core will not be able to be seen when the car is moving. In that case, assess the spray distribution by looking at which

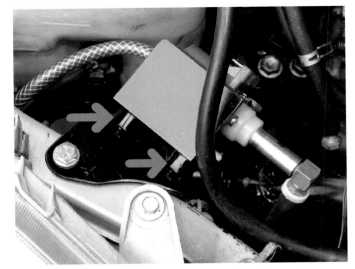

The pump was mounted on a rubber sleeve (arrowed) to suppress vibration transmission to the car's body.

WORKSHOP PRO — MODIFYING THE ELECTRONICS OF MODERN CLASSIC CARS

parts of the core get wet. It's important to again note that in most applications, the spray nozzle will not be aimed at the intercooler core, but instead will be positioned relative to the airflow reaching the core.

The next step is to do some temperature measurement. When measuring either intercooler core temperature or intake air temperature after the intercooler, the on-boost temperatures should be lower with the spray working. If there is little change – and the intercooler is getting hot – look at the spray direction and also possibly the size of the nozzle. In the guinea pig car, when running off boost at 100km/h (~60mph), by using the spray it was possible to drop the measured intake air temperature to a little *below* ambient! In cases where the intercooler core gets really hot in normal use, and especially when the ambient temperature is high, the high-pressure spray will make a substantial difference to intake air temps.

ADDING A TWO-SPEED FAN TO YOUR INTERCOOLER

In the above section I added a water spray to an intercooler. But what if it's not enough – for example, you have an intercooler that's mounted in the guard (fender) or in the engine bay? Especially if the intake air gets hot at slow road speed (or when the car is stationary), you might need to add an intercooler fan. And it's even better if it's speed controlled and operates only when needed. First – sourcing and fitting a fan.

THE FAN

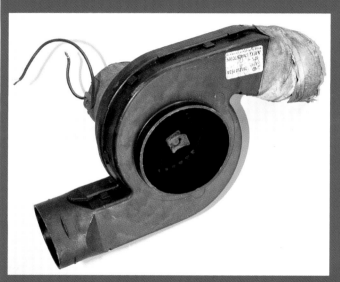

It was the purchase of this cabin ventilation fan – or more correctly, centrifugal blower – that opened-up the possibility of high pressure forced-air cooling of the engine bay intercooler in my Nissan Maxima. The fan was marked with VW and Audi brands and was made by AEG. It had two outlets (each of about 50mm in diameter) and a single central inlet (about 90mm in diameter). I later learned the fan was a 1974-79 Volkswagen Kombi cabin blower, as fitted to cars with the Type 4 1700, 1800 or 2000 engines. It is found in the engine bay above the motor.

The motor was also pretty sizable – a good clue to the power of the blower, which in this case was huge. Attach this blower to a battery, and, firstly, you'll need to hold on to the thing, or the reaction of the fan rotation will cause the motor to spin the other way and jump all over the floor. Secondly, the air coming out of the two outlets will feel like some kind of jet blast. Well, I exaggerate – but not much. You could certainly feel the airflow from the two outlets at 3-4 metres (yards) away. With a current draw of 15.5 amps at 12V (that's 180 watts!), this is one powerful blower. Incidentally, the fans in these designs aren't shaped like radiator cooling fans; instead they're centrifugal designs. Air is drawn into the 'eye' of the hurricane and then thrown out the edges. The housing in which the fan sits directs and constrains this flow – most designs have just the one outlet, rather than the two of the blower shown here. A centrifugal fan is capable of developing a much higher pressure than an axial (ie propeller-like) fan. That makes it ideal for the intercooler application, where it has to move air through a resistance.

When sourcing an intercooler fan, designs to look for include cabin ventilation fans fitted to people movers having rear vent outlets, those fitted to trucks and buses (but check that the fans are 12V not 24V), and blowers on prestige European cars. If you have a chance to try before buying, place your hand over the inlet and see how much suction it can develop – the more the better – because the fan will be located so that it draws air through the restriction that is formed by the intercooler fins. Note that I don't recommend small PC fans and similar – they're just not powerful enough.

110

5. OTHER ENGINE BAY MODIFICATIONS

SHROUDS

Heat exchangers like radiators and intercoolers need a lot of surface area to carry out their heat exchange function – that's why radiators are always large and thin. But it also means that fans working on radiators need to be shrouded so that all the air they are moving is drawn through the radiator core. In terms of the area it covers, an unshrouded fan might be 'missing' over half the radiator. (There's another reason for a shroud – an unshrouded fan will also have a lower flow efficiency … it will be whizzing air around its blade tips, rather than pushing it forwards or backwards.)

The centrifugal blower type fans being used here already have an effective shroud – it's the housing in which the fan sits. However, to make sure that all air being moved by the fan is being drawn through the intercooler, another shroud is needed at the intercooler end of the system.

In the case of the installation shown here (where the blower is mounted against the intercooler), the blower and the intercooler shroud are one assembly – but it doesn't have to be like that. The blower and the shroud can be connected by a long flexible tube, allowing the blower to be mounted remotely. Let's look now at some of the shrouding approaches that are possible.

TYPES OF SHROUDS

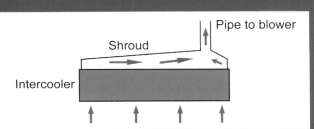

An intercooler shroud can be organised in a few different ways. In the approach shown in this diagram – called a full shroud – all the air that passes through the intercooler is gathered by the shroud and fed through one outlet, which is connected to the suction side of the blower. The advantage of this approach is high cooling efficiency – when the blower is operating, all the air is being pulled through the intercooler across its full area. The disadvantage is lower cooler efficiency when the blower isn't operating, because the shroud and blower will restrict flow when normal forward car movement is pushing air through the core. One way of overcoming the disadvantage is to keep the fan running all the time, perhaps only at slow speed when at low loads. In fact, that's the approach I took on the Maxima. But there's another way of doing it too – a partial shroud.

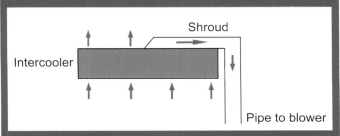

Even if you have an intercooler with very restricted space, you can still use a full or partial shroud. The trick is to make the shroud a different shape. For example, the shroud can be relatively thin, with the suction hose joining the shroud at one end, and exiting past the side of the intercooler. Or you could make the exit hose parallel with the intercooler surface – there's a variety of ways of doing it.

The easiest way that I've found of making an intercooler fan shroud is to use sheet aluminium. If you buy from a scrap metal dealer, you'll find that aluminium is extremely cheap. Aluminium is also easy to work with hand tools (cutting, filing, bending, drilling), doesn't rust, can be polished, and is light in weight. This photo shows the design of another intercooler shroud. It used a cut-down small radiator fan and the shroud covered the full intercooler area.

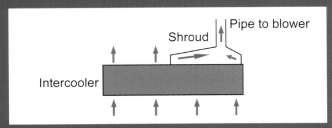

When using a partial shroud, half (or whatever proportion you want) of the intercooler core keeps functioning as it did before, while the other half operates as a forced-air design. In a mostly urban driven car, the highest intake air temperatures you're likely to record are in stop-start conditions, especially after being stationary for a while. In that case, I'd suggest the full shroud. On the other hand, a car mostly driven on freeways or country roads could use the partial shroud.

111

WORKSHOP PRO: MODIFYING THE ELECTRONICS OF MODERN CLASSIC CARS

TYPES OF SHROUDS

Because the blower was to be mounted right up against the shroud, I built both the blower mount and the shroud at the same time. For the blower mount, a piece of aluminium sheet was cut out with an electric jigsaw and then folded into shape using a homebuilt sheet metal folder. (You could just as easily use a couple of bits of timber and a vice.) A holesaw was then used to make the openings opposite each of the air outlets – you don't want these obstructed if the blower is to work as well as it should.

The blower was attached to the bracket using the already existing mounting screws that mount the plastic part of the fan to the motor.

Next, the shroud was folded-up out of sheet aluminium. A hole was cut for the 'eye' of the blower to project into and the two assemblies – the blower support and the shroud – were bolted together.

To smooth the transition from the shroud into the blower, a cut-down flared plastic speaker port was used. If you are mounting the blower remote to the intercooler, the speaker port also makes an ideal way of providing a mounting for the hose that will link the shroud to the blower.

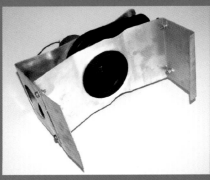

All this work resulted in this assembly – which looks nothing like it should be associated with an intercooler!

Here's how it all looks when it's mounted on the underside of the intercooler. Note that this intercooler had enough mounting lugs and brackets to allow the shroud to be bolted to the intercooler – if this isn't the case, you may have to fold lips that allow the shroud to clip into place. At the sheet metal folding stage don't worry too much about getting an airtight seal between the shroud and the intercooler because ...

... the seal can be easily formed from strips of foam rubber, painted black with a spray can and held in place with contact adhesive. It's very important that the shroud is sealed to the intercooler – air will flow to the fan by the easiest means, and if it can avoid having to pass through the restriction of the intercooler, that's just what it will do.

5. OTHER ENGINE BAY MODIFICATIONS

TESTING

The simplest test is to connect power to the blower motor and see what happens. However, a word of warning. Make sure that you try the power connections both ways around – these fans are directional, *although they will blow air whichever way the fan is spinning*. However, one direction of rotation should flow a lot more air than the other – normally the fan speed is slower in the 'right' direction.

Placing a sheet of paper over the top of the intercooler is a good way of seeing how much air is flowing – the paper sheet should be immediately sucked flat against the top of the core. If the airflow is less than expected, check that the air outlet(s) of the blower are unrestricted, and that all the sealing between the shroud and the intercooler is airtight.

In the case of the design shown here, I was initially disappointed by the airflow – until I swapped the polarity of the fan leads ... Then, with the fan rotating in the correct direction, a *massive* amount of air was drawn through the intercooler. In fact, the fan at full speed was working too hard for continuous use – after running for 10 or 15 minutes, the motor was getting quite hot (not really surprising when you look at its power) and the noise of the blower working at full power was quite loud. That's no problem though – it's very easy to slow down a brush-type DC motor so that you have high and low speeds available – as shown below.

VARYING FAN SPEED

In many conditions, the full power of the intercooler fan is not needed. Instead, it is better to vary fan speed to suit the conditions.

There are two primary ways of altering fan speed. One of the most elegant techniques is to use a PWM controller. This is the approach I would have taken, had I programmable management fitted to the car. Such an approach would have allowed steplessly variable fan speed control, with fan speed mapped against road speed – or even, using a 3D table, against road speed and engine load. (Incidentally, in this situation use a solid-state relay that's pulsed by the programmable ECU – then you'll have the required power handling.)

However, there is another technique that can be used. It will give you only high and low speeds, but it's easier and cheaper. This approach uses a dropping resistor so that in slow mode, the voltage that the fan sees is much lower than full battery voltage. This technique is often used by original equipment manufacturers to control radiator fan speed or water/air intercooler pump speed.

So how does it work?

USING A DROPPING RESISTOR

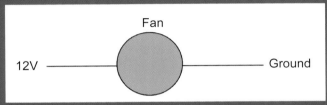

This is the starting point – the fan is connected to 12V and ground. It runs at full speed all of the time.

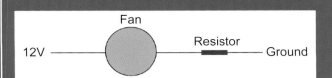

Now we place a resistor in the circuit. This lowers the voltage that the fan sees, which slows it down.

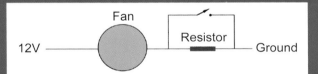

Placing a switch in parallel with the resistor bypasses it – so this becomes the fast/slow speed switch. Close the switch and the fan speed is fast. Open it and it's slow.

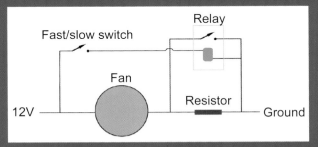

Now let's add a relay to do the speed switching. When the new fast/slow switch is closed, power is fed to the relay's coil (green) and the relay closes, bypassing the resistor. The fan therefore runs at full speed. Open the switch and the fan speed slows. The advantage of using the relay is that the slow/fast switch can now be a low current switch. However, in this design the fan would be running all of the time – even with the car off! That's because the current draw of the fan is so great that it should be fed straight from the battery. We therefore need to use another relay to turn the fan on and off.

WORKSHOP PRO — MODIFYING THE ELECTRONICS OF MODERN CLASSIC CARS

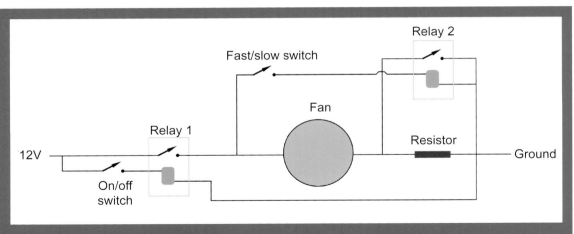

Now we have the final circuit. Relay 1 is the main on/off control, operated by an on/off switch, that can now be a low current design. Relay 2 is operated by the fast/slow switch (another low current design) that controls fan speed. For example, you could use an ignition-switched source to power Relay 1, so the fan operated only when the car was running. The fast/slow switch could be a normally open temperature switch that closes when the intercooler core temperature reaches (say) 45°C (113°F). That way, the fan would run at slow speed all the time the car was running, but fan speed would increase if the intercooler temperature was high. But what about the dropping resistor? Can that be any old resistor? The answer to that is an emphatic no!

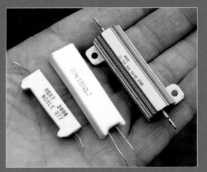

The resistor needs to absorb the power that's no longer being fed to the motor – the power has to go somewhere, and it is dissipated by the resistor as heat. This means that high power resistors like those shown here will be needed. It's difficult to be specific about the values (these will depend on the exact blower that you are using, and what speed you want the slow setting to be), but as a guide in the Maxima application, two 0.5Ω 50W resistors were mounted in series, giving a total resistance of 1Ω, with 100W power dissipation. This decreased the voltage that the fan motor saw to about 7V, which slowed it down considerably. (However, even at the slow speed, the airflow through the intercooler could still be easily felt by hand.)

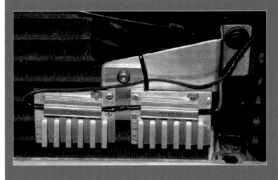

The resistors, that already came with finned heat-dissipating alloy bodies, were mounted on two heatsinks (salvaged from a defective PC power supply), with the complete assembly located in the airstream in front of the radiator.

Here are the two relays that control intercooler fan speed, mounted on an aluminium bracket. The fuse box behind the relays was salvaged from another car, and contains the fuses (high current and low current) for the new fan system.

5. OTHER ENGINE BAY MODIFICATIONS

ELECTRONIC RADIATOR FAN SWITCH

There are a couple of reasons why you might need an electric radiator fan temperature trigger. You could have needed a cooling system upgrade and have sourced a bare fan from a car dismantler, installed it – and then wondered how you're going to have it automatically turn on and off. Or the temperature switch in your current car's cooling system may have failed – and you're not too rapt with the cost of a genuine replacement. (And it would also be nice to have control over when the fan actually comes on, too.) Or maybe you've replaced the radiator with an upsized design that cannot take the original radiator temperature switch.

Of course, there are lots of solutions (eg commercially available adjustable radiator temperature switches) but the approach in this story has significant advantages over other techniques. So what advantages then?

First, it's easy to alter the temperature at which the fans will cut in. In fact, you just turn a 'pot' (potentiometer) on the circuit board, or change the setting on the digital display by 'up' and 'down' keys. Second, you can adjust that point very finely indeed – in some systems, to the individual degree. Third, the temperature at which the fan switches off again (ie the hysteresis – the difference between the switch-on and switch-off temperatures) is also adjustable. And finally, you don't need to install a new temperature sensor. Instead you can take the temperature signal straight off the engine management ECU.

HOW IT'S DONE

The approach uses a simple voltage switch, such as those widely available on eBay. These are available in both pot-adjustable and digital adjustable forms.

This adjustable voltage switch can be used to monitor the output of the standard coolant temperature sensor, switching on the radiator fans (via an additional relay) as required. This design is pushbutton adjustable, and the display shows the monitored voltage.

So how does the system work? Well, the ECU sends out a regulated voltage to its coolant temperature sensor, which is a device that changes in resistance with temperature. If the resistance is low, the ECU sensor voltage is pulled lower. If the resistance is high, the ECU sensor voltage stays higher. Since most sensors have a resistance that gets lower as the temperature gets higher, the voltage sensed by the ECU gets lower with increasing temperature.

By tapping into the wire with the voltage switch, we can then additionally use the sensor to switch the radiator fans. Because of the high input impedance of the voltage switch, the signal is not 'loaded down' and so it continues to work fine with the ECU.

FITTING AND SETUP

In this case the guinea pig car was my Maxima V6 Turbo with twin factory fitted electric fans. A change in radiator meant that the original temperature switches could no longer be used.

The first step in the installation is to back-probe the connections to the coolant temperature sensor, either at the ECU or near the sensor itself. (On the Maxima I did it near the sensor because it was quicker and easier to find the right wire without having a dedicated workshop manual). As mentioned, you're looking for a voltage (normally between 0-5V) that decreases as the car warms up. So you can easily see this change, start off the measuring process with the engine cold.

Once you've found the right signal wire, solder a new wire to it and then run it back into the cabin. Provide power and ground to the voltage switch, and connect the wire from the sensor to the switch input. Set the switch parameters so that it trips on a falling voltage.

Start the car and set the voltage adjustment pot until the switch clicks over. If you had the radiator fan connected, it would be on now too. Let the car warm up and then adjust the voltage switch until it just turns off. That is, you've set the switch so if the car gets any warmer, the switch will activate.

With the hysteresis set to its minimum, the switch will go off quickly once the temperature starts to drop. If the switch goes off too early, increase the hysteresis. Note that the set-up process can require some trial-and-error changes, so leave the switch accessible for a day or two of normal driving so that the fine-tuning of the switch behaviour can be adjusted. Make sure that you don't set the temperature threshold too low or the fans will be on all the time – in most cars you want them trip when the temperature needle gets to say 60 or 65 per cent of full gauge movement. (Of course, you can also directly measure the coolant temperature and set the switch up that way.)

WORKSHOP PRO: MODIFYING THE ELECTRONICS OF MODERN CLASSIC CARS

Finding the correct wire on the temperature sensor – one that falls in voltage as the engine warms up.

CONNECTING THE FAN

The next step is to connect up the fan. While the voltage switch has an existing onboard relay, all radiator fans should be driven using a separate, heavy-duty automotive relay.

If the voltage switch is being used to replace an existing temperature switch, you can use the relay that probably already exists in the car's wiring. In that case, just wire the NO (normally open) and COM (common) terminals of the voltage switch relay to the connections that previously led to the temperature switch. In fact, that's exactly what I did on the Maxima, an approach that saved having to buy a new relay and run all its associated wiring.

With surprisingly little set-up time I soon had the radiator fans on the Maxima working as I wanted. I left the hysteresis at its minimum level and found that I could soon tell by closely watching the temperature gauge when the fans were about to come on, and when they were about to go off again. In short, the system worked perfectly.

VOLTAGE BOOSTERS

Voltage-boosting modules are now widely available online. They accept normal battery voltage, and output a voltage that is higher. How much higher depends on the specification of the module and the adjustment of an onboard pot. Small booster modules will handle around 50W power, while larger ones are good for about 200W. Note: for durability and to keep the heatsink temperatures under control, always de-rate the claimed power output of these modules by about 50 per cent.

A small voltage booster good for about 50W (ie a current of about 3-4A).

116

5. OTHER ENGINE BAY MODIFICATIONS

The best use for these modules is to boost the voltage going to items that provide much higher output with only slightly higher voltages.

For example, you can use a 50W voltage booster to easily increase the pressure and output of windscreen washer pumps commonly used for intercooler water sprays. You can use it to boost the output of filament interior lights, brake lights and reversing lights. Depending on the measured current draw, you can use it to boost the flow of water/air intercooler pumps. You could also use it with 50W headlights in order to brighten your main beam (one unit per light). The larger 200W unit is suited to increasing fuel pump output, or boosting the output of a pair of headlights.

But what happens to the device you're powering when you increase the voltage going to it? In short, not only will its performance improve, but its life will be reduced. In many cases, that's of little concern – something like an intercooler water spray pump is used so little (in relative terms) that its life will still be fine. Incandescent light bulbs will have a shorter life, but as they're a replacement item, again it's not a huge problem. However, you should select the increase in voltage with care. A motor (eg a pump) used infrequently in short bursts could be run at 18V without many issues, but a filament lamp being used for long periods shouldn't be fed much over 15V. In general, make sure that items don't get too hot!

In some cases, you can switch in the voltage booster only when the higher output is required. For example, this approach can be taken with a fuel pump.

Note that devices that use internal voltage regulators, or are current limited, shouldn't be run at higher than standard voltages. Normally, there will simply be no difference in the performance of the device but in some cases (eg LEDs using dropping resistors), the device may be damaged by over-current. So LEDs and electronic bits and pieces like car radios and other electronic modules aren't suitable for running at higher than their design voltage.

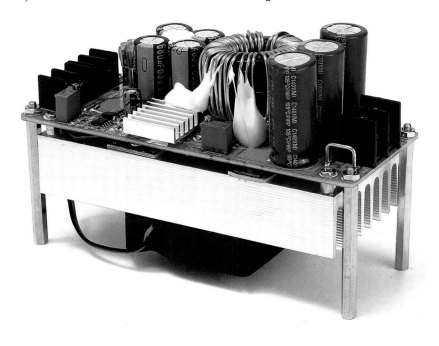

A larger voltage booster suitable for 200W loads (14-15A). This unit can be used to increase the output of a fuel pump.

www.velocebooks.com / www.veloce.co.uk
All current books • New book news • Special offers • Gift vouchers

WORKSHOP PRO MODIFYING THE ELECTRONICS OF MODERN CLASSIC CARS

Chapter 6
Modifying other electronic control systems

- Modifying the weight of power steering
- Modifying regen braking
- Switching off an airbag
- Adjustable stability control
- Switching off traction control but not stability control
- Developing a torque split controller
- Improving auto transmission shifting
- Fully programmable automotive controllers

WORKSHOP PRO: MODIFYING THE ELECTRONICS OF MODERN CLASSIC CARS

MODIFYING THE WEIGHT OF POWER STEERING – THE LEXUS

This section describes how I modified the electronically variable power steering system in the Lexus LS400 I then owned to make the steering weight adjustable. And, while the detail of the modification is specific to the Lexus, the same procedure can probably be followed in other cars that use a similar approach to electronically varying the steering assistance.

To give the best overview, I'll describe the step-by-step process that I undertook to identify how the variable power steering works, how the weight change was made, the development of the prototype and its testing, and the final result.

How the system works

The first thing that you need to do to undertake any modification of this sort is to have a thorough knowledge of how the electronic control of the power steering works. I think that this can be achieved in only one way – by reading the factory workshop manual.

The LS400 has what Lexus term 'Progressive Power Steering' – PPS. This uses a solenoid valve to vary the flow of hydraulic fluid to a reaction chamber – a fluid force that actually resists the power assistance. If a lot of fluid is allowed to flow to the reaction chamber, the steering effort is higher. If little fluid flows to the reaction chamber, then the steering effort is lower.

The key point: *in this steering system, more fluid flow into the reaction chamber equals a higher steering effort.*

A solenoid controls the amount of fluid flowing to the reaction chamber. The solenoid consists of a coil, a return spring and the valve. When no current is applied to the solenoid, it opens, allowing more fluid to flow to the reaction chamber and so the steering to become heavier.

The current to the solenoid is varied by means of pulse width modulation – the current is pulsed on and off quickly.

I altered the power steering assistance on my Lexus LS400 with just a simple, high-power pot.

The Lexus power steering ECU.

If it is on for only half of the time (ie it has a duty cycle of 50 per cent) the coil will 'see' only half battery voltage, and so will not close fully. If the duty cycle is reduced to, say, 30 per cent, then the valve opening will change. Note that unlike an injector, the frequency of the pulsing is so quick that the valve doesn't open and shut to the individual pulses – instead the plunger hovers at mid-points.

The PPS is controlled by its own dedicated ECU – a little box. It's a relatively simple box, too, with just one input – road speed. As speed goes up, the duty cycle with which it feeds the solenoid goes down, and so the valve opens further, increasing steering heaviness. Or, to put it the other way, as the road speed falls, the solenoid duty cycle increases, closing the valve and so lightening the steering.

Key point: *in this steering system, a lower valve duty cycle equals a higher steering effort.*

Well, with a bit of reading between the lines and some examination of the diagrams, that was what the workshop manual told me.

Testing

The first test that I undertook to prove that this information was correct was to pull the fuse that feeds the power steering ECU. This would both disable the variable speed sensitivity control (the power assist hydraulics would keep on working), and also cause the solenoid valve to fully open, feeding lots of fluid to the reaction chamber and so making the steering heavier. I removed the fuse and went for a cautious drive.

Yes, the steering was much heavier at all speeds – especially at low speeds. In fact, while at first I had considered increasing the steering weight simply by disabling the PPS, the resulting steering was heavier than I wanted. Not nearly as heavy as a non-powered system, but a long way heavier than the overly light standard system.

The requirement

So, having the equivalent of zero per cent duty cycle (ie no power at all) going to the valve gave too heavy a steering system, while the standard (varying) amount of duty cycle

6. MODIFYING OTHER ELECTRONIC CONTROL SYSTEMS

gave steering that I perceived as too light. What was needed was a system that reduced the duty cycle coming out of the PPS computer, so that at all speeds, the valve would allow more fluid to flow to the reaction chamber.

(Or, what about increasing the speed signal going to the ECU? That way the steering would also be heavier. However, I discarded this idea, partly because the transmission/engine ECU, the trip computer, the ABS and Stability Control Systems – and some other systems – also use the speed signal input. Even if it were modified after the signal had been taken for these other systems, best to leave it alone. And furthermore, a signal would also have to be generated when the car was stopped – telling the ECU that in fact the car was moving!)

So how to reduce the duty cycle output of the power steering ECU? And how to give it a variable adjustment, so that steering weight could be set to exactly suit my preference? Such a modification of the output signal duty cycle could be made using a micro-controller, power output transistor, other circuitry – and lots of work. But was there a simpler way?

Could the modification of the output signal be done using just passive components? After all, the coil of the solenoid sees the average voltage being applied to it – it doesn't open and shut to the individual pulses, because they're being applied in too quick a succession. So, the coil doesn't know that it's getting sent a 50 or 60 or 70 per cent duty cycle – it just sees the average voltage resulting from that quick succession of on/offs.

So, what if the duty cycle remained unaltered – but the actual voltage of the 'on' pulses was reduced? This suggests that a resistor placed in series with the solenoid pulse-train feed would have the same effect on the solenoid plunger position as reducing the duty cycle of the pulses. I decided to carry out some experiments.

A paralleled test bundle of large, old resistors, used to provide a trial voltage drop.

Experiments

The resistor that was going to be placed in series with the coil would have to dissipate the power that the coil would no longer be seeing. How much power was the question – and that was easier to find out by trial and error than any other way. (Note that placing a resistor in series with the solenoid coil in this way means that less current will flow in the circuit, so a heavier load won't be being placed on the power steering ECU.) The first experimental resistor – a standard ¼W 5.6Ω resistor – got too hot to touch in about 1 second, literally. So obviously a lot more than a quarter of a watt was going to be generated as heat!

Next up, some large old resistors were used, wired in parallel until an array that gave 4.8Ω resistance and 3W power handling was obtained. These grew only warm to touch, however the result on the steering was less encouraging – there was no apparent change in weight. I then went up in resistance, still binding lots of large resistors together to give the power handling, but again this didn't seem to have an effect. Getting bored of going up by small increments, I then jumped to 100Ω – and immediately could feel a major difference in the steering: it was much heavier. Not as heavy as with the fuse out, but still heavier than I'd prefer in normal use.

So, on the basis of that testing, a 0-100Ω variable resistor of about 2.5W power dissipation would give me a dial-it-up steering weight variation from normal through to heavy.

Potentiometers

The easiest way to gain a variable resistor is to use a potentiometer, wired as a variable resistor. Wire-wound potentiometers with 3 watts power handling are available from electronics stores – so I went and bought one. (Now it would be easier to get one online.) With this pot wired in series with the solenoid I could, for the first time, change steering weight on the move by twiddling the knob. It's an uncanny feeling putting in rapid small steering inputs, and at the same time turning the steering weight adjustment knob.

However, I found that to get the steering near to my weight preference, I was right up at one end of the pot's range. I attached to the pot a piece of cardboard with a 0-10 scale on it and added a knob. I then went out test driving, assessing the full range of adjustment that I'd need – from the heaviest I ever wanted the steering to be, right through to the lightest. When this range of knob adjustments was compared with the scale on the cardboard, I found I was using resistances only from 0Ω (ie standard steering weight) to '2' on the scale (ie 20Ω). So yes, that earlier test with a 4.8Ω bunch of resistors (where I'd felt no change in the steering) was wrong – when the alteration could be suddenly input with a knob, you could easily feel the difference.

So, a pot that had about a 0-20Ω range would be suitable for the final installation. However, it would also need to have adequate power handling – and here's where I had a problem. The 3W test pot was getting very hot.

However, that pot had been only a cheaply made one, so I then bought a better quality 20Ω pot.

This gave a very good adjustment range (now I could use the full sweep of the pot rather than being up at one end), but the pot was still hot to touch. In most electronics applications, resistors do get very hot (ie much too hot to keep your fingers on them) when they are working near their maximum power dissipation. However, in this application, I wanted the temperature of the pot at its maximum to be just warm. I added a smallish heatsink (about 40 x 30mm – 1.6 x 1.2in) and that settled it down a bit – but I then swapped to a larger heat sink (about 75mm – 3in square) and the temperature of the assembly rose only a little above ambient.

The heatsink was mounted with the heatsink fins aligned vertically (to give proper airflow), and then stuck with quality double-sided tape to the side of the body control ECU that lives under the driver's side of the dash. While I wanted access for setting-up adjustment (and perhaps long-term tweaking as the tyres wear) I didn't want or need on-dash control. Instead, with the car stopped, the pot was easy to get to.

The final pot used, a good quality 20Ω unit with a power of 3W. It was mounted …

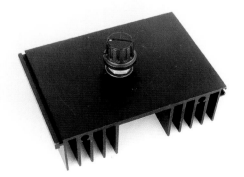

… on a heatsink that kept it quite cool.

Results

So what difference did the adjustment make? To quantify the change, I attached a spring balance to the outer end of a steering wheel spoke and measured the force required to turn the wheel, with the car stationary and parked on smooth concrete. With the pot set to the 'standard' position (ie no change over the normal Lexus steering weight) I measured a required pull of 4.5kg to turn the wheel at a constant rate. With the pot set to its 'heaviest' end, the required pulling effort doubled to 9kg. In normal road use, I was usually about halfway through the pot's range.

On the road, the heavier steering made an amazing difference. The actual feedback of the road surface was unchanged, the steering precision was unchanged, but the extra 'meat' of each input altered the driver perceptions a lot. I think perhaps it was the feedback to the steering input which was the big difference.

No longer could you just heft on some lock when entering a mid-speed corner. Instead, you fed it in much more progressively. It was a little like if you have driven in an arcade game that has zero steering resistance – you don't feel that you can steer very well at all. That's not to say that the Lexus previously had arcade game steering, but in the same way as resistance in the steering wheel of a game makes the steering better, so a lift in weight improved the steering of the Lexus.

In addition to mid-speed corners, the difference was also very noticeable in freeway sweepers, where the required amount of lock could be apportioned much more accurately. Also, the chances of making an inadvertent steering input were lessened. For example, if you shifted in your seat, or sneezed, or glanced down at the instruments or at a roadside sign, the steering wheel more strongly resisted the hand movements that you didn't intend to make.

Note, however, that even with this modification, the steering still altered in weight as it did when standard – more assistance at parking speeds and less when moving quicker. It's just that it was heavier all the way through.

But perhaps the biggest eye-opener was to snake down an empty road at 50km/h (30mph), swinging the car from side to side as another person turned the steering weight pot – you suddenly realised with startling clarity that the amount of steering weight makes a huge and instant difference to how a car feels on the road.

OTHER CARS?

If you are thinking of applying this technique to other cars, there are two key factors to ascertain:
- Does the car use a pulse width modulated solenoid to control the power steering weight?
- If so, does the steering get heavier as the duty cycle of the control signal decreases?

If the answer to both questions is 'yes,' start experimenting!

And of course, any car system that uses PWM control (other than injectors) where a smaller duty cycle gives the behaviour you want, is likely to be able to be modified with just the simple, high-power pot.

6. MODIFYING OTHER ELECTRONIC CONTROL SYSTEMS

MODIFYING THE WEIGHT OF POWER STEERING – THE PRIUS

Above I showed how electronically-controlled hydraulic power steering assistance was modified on a Lexus LS400. Now, I want to show a completely different technique I used on an NHW10 Toyota Prius fully electric steering system. Again, I'll take you through step-by-step how I devised the modification.

Typically, an electric power steering system consists of:
- a powerful electric motor geared to the steering shaft
- torque sensor(s) that detect how much effort is being put into the steering
- an electric power steering Electronic Control Unit (ECU)
- road speed input to the ECU

The ECU looks at the steering torque and steering direction being applied by the driver, as well as the road speed, and directs the electric motor to provide the required amount of assistance in the correct direction. (Torque refers to the strength of twist being applied to a shaft. The higher the twisting force, the greater the torque.)

Since the key ingredient in modification is the torque sensor, let's take a closer look at it. As with some conventional hydraulic power-assisted steering systems, a torsion bar is used to measure the relationship between the torque being applied to the steering wheel by the driver, and the resistance being posed by the tyres.

It's important to realise that this measured torque is a two-way process – if, for example, the front wheels are on wet grass, they'll turn very easily. So, despite the driver turning the steering wheel hard, not much torque will need to be applied to alter the steering angle of the tyres. However, if the front tyres are on coarse bitumen, they will resist turning, and so the torque applied by the driver will

Back-probing the Prius power steering ECU to see what signals occurred with left and right steering movements.

need to be much greater to get the tyres to turn. In other words, the torque sensor indicates both the driver's input of torque and the torque reaction of the tyres. The use of the torsion bar therefore takes into account the real steering effort needed – irrespective of road surface, tyre inflation pressure, and road speed.

So, how does this torsion bar system work? The torsion bar forms part of the steering column – it twists when subjected to both high input torque and high tyre reaction torque. Two sensors are used. Each measures the amount of twist and outputs a voltage that is proportional to this. When no twist occurs, the voltage output of each sensor is in the middle of its range. So, with sensors with an output range of 0-5V, each sensor reads close to 2.5V when no steering torque is applied.

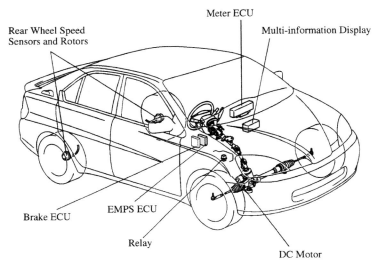

The Prius electric power steering system. Note that Toyota calls the system 'EMPS.' (Courtesy Toyota)

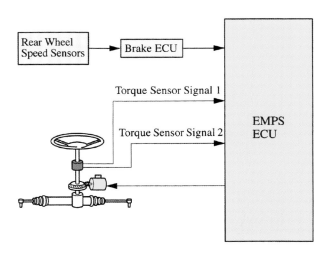

Two torque sensor inputs are used in the Toyota Prius system. (Courtesy Toyota)

WORKSHOP PRO: MODIFYING THE ELECTRONICS OF MODERN CLASSIC CARS

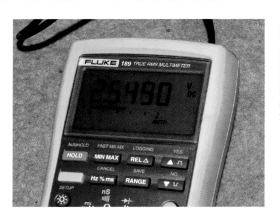

Each torque sensor outputs near 2.5V when no torque is being applied to the steering.

However, when subjected to torsion, the sensors' output voltages change. When there's increasing left-turn steering torque being applied, one sensor increases in its output voltage while the other sensor decreases in its output. The opposite occurs on right-hand corners.

Therefore, the larger the difference between the output voltages of the two sensors, the more steering effort that is occurring.

Because of the way the output signals of the torque sensors are configured, the ECU knows both the direction that the torque is being applied in and how great it is. The ECU then instructs the electric motor to assist appropriately, and as a result, the required steering torque effort by the driver decreases, resulting in a lower difference in the output voltages of the torque sensors. The assistance provided by the motor is therefore reduced.

One of the hard points to grasp about these torque-measuring systems is that the output difference between the two sensors is not proportional to the amount of steering lock applied. This is because the tyres have most resistance while they are being turned – once a certain amount of tyre angle has been adopted, the effort required to maintain that steering lock is much less than the effort required to first gain it. Instead, the greatest difference between the two sensor outputs occurs when steering input is being rapidly applied on a grippy surface at low speed … which is fine, because that's when you most need the assistance!

Modifying the system

So to summarise the above paragraphs for those just skipping along: the greater the difference in the output voltages of the two torque sensors, the greater the amount of steering torque that the ECU knows is being applied to the steering.

In nearly all cases, the desired outcome of modified electric power steering will be more steering feel – or in other words, you want less power assistance. So to achieve the outcome of less power assistance, the ECU needs to be fooled into thinking that there is less steering input effort than is really occurring. To achieve this, all that we need to do is reduce the difference between the voltages of the torque sensors. This can be achieved very simply by the use of just two multi-turn potentiometers (pots). Even including the cost of a box to mount the pots in, the total bill will be less than a cheap pizza.

When assessing how much weight change there will be in the steering by altering its electronic control, first pull out the power steering fuse, and see how heavy the steering gets.

How to do it

The first step with any electric power steering system is to disable the system and go for a drive. Usually, switching off the system is just a case of pulling out the electric power steering fuse or relay. Of course, the steering will be much heavier when moving slowly, but the car will still be driveable. What you are looking for is the change in steering weight at speed – say 80km/h (50mph). Is the steering much heavier, or the same as usual?

If it's the same as usual, the amount of power assistance being applied at this speed must normally be zero. (In that case, you're not going to be able to improve steering weight by modifying the system!) However, if you notice a firmer, meatier steering weight, you can be sure that there's too much assistance normally being given at this speed – and so there's room to make improvements.

The next step is to find some of the functions of the power steering ECU pins. At a pinch you can get away without a workshop manual but it's always best to have one. Ground one lead of a multimeter, and then use the other to back-probe the plugged-in power steering ECU. Have the car running, and ask an assistant to waggle the steering while you're taking the measurements.

6. MODIFYING OTHER ELECTRONIC CONTROL SYSTEMS

THE ELECTRONIC MODIFICATION

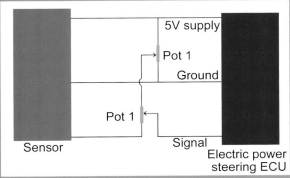

Let's take it step-by-step. With lots of the other connections left out, this diagram shows how one of the steering torque sensors connected to the electric power steering ECU.

Two pots are then added. Pot 1 is placed across the 5V-to-ground connections. If this pot is set to its middle position, 2.5V will be available on its wiper. Pot 2 is wired with one end connecting to this 2.5V supply and the other to the sensor output. This pot's wiper goes to the ECU. If the wiper of Pot 2 is placed closer towards Pot 1, the signal the ECU sees will be held more and more at 2.5V – that is, no torque change. On the other hand, if the wiper of Pot 2 is placed closer to its other end, the ECU will see more and more of the unaltered signal. So with Pot 1 set to provide 2.5V on its output, by adjusting Pot 2 you can alter the signal from being always held at 2.5V at one extreme, to being dead standard at the other extreme. Set Pot 2 to 'in-between' positions and you can get 'in-between' values. The two pots used are 10kΩ multi-turn designs. If you use small trimpots, these are very cheap, or if you use full-size multi-turn units, more expensive. I set the system up with the latter, simply because I already had them on the shelf. (Always use multi-turn – eg 10-turn – pots as this makes the setting-up much easier.)

Adjust Pot 2 so that its wiper is fully at the end closest to the signal input. Start the car and drive it – it should drive normally. If it doesn't, check your wiring. Then adjust Pot 2 so that the wiper starts to move towards the other end. The steering should now get heavier. If you go too far, it's likely that you'll trigger a fault condition – when setting this pot, drive the car lots to make sure that (a) the weight is good across a variety of driving situations, and (b) no fault condition is triggered.

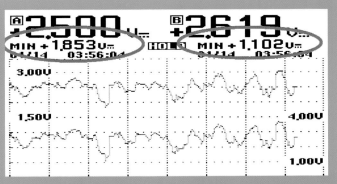

Using a Fluke 123 Scopemeter to data-log both the input signal from the sensor and the modified output shows the changes that have been made.

As can be seen, the input trace (bottom) and the output trace (top) appear to have the same shape. However, close inspection shows that the upper trace always moves less distance from the midpoint of about 2.5V. In fact the recorded minima (circled) show that the output dropped only as low as 1.853V, compared with 1.102V for the input. Other data (not shown here) indicates that the maximum voltage recorded on this drive from the sensor was 3.673V, versus 3.029V on the modified output.

In other words, the output voltage holds closer to the 'no-torque' value of about 2.5V, telling the ECU that there was less steering torque being input than there really was. The result is less power assist, and so greater road feel.

For the Toyota Prius on which this modification was performed, the following important voltages were found:
- Torque sensor #1: 2.5V output with no torque input, varying downwards with left-hand torque and upwards with right-hand torque
- Torque sensor #2: 2.5V output with no torque input, varying upwards with left-hand torque and downwards with right-hand torque
- 5V regulated output

Either of the two sensors can be intercepted – the ECU is just looking for the difference between the output voltages. So how is the modification done? The accompanying box shows just how easy it is.

MODIFYING REGEN BRAKING

The availability of cars using hybrid petrol/electric drivetrains creates the possibility of doing unique modifications – ones that in the history of cars, have literally never before been done. In this 'world-first,' I modified the regenerative braking capability on a Toyota Prius to give greater braking capability in light braking applications. It's a modification that needed no brake fluid, no rags, created no brake dust and yet made a major difference to how the car braked!

So what is 'regenerative braking,' or 'regen'? Regenerative braking occurs only in vehicles that use

MODIFYING THE ELECTRONICS OF MODERN CLASSIC CARS

electric power. During braking, the vehicle's electric motor is operated as a generator, pushing power back into the battery. By using regen rather than friction brakes, the power that's usually wasted in heating the brakes gets turned into usable energy that is stored for later use. In a pure electric car, the use of regen braking extends the range, while in hybrid petrol/electric cars, it improves fuel consumption and emissions because the petrol engine doesn't need to be used as much. However, regen braking isn't strong enough to do all the braking, so cars with regen braking use conventional hydraulic friction brakes as well. And this is where it starts getting tricky. If there are two completely separate braking systems at work, how are they combined in their outcome so the driver has to press only the one brake pedal?

To keep the cars feeling as much like other cars as possible, the hybrids have both the regen and conventional brakes controlled by the one brake pedal. In the first part of its travel, the brake pedal operates the regen brakes alone, then as further pressure is placed on the pedal, the friction brakes come into play as well. And it's not only when the brake pedal is being pushed that the hybrids (and pure electric cars) regen. When the throttle has been fully lifted, a gentle regen automatically occurs.

The regen in my NHW10 Prius had always seemed to me to be less than fantastic. Firstly – and perhaps because I was running Kevlar pads and slotted discs – the friction brakes seemed to do nearly all the work. When braking at slow speeds, the friction brakes could be heard working (the high-performance pads had a slightly graunchy sound), and at higher speeds, the pedal pressures were very much of the sort that occur with standard hydraulic-only braking systems. In short, it never felt much like the regen was doing a lot.

The Prius ran a centre dash LCD that showed how many watt-hours of regen had occurred each 5 minutes. These were indicated by small 'suns' that appeared on the display – each sun being indicative of 50 watt-hours. (So when a sun appeared, enough power has been put back into the battery to run a 50-watt light bulb for an hour – it's not a trivial amount.) But in my driving, seeing a lot of 'suns' was rare – indicative that the regen wasn't contributing much.

So it seemed to me that if the regen could be tweaked to do a greater proportion of the braking work (especially in light braking), fuel economy would benefit, the braking would be smoother, and it would take lighter pedal pressures.

The system
The Prius was – and is – a car with stunning engineering. So it comes as no surprise to find that, even on the first Prius, the system that integrated regen and hydraulic braking was complex. In fact, it was the ABS ECU that handled regen braking as well as ABS functions, sending a signal to the hybrid ECU to tell it how much regen to impose. But how did the ABS ECU know what to do?

Rather than measuring brake pedal travel (which could vary with pad wear, etc), the system used pressure measuring sensors to detect master cylinder pressure. The higher the master cylinder pressure, the harder the driver was pushing on the brake pedal. If the driver was pushing only gently, the piston displacement would be small, and so the hydraulic brakes would be only gently applied. In that situation, the ECU knew that the driver wanted only gentle deceleration, and so instructed the hybrid ECU to apply only a small amount of regen. However, as master cylinder pressure increased, so did the amount of regen that could automatically be applied.

(In fact, there were four pressure sensors in the braking system and two pressure switches – but it's the master cylinder pressure sensor that was most important.)

Modifying the system
So if the amount of regen that occurs was largely dictated by the output of the master cylinder pressure sensor, what about intercepting and altering this signal? That way, the ABS ECU would think that there was more master cylinder pressure than was actually occurring, so resulting in more regen being applied. Since the actual hydraulic pressure going to the brakes would be unchanged, there'd be a greater proportion of regen braking in the mix.

The voltage output of the pressure sensor ranged from about 0.4-3 volts, rising with increasing pressure. So if a small voltage could be added to this signal, the ECU should respond with more regen braking. But if this was done, would it detect a fault condition? The workshop manual stated that a fault would be detected if the voltage from the sensor was outside of the range of 0.14-4.4V, or if the voltage output of the sensor was outside a certain ratio to its nominally 5V supply voltage. Further, the latter was checked when the brake switch was off (ie brake pedal was lifted). In other words, the voltage needed to be within a certain range, and in some cases this was checked with the brake pedal not being used.

Wiring it up
With a workshop manual it was straightforward to find the right wires – but always double-check with a multimeter and make sure the colour-codes of the wires match those shown in the manual. The pot and relay (the latter salvaged from an old ABS controller!) were mounted on pre-punched board.

Testing
The first step was to make sure that the relay clicked in

6. MODIFYING OTHER ELECTRONIC CONTROL SYSTEMS

THE CIRCUIT

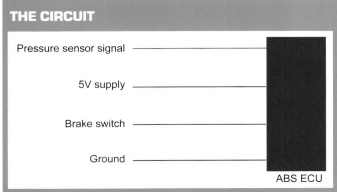

Leaving out a lot of the ECU connections, here's what the master cylinder pressure measuring system looked like. From top, there was the voltage signal from the sensor, the regulated 5V supply to it, the input from the brake light switch (12V when the brakes were on), and the ground connection.

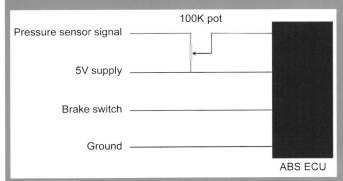

As indicated in the main text, what I wanted to do was to lift the voltage output of the master cylinder pressure sensor, especially at low sensor output levels. This was easily achieved with a single 100 kilo-ohm pot. If the pot was wired between the output of the sensor and a regulated 5V, and the wiper of the pot was then connected to the ECU, the voltage that the ECU saw could be varied from a constant 5V (not wanted!) right through to the standard signal. If the wiper was adjusted so that it was just a little way towards the 5V end, a small voltage would be added to the signal. (This is exactly the same as we saw in Chapter 3 when changing the output of the MAP sensor.) Note that to allow for the required fine adjustments, a multi-turn pot should be used.

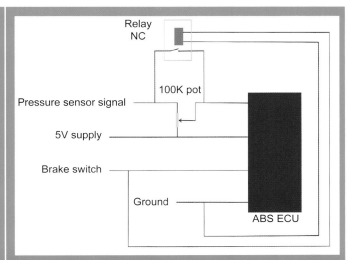

But what about the way the ECU checks the sensor output voltage when the brakes are off? In that case, it might spot that the output voltage of the sensor always appeared to be a bit high. The easy way around this was to add a relay that bypassed the pot whenever the brake pedal was released. This was achieved with a low-current SPDT 12V relay. As shown here, whenever 12V was available on the brake light circuit, the relay opens, sending the signal through the pot. But when the brake lights were off, the relay closed and so the sensor input voltage to the ECU was effectively standard. As a result, the system worked as standard until the brake light switch came on, whereupon whatever adjustment had been set on the pot immediately came into effect.

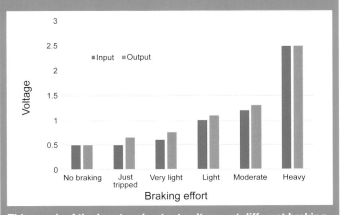

This graph of the input and output voltages at different braking efforts shows what happens. When the brake pedal was not pushed, the input and output voltages were the same. When the brake pedal had just been tripped, the output voltage rose. The amount that the output was greater than the input progressively reduced as braking effort increased, until at heavy braking load the signal was back to standard.

and out with movement of the brake pedal. Next, the pot was set so that its wiper was right at the 'sensor' end, and a multimeter was used to check that the signal going into the ECU was the same whether the relay was tripped or not. The car was test-driven and the system worked exactly as standard – at this stage, just what was wanted.

Next, the pot was adjusted a little so that the output signal of the sensor was dragged upwards. Driving of

The completed modification circuit board – one multi-turn PCB-mount pot (it's a screwdriver-adjustable one) and one relay.

the car then showed a fascinating outcome. When the brake pedal was pressed just far enough that the brake light switch tripped, the car would immediately decelerate with regen braking alone. It was quite a clear and distinct feeling – no friction brakes operating, but the car rapidly slowing. In fact, at this level of pot adjustment, too rapidly slowing … The pot was then adjusted back a little and further driving undertaken.

As finally set, the very light pedal brake pressure voltage at the ECU input was lifted from 1V to about 1.15V – just a 15 per cent increase at this end of the sensor's output. This resulted in a clear deceleration when the pedal was lightly pressed, and much stronger regen than normal as the pedal was pressed harder. At high braking efforts, the behaviour of the car was near standard – it was in very light braking where there was a clear difference.

But I'll be frank – I am not sure that everyone would like the end result. Why? Well, that takes some explanation. With a purely hydraulic braking system, if they so desire, the driver can brake extremely gently. In fact, perhaps only imperceptibly slowing the car. However, with the modified regen system, the minimum braking deceleration that could be achieved was noticeably more than the minimum with a purely hydraulic system. That is, if you wanted to slow down, you put your foot on the brake pedal. If you didn't want to slow down, don't brake – because if you did, for the same light pedal pressure, you would slow at a quicker rate than in a conventional car.

So extremely gentle braking, of the sort perhaps where in a conventional system you'd just barely be touching your foot on the brake pedal, was no longer possible. Once that switch had tripped, braking was occurring. This is because when the relay was activated by the action of the brake pedal, the ECU immediately thought that the brakes are being applied more strongly than they really were – and so gave you the extra regen braking.

But it's important that this minimum braking threshold not be overstated. Testing with an accelerometer showed that when the throttle was lifted (without any braking occurring), the deceleration from 100km/h (~60mph) was about 0.1g. When the brake pedal was pressed at its modified minimum value (that is, the relay was just tripped), the deceleration increased to 0.11g. At 60km/h (37mph), a throttle lift resulted in 0.08g deceleration, while minimum braking resulted in 0.12g deceleration. (As a guide, about 0.8g indicates very hard braking.)

And the benefits? First, the regen braking was clearly doing much more of the braking work. Or, to put it another way, much less energy was being wasted in the brakes. This could be both seen in the display of watt-hours regenerated (there were more 'suns' than usually achieved), and also in the feel of the car. The regen braking was smooth and effortless, slowing the car substantially before the brake pedal was moved a little further to activate the hydraulic brakes and bring the car to a halt.

It was hard to assess urban fuel consumption, but with more energy going back into the battery (and not into heating the brakes), the fuel economy was also sure to be improved. The wear of the hydraulic brakes and pads was also substantially lessened. (These brakes still got used in anger down long hills, though – as the battery filled, regen

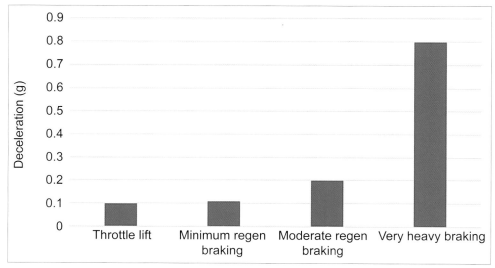

This graph compares the figures – a throttle lift typically gave about 0.1g deceleration, minimum regen braking about 0.11g, moderate regen braking about 0.2g, and a hard emergency stop resulting in braking of (at least) 0.8g.

6. MODIFYING OTHER ELECTRONIC CONTROL SYSTEMS

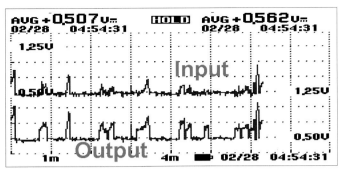

Logging the input signal from the sensor and the modified output signal to the ECU showed that over a 6-minute hilly urban drive, the average value of the sensor signal was 0.507V, and the average value of the modified output was 0.562V. However, as the lower trace shows, the modified output of the sensor had a lot more 'area under the curve,' a better indication of the changed feel on the road.

progressively reduced its braking action and the hydraulic brakes then did more and more of the work.) The regen also switched off at very low speeds, so the discs still stayed shiny.

This modification shows clearly that no matter the system, if it uses straightforward analog control voltages, it is easy to change.

SWITCHING OFF AN AIRBAG

> **WARNING!**
> Unwanted airbag inflation can cause severe injury or death.
> The technique described in this story may be illegal in some jurisdictions.
> Modifications to airbag systems should only be carried out by experienced and qualified persons with access to the full factory workshop manual and information.
> Inappropriate modifications and/or techniques can:
> - cause unwanted airbag inflation
> - prevent appropriate airbag inflation in a crash
> - or require the replacement of expensive parts

The main reason that you might want to switch off an airbag is if you have a young child riding in a front seat of an airbag-equipped car. Children under the age of about 12 should not ride in the front seat of a 1990s/2000s airbag-equipped car. That's not something you'll find trumpeted widely, but if you talk to engineers involved in crash testing, they'll quietly make the point that anyone smaller than about a typical 12-year-old will be badly placed should the airbag inflate. The primary problem is in the location and height of the head. It's for this reason that some cars have,

as standard, a passenger side airbag deactivation switch. But what if your car doesn't have such a switch? How can you switch off the airbag without triggering a fault code?

Information
First, it is absolutely vital that you have access to the full manufacturer's workshop manual. This is the case for two reasons: (1) the manual contains vital safety information including the appropriate measures that need to be taken before working on the airbag system, (2) the electrical characteristics of that system need to be known if the airbag is to be disabled without a fault code being logged. In many car modifications, technical systems (eg the engine management system) can be modified without full information being available. The modification described here does not fit into this category!

In the case of the Honda Insight on which the passenger side airbag disabling modification was undertaken, the primary 'safing' procedure was to disconnect the negative terminal of the battery and wait three minutes for the internal energy storage capacitors to discharge.

Airbag system sensing
The airbag ECU in all cars is designed so that it can sense when there is a problem with the integrity of the wiring leading to the airbag, or the airbag itself. The ECU obviously can't fire the airbag to see if it works, so it looks instead for two main fault conditions: the airbag is showing zero resistance (ie there is a short circuit across the airbag and its wiring), or there is infinite resistance across the airbag and its wiring (ie the wiring or the airbag has a break).

So, in between the values of infinite and zero resistance, there must be a resistance that the ECU sees as correct. In the case of the Honda, this is 2Ω. This value can be deduced from the fact that, as shown in one of the workshop manual fault-finding sequences, when testing the integrity of the wiring leading to the airbag, the airbag is unplugged and a 2Ω resistance substituted. If the fault code disappears, there's no problem with the wiring between the 2Ω resistance and the ECU.

However, in some workshop manuals that I have looked at, such information isn't readily apparent. That is, a substitute resistance is not clearly stated. The apparent solution would be to measure the airbag resistance directly with a multimeter. But be very careful if doing this! If the procedure of measuring airbag resistance with a multimeter is not shown in the workshop manual, don't do it! This is because multimeters test resistance by applying a current to the tested object and then measuring the voltage drop. In other words, when you measure resistance with a multimeter, you are applying some power to whatever you are measuring. In the case of an airbag, if that current is too high, the airbag may inflate. The Honda workshop manual states that the test multimeter should output a current of less than 10 milliamps on the lowest value resistance scale; however, it also states that you shouldn't measure airbag resistance directly.

So why do you want to know the resistance the airbag poses to the ECU? The answer is that if the airbag is disconnected and a substitute resistance supplied instead, the ECU won't see any problems with the system.

Substituting a resistance

If the resistance that the airbag normally poses to the ECU is known, a double pole double throw (DPDT) switch can be used to switch the resistance in as the airbag is switched out. If following this route, maintain the correct polarity of the airbag and ECU wiring; that is, make sure that when the airbag is connected to the ECU by the switch, the polarity of the connections remains unchanged.

The appropriate resistance can be created by using a normal electronics ¼W resistor (which will probably then burn out should the airbag inflation signal be sent during a crash!) or by using 5W or 10W resistors (which will probably withstand the current burst).

Switching to select or deselect the airbag should only be done with the ignition off, and as a matter of course, no-one should be positioned in front of the airbag when the switch is operated. The wiring diagram for the modification is deliberately not shown here: if you don't have the skills to wire in a DPDT switch, I don't want you doing it!

Switch

A good quality switch must be employed. This is because when the airbag is switched on, the integrity of the switch is important in making sure the airbag fires in a crash. And, when the switch is in the off, position, you need to be sure it is in fact off. Switches with silver-plated contacts are available from quality suppliers; in this instance, I don't suggest you use the cheap switches available online. A good approach is to use a key-operated switch. Don't use a switch designed for just small currents – a 5A rating should be a minimum.

The best approach is to attach all the wires to the switch on the bench. Use a multimeter to measure continuities and resistances, making sure that when the switch is in one position, the appropriate substitute resistance is shown on the ECU connection wires, and when the switch is in the other position, the ECU wires are connected to the airbag wires. Make absolutely certain the 'airbag on' and airbag off' switch positions are clearly marked.

Special crimp terminals are available for when working on airbag wiring. However, at the end of the day, the wires are just insulated copper wires so other, more traditional techniques, can also be employed. If soldering, make absolutely certain that the wiring cannot flex anywhere near the solder joint, where it will be more brittle than elsewhere.

Note that even with the airbag on/off switch in place, both short and open circuits will still trigger the fault warning light as it did with the system standard.

BLACK BOX ADJUSTABLE STABILITY CONTROL

> I have chosen to present this section largely as I did when I tested the product in 2009 for an online magazine. I thought then that the product was probably 20 years ahead of is time – and perhaps it still is the same distance into the future! Was it good? Yes! Is it worthy of further thought? – absolutely yes. Is it still available? No – sales were poor. Can the same result be achieved with the simple interceptor techniques covered in this chapter? I think it's very likely they can be.

Walk into a suspension workshop and you could be forgiven for thinking nothing's changed in 50 years. You'd like better handling, Sir? No problem – here we have new, shortened and stiffer springs. And you'll need new dampers – yes, here are our better quality, adjustable dampers. Improved bushes and stiffer anti-roll bars? Yes, they'll also need to be on your list …

But snap back to current reality and it doesn't take much nous to realise that confining handling changes to just the mechanical bits is fundamentally flawed. What's been missing until now is an ability to adjust the standard electronic systems that, in many cars, make more of a difference to real-world handling outcomes than pretty much anything you'd normally do with springs or sway bars. Enter Whiteline's Black Box – a driver-adjustable control system for factory-fitted, electronic stability control systems.

The Black Box gives the driver infinite adjustment over the way the electronic stability control system operates. Want more oversteer? Turn the knob. Want the car's ability control system to have the same handling approach as standard – but to intervene much later, when the car is more out of shape? Turn the knob. Want to have instantly accessible pre-sets for wet weather, beginner driver and the track? Turn the knob …

This is a revolution in handling modification that cannot be understated in its importance. Not only can you adjust the handling balance of the standard car, you can also modify the suspension (with those springs, bars and dampers) – and then adjust the stability control to match the mechanical modifications!

Or what if you boost engine power and find that the car is more inclined to power either oversteer or understeer? Again, simply adjust the Black Box so that the stability

6. MODIFYING OTHER ELECTRONIC CONTROL SYSTEMS

control system works perfectly with the increased power.

The Whiteline Black Box can never be outdated, no matter what changes you later make to the car. If you want improved handling, that makes the Black Box the first modification to buy.

Electronic stability control

Electronic stability control (ESP) uses an electronic sensor to detect the rotation of the car around its vertical axis. This sensor, called a yaw sensor, can detect whether the car is turning clockwise or anti-clockwise, and how fast it is doing so. The other major input to ESP is a steering angle sensor, that – as the name suggests – senses how much steering lock you have dialled-in, and the direction of that steering lock.

In normal driving, the system watches the steering lock and checks that the car is turning in the right direction, and at the right rate, to match the steering input. If the system detects that the car is not following the path dictated by the steering input, it knows the car is sliding. If the car is not turning as much as the steering indicates it should be, the car must be understeering. If the car is turning more than the steering input, it must be oversteering.

If either of these conditions is detected, ESP brakes individual wheels and/or reduces engine power. These interventions are designed to return the car's path to that being requested (via the steering) by the driver.

From this (very quick!) overview you can see that the system makes a judgement about how much the yaw angle of the car can deviate from the steering input before ESP intervenes. In other words, how much sliding (ie understeer or oversteer) is permitted.

What a lot of people don't realise is that the judgement about when ESP should intervene varies a lot from model to model. Some manufacturers program the ESP software to allow the car to develop quite a lot of 'attitude' before the ESP system intervenes. Other manufacturers have the brakes hammering away and the power being decreased before the driver can detect any sliding at all! Some cars even have switchable levels of ESP, where for example the car has to be sideways almost beyond the point of recovery before the ESP comes in. The sensitivity of the ESP, and whether it lets a car have more oversteer (or more understeer) before it intervenes, is entirely up to the car company's policies and its engineers.

And that takes us to a really important point. Whiteline's Black Box allows you to make the judgement as to how early ESP should intervene. It also allows you to make the judgement as to the driving conditions, driver experience, modifications made to the mechanical parts of the suspension, or to engine power. In short, it allows the custom tailoring of the ESP system for your specific application.

The Whiteline Black Box – an unsuccessful commercial product, but one that really showed the possibilities of modifying stability control.

Black Box

So what does this wondrous modification device look like? As the pictures show, the Black Box is, well, black, and, um, a box. It's about the size of a small paperback novel and has on its front face two pushbuttons (accessed through the flexible membrane faceplate), a two-line back-lit LCD and a knob. There are also two LEDs visible – a red and a green.

It is recommended an auto electrician does the installation, although with only a few simple connections and with full fitting instructions provided, it's easily within the realm of someone working at home.

On the tested Australian VE Holden Commodore, the connections are:
- Direct plug-in via an adaptor to the vehicle yaw sensor
- Spliced connections to the wiring at the back of the OBD port

There are three modes of operation. These are:
- OEM – ESP operates as standard; Mode LED is off
- Active – ESP behaviour is altered on the basis of the selected Black Box pre-set; Mode LED is green
- Tuning – the action of the ESP can be altered real time (this mode is also used to set the parameters for each of the pre-sets); Mode LED is red

After starting the car, the user has to scroll through a warning and product disclaimer notice before pressing OK. Because at this stage no pre-sets will have been set, the next step is to enter Active (tuning) mode, a process that requires a tricky simultaneous pushing of the two buttons. In tuning mode, two controls are available. Because the Black Box is a world first, the developers have had to invent new terms to describe the action of the tuning adjustments.

'Volume' describes how much action is permitted by the ESP. When Volume is set to 100, the ESP system is as sensitive as standard. When Volume is set to 0, the ESP has no input. So for example, if you have a car that

WORKSHOP PRO — MODIFYING THE ELECTRONICS OF MODERN CLASSIC CARS

has ESP that intervenes too early, all you need to do is to dial-back the Volume a bit. (In engineering terms, think of volume as 'gain.')

'Bias' describes how much understeer or oversteer occurs. So if you want the car to oversteer a lot, you set the Bias control to +40. If you want the car to understeer a lot, you set the Bias control to -40. If you're happy with the factory understeer/oversteer, set it to 0.

In Tuning mode you can set both Volume and Bias independently and across the full range of values. Once you have selected a combination that suits, this can be memorised to become one of the five available pre-sets. Therefore, you can have up to five (plus standard) different ESP settings available. To state this again, by using the pre-sets you can have the ESP set perfectly for wet conditions, for dry conditions, for your friend driving your car, for the track – and can even alter the 'tune' of the pre-sets as your tyres wear, your dampers start to get tired, and so on!

With the pre-sets previously calibrated, you'd normally get into the car, OK the warning and then select the right pre-set for the conditions.

Testing

I was able to drive a VE Commodore SV6 fitted with the Black Box. The main venue was a racetrack, but I was also able to drive the Commodore on both bitumen and dirt roads. To be honest, the system is the type that would be best experienced over a few weeks of driving in all sorts of conditions, but even in my brief sampling, it was obvious that the Black Box does what its maker claims.

By altering the settings, I was able to change the action of the Commodore's stability control to give more composed and faster cornering: in comparison, the standard settings felt quite lame, with bags of understeer and overly early intervention. On dirt, the traction control and ABS continued to work exactly as standard, while altering the Black Box settings gave the opportunity for lots of sliding without the car ever getting away.

How it works

Black Box development has so far cost Whiteline a lot over three years – so the company is not about to divulge all the details on how the operating system works. However, the system clearly intercepts and alters the output of the yaw rate sensor – which is easy enough to do with an analog sensor. But much trickier is the way in which it intercepts CAN bus data so as to alter the instructions being sent between vehicle ECUs. [Years later, I learned that in fact it was just a simple interceptor of the voltage signals – no CAN bus interception needed. That makes it achievable with the techniques I've outlined earlier in this chapter.]

By working with the standard car electronics rather than fighting against them, the Black Box takes car handling modification straight into this century. I think it's a brilliant product. ...

SWITCHING OFF TRACTION CONTROL WHILE LEAVING STABILITY CONTROL ACTIVE

I'll now show you a technique that lets you disable traction control while still leaving stability control operating. This will allow you to spin the drive wheels in a straight line as much as you want, and will also allow you to move the car around with power when cornering, without the traction control system always intervening. However, the 'safety net' of stability control will remain – that is, if you get too far sideways, the stability control will still reassert itself, helping to prevent the car leaving the road. It's a brilliantly simple way of reinvesting your car with fun while still having the electronics available to help in times of major crisis.

> **WARNING!**
> As with any car modifications, altering the electronics of traction control and stability control systems can be dangerous. If you make incorrect wiring connections, your car may cease to have ABS, traction control and/or stability control. When modifications have been undertaken, the car may handle in a very different way to standard, and the driver may need higher skills. As with any modification of the brakes and suspension, good quality workmanship must always be used. Further, very careful test driving should always be undertaken until you are confident with the altered handling characteristics of the car.

Testing the stability control modifier on the track. On this car, the brake lights are illuminated when the stability control is operating.

The modification requires no difficult electronics construction, although it must be said there are plenty of

6. MODIFYING OTHER ELECTRONIC CONTROL SYSTEMS

The modification to the traction control was performed on this Lexus LS400. The approach taken meant that stability control and ABS worked normally.

surrounded by a coil of wire. A toothed cog, which is attached to the hub, rotates past this sensor. This type of sensor is called 'inductive.' The output of an inductive sensor is a voltage that rises and falls as a sine wave – it's like an AC (alternating current) waveform because, essentially, the sensor is like a little alternator.

As the wheel rotates faster, the frequency (the number of up/down movements of the voltage per second) also becomes faster. The Electronic Control Unit (ECU) is looking at this frequency, and from it can calculate the speeds of each wheel equipped with a sensor (that's usually all four wheels).

An important point about an inductive sensor is that it has a lot of 'grunt': because you effectively have four little AC generators going – one in each wheel – there is plenty of signal strength available. This makes the next stage of this modification really easy.

Telling the ECU lies

Let's concentrate on a rear-wheel-drive car. (For a front-wheel-drive, just reverse the obvious in the following discussion.) The traction control system knows that the rear-left wheel is slipping when it is spinning faster than the front wheel – let's say, faster than the front-left wheel.

It stands to reason, then, that if we tell the ECU that the front-left and rear-left wheels are spinning at the same rate, and that the front-right and rear-right wheels are spinning at the same rate, the ECU will see no wheelspin by the drive wheels. It doesn't matter if the rear drive wheels are smoking up a storm, if the ECU thinks that each of the rear wheels is rotating at the same rate as the wheel directly ahead of it, the traction control system won't intervene. After all, it can't know that anything odd is happening (unless the system measures road speed via the undriven wheels and then matches that against selected gear and engine revs – fairly unlikely).

And it's very easy to make the ECU think that the rear wheels are spinning just as fast as the front wheels. All that we do is disconnect the rear-left wheel ECU input from the rear-left speed sensor, and instead connect it to the front-left sensor. And, likewise, the rear-right wheel ECU input to the front-right sensor. So both left-hand wheel speed inputs get their information from the front-left wheel, and both right-hand wheel speed inputs get theirs from the front-right wheel.

If we use a couple of DPDT electrical relays to do this, we can easily switch between the standard wheel speed inputs and the no-traction-control setup. You could, of course, just use manual DPDT switches, but this would make on-the-go changes difficult, and would also prevent the ABS from working. The ABS? OK, take a deep breath.

If the system is set up so that the ECU cannot detect wheelspin, it also won't be able to detect wheel lock-up.

Sensors

The traction control system determines if wheelspin is occurring if one or other of the driven wheels is rotating faster than the undriven wheels. Wheel speeds are sensed by ABS sensors; these almost always consist of a magnet

wires to connect up. This makes having access to a factory workshop manual of the traction control ECU very useful – although not vital.

The test car on which the modification was performed was a 1998 Lexus LS400. Obviously, I can guarantee that the modification works well on this car, and I can't think of any reason why it wouldn't work on other cars as well. (In fact, as you'll read later, you can test if the modification works on your car without having to spend a single cent!) Traction and stability control systems operate in different ways on different cars, so I can't make a blanket statement that this approach will definitely work on every car, but I think it's likely that it will …

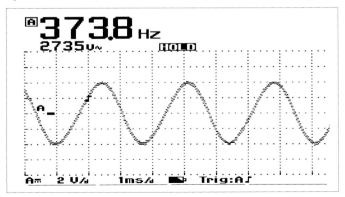

A scope screen grab of the output signal of one of the ABS sensors.

133

WORKSHOP PRO: MODIFYING THE ELECTRONICS OF MODERN CLASSIC CARS

So, the ABS system will turn off at the same time as you switch the system into no-traction-control mode. Since I am a believer in having ABS operating all of the time, that outcome needs to be got around. However, if you use relays to do the switching, it's easy to switch back into normal mode whenever you hit the brake pedal, using the signal input from the brake lights. To do this, we use a relay triggered by the brake lights to switch the signal interception relays back to standard. To provide a manual Traction Control on/off switch, all that's needed is a switch in series with this 'brake light' relay. The circuit has also been designed so that if power is lost to the system, the traction control returns to standard.

Let's look at that now in terms of circuit diagrams.

DISABLING TRACTION CONTROL

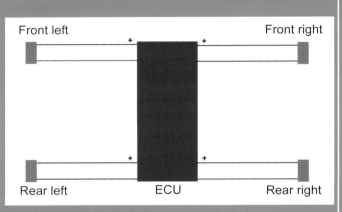

This is basic traction control system – four wheel-speed inputs.

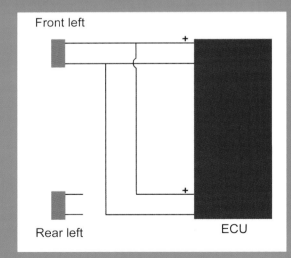

This shows only one side of the car – obviously, the other side is duplicated. Here the front-left wheel speed output has been additionally connected to the rear-left ECU input. The result is that, in this rear-wheel-drive car, the ECU cannot detect wheelspin.

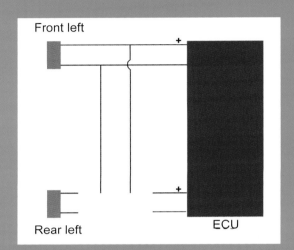

You can see here that if we connect the ECU rear-left inputs to just the rear-left sensor, the system will act normally. However, if we instead connect this ECU input to the front-left speed sensor, rear wheelspin won't be recognised. We can do this changeover with a DPDT switch or relay.

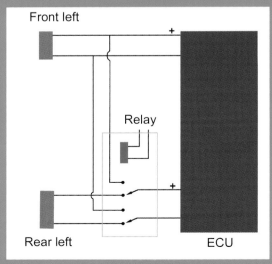

Here a relay has been used to make the changeover. Note that when power is not applied to the relay's coil, the system is in standard mode.

6. MODIFYING OTHER ELECTRONIC CONTROL SYSTEMS

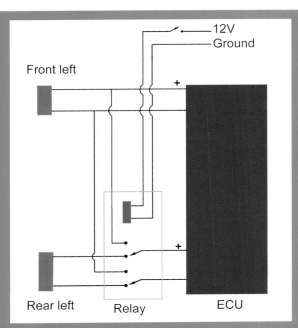

By adding a switch that operates the relay, we can manually switch the traction control system on and off.

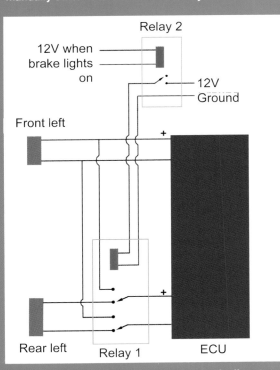

By adding another relay, we can automatically return the system to standard when the brakes are applied, so that ABS continues to operate normally. Relay 2 is switched on when the brake lights are on, so cutting off power to Relay 1 and returning the inputs to standard.

Testing

As mentioned earlier, it's easy to see if the approach will work before actually starting to install switches and relays. You will need to find the pairs of wires that come from each wheel speed sensor. (At a pinch this may be able to be done at the wheel speed sensors themselves – which would mean you could do it without a workshop manual.) Cut the wires leading to the driven wheels, and connect each of these pairs to the wheel at the opposite end of the car.

Note that there's one other tricky thing to know about. You'll find that one side of each of the sensors is grounded inside the ECU. This wire can be found by using a multimeter to measure for continuity between each ECU wheel speed sensor input and ground. The sensor input that has continuity to ground is the ground wire – so the other in the pair must be the positive, or signal wire. Make sure that you connect negative to negative and positive to positive. Once you have made these connections, start the car and go for a gentle drive.

Note that during this testing your car will NOT have ABS and will NOT have traction control!

However, no warning lights should come up on the dash (eg the ABS, traction control and stability control warning lights shouldn't glow). Additionally, the stability control system should still operate while the traction control doesn't. Now, obviously, take great care when finding out what the car's behaviour is – a large expanse of grass is perfect, as you can do the testing at very slow speeds.

Wheelspin in a straight line should be uncontrolled (except by your right foot!), while you should be able to push the car more into a corner under power without the traction control intervening. However, when the car is sliding in a direction different to where you're pointing the steering, the stability control system should still operate much as it did previously. In other words, you should be able to throttle-steer much more than the standard system allowed.

If warning lights come up when you first start the car, either you have made a wiring error (check polarity as mentioned above) or the ECU is smart enough to detect that there is one sensor connected to two inputs.

In the guinea pig Lexus, the (non-LSD) rear-end would spin as much as you liked in a straight line, and would also allow far more cornering attitude changes with the accelerator than could be achieved in unmodified form. However, start whipping-on opposite lock to control the rear-end powerslide, and the stability control would rapidly calm the situation. In short, it reduced the early intervention of the traction control so you could have a lot of fun without switching off all the electronic assists.

WORKSHOP PRO: MODIFYING THE ELECTRONICS OF MODERN CLASSIC CARS

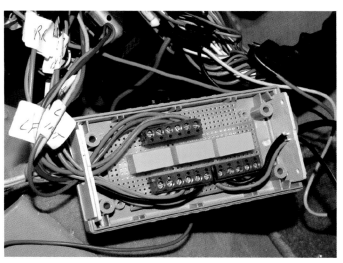

The relay board wired into place. Note the temporary paper labels attached to the wires on the left – it's vital that you don't get confused as to which wire is which.

WIRING

As covered in the box on the previous page, the switchover system uses three relays. The changeover relays need to be Double Pole Double Throw (DPDT) 12V designs, while the switching relay can be a Single Pole Double Throw (SPDT) 12V design. At the risk of confusing you, you can also make the switching relay a DPDT design (and then use only half of the contacts). Using all-DPDT designs makes it easier to source the relays, as you can just buy three that are the same. In addition to the three relays, you'll also need some pre-punched circuit board, terminal strips and a box.

The key part of the system is making sure that the right relay contacts connect to the right parts of the car. Easiest is to wire the relays on the pre-punched piece of board, mounting both the terminals and the relay on the board. That way, you need only connect from the screw terminals to the car, rather than trying to connect directly to the relay terminals themselves. The way in which you connect the pins to the terminal strips is up to you, but I chose to use a short length of insulated wire. That way, it's easy to arrange the wires without concern for short-circuits.

As far as I know, this modification is unique – but it worked very well.

DEVELOPING A TORQUE SPLIT CONTROLLER

> The following material was first published in December 1998. At the time, the idea that the handling of the Nissan Skyline GT-R was not perfect was seen as absurd: according to many, this story was simply indicative of my hopeless driving ability. But in the years since, the use of aftermarket torque split controllers on GT-Rs has become common – and of course, Nissan itself improved the logic of the four-wheel drive systems on subsequent GT-R models by dramatically decreasing the silly amounts of power oversteer available on the R32 GTR.

After owning a turbo 4WD Subaru Liberty (Legacy) RS for four years, I'd decided that I wanted a Skyline GT-R. The famous winner of the Australian Bathurst Touring Car Races (until it was banned!), the Skyline is the car that every person into turbo four-wheel drive cars simply lusts after. Not only has it great grunt, but also the brakes and handling to match.

And it's the topic of handling that needs some dwelling on. I remembered once doing a GT-R story for an American magazine. The conversation with the car's owner had gone something like this: *(continues on page 138)*

FINDING THE RIGHT WIRE – AND REPLICATING ITS SIGNAL!

This is a short story of electronic detection: of using clues and tools to work out something which, at first glance, looks impossible. The outcome is specific to one make and model, but the approach is universal.

Many 'grey market' Japanese imports ('grey market' means the car was never imported officially) come with navigation systems. That's great, but not so good are the Japanese software and Japanese language directions. Initially it seems a good idea to convert the navigation system to English and local mapping software, but this is very hard to do. Much easier is to simply pull the navigation computer out of the car, so saving a bit of weight and at the same time, freeing-up some space.

But that move can have some unexpected outcomes. Firstly, in some cars, the navigation system operates the car's clock. So what happens to the clock when the navigation computer is pulled? In the case of the guinea pig car – an NHW10 Toyota Prius – the result was good. Whereas previously there had been no way of setting the clock (and it had always been an hour out), with the navigation computer unplugged, up came clock setting buttons on the colour dashboard LCD screen – excellent.

But not so good was that at the same time the driver's side door speaker stopped working! There wouldn't seem to be any connection between a navigation computer and a car radio, but there is. What happens is that the navigation computer sends its oral instructions through the driver's side door speaker. Because it doesn't want to be yelling its instructions over the top of the radio, when the instructions come through, the speaker is disconnected from the radio. That's fine – but when the navigation computer is unplugged, the speaker goes silent all the time.

And how do you fix that?

Thoughts

As always, the first thing to do is to consult a wiring manual for the car. But unfortunately, while I have a manual, it's for the wrong model. The manual I have is for an NHW11 model, not my NHW10. And the wiring changed substantially between these models. But thinking about the way the system could work, there were at least two possible approaches.
1. The front-right speaker signal from the radio could be fed to the navigation computer. The navigation computer could either pass this signal to the speaker or, when it wanted to voice instructions, disconnect the radio and switch the speaker output to the navigation system.
2. Alternatively, the navigation computer could signal the radio when it wanted the radio to cut off the speaker. The navigation computer would always be wired to the speaker, but it would voice instructions only after the radio had been instructed to turn off its feed to that speaker.

And here's where the wiring diagram was useful – as shown in the diagram, the radio was always connected to the four speakers (approach 2), rather than having one of the speakers fed through the navigation system (approach 1). So the navigation computer must signal the radio to shut down the feed to the front-right speaker. With the computer unplugged, this signal was present all the time, so switching off this speaker.

But how on ground could the shape and form of this signal be found, and then recreated and fed into the right wires? It seemed impossible ...

Testing

The navigation computer was connected to the wiring loom with three plugs – large, medium, and small. The small plug carried only one shielded wiring connection – almost certainly to the GPS aerial. The medium-sized plug contains eight connections, and the large plug nine connections.

The wires that signal the radio to shut down the speaker must be in one of these plugs – but which? The first step was to unplug each in turn, so finding which plug contained the relevant wires. Pulling the medium plug caused the clock to come good, and pulling the large plug caused the speaker to switch off. So the relevant wires were in the large plug – that simple move had halved the number of possibles from 18 to 9.

But the nine wires that were in the large plug could be for anything: power, ground, video feeds to the central colour LCD, speed or braking inputs. (But, on reflection, probably not the video feed to the LCD. Because the LCD screen clock came good when the small plug was pulled, this plug probably contains the LCD feeds.) But how do you sort out which of the wires in the remaining large plug are which?

Thinking about it further, most of these wires are going to inputs to the navigation computer. The only outputs are likely to be to the radio, to tell it to shut down the speaker. So the inputs will be measurable whether the plug is in the navigation computer or is pulled, but the outputs will be present only when the plug is in the navigation computer. If that's the case, some measurements with a multimeter would reveal all.

I grounded one lead of a multimeter and, with the ignition switched on, measured the voltages on each of the plug wires with the plug pushed into the navigation computer. I then pulled the navigation system plug, and again measured the voltages. Two of the pins which previously had 6.2V on them had dropped to zero.

It therefore seems likely that the radio looked for 6.2V on these two pins before it activated the front-right speaker. I grabbed two short lengths of hook-up wire and connected the 6.2V supplies to their adjoining pins. The speaker came to life ... It was then a simple case to cut and solder together the correct wires, covering the joins with insulating tape.

In this situation I was a little lucky – if the car had a CAN communications bus and looked for signals communicated on the bus before it activated the speaker, I'd have been stuck. But by the same token, finding the right wires and then feeding-in the correct voltages was no easy task.

In this type of situation:
1. Think about the different ways in which the system might operate.
2. Isolate the relevant plug by pulling the plugs in turn.
3. Remember that input signals will be present irrespective of whether the plug is pulled or not, but output signals will only be present when the plug is inserted.*
4. Recreate the required output signal and feed it to the appropriate pins.

An exception to this is where the sensor needs a power supply, or pull-up or pull-down resistor, to function.

See, even what initially appears impossible can be done without blowing anything up!

WORKSHOP PRO: MODIFYING THE ELECTRONICS OF MODERN CLASSIC CARS

"What's she like around corners?" I'd asked.

The owner had shaken his head. "It's just fantastic – it never even moves. Once I went into a corner way too fast and the torque gauge on the dash flicked for a moment – that was it. It's just incredible ..."

I've always been a person who prefers a good handling car to one that is fast but can't go around corners. So to have both blistering straight line speed, and what many regard as one of the all-time best handlers in the one package – wow!

Buying

At that time, Skyline GT-Rs in Australia fell into two categories – the 100-odd R32 model cars that were imported by Nissan Australia in 1991, and the R32 and R33 models privately imported direct from Japan.

The latter R32's were a lot cheaper, but at the time I was looking, had some major problems. Problems like having been burned or crashed in Japan to the point where the Japanese owner would rather get rid of it than persevere with repairs ... This put me off the imports, and so I turned my attention to the Australian-delivered cars. I fully expected to have to travel widely in Australia to find the car, but in the end, I located it in my home city of just a million people, Adelaide. Even better, the car was effectively brand new, having only 3500km (2200mi) on the clock. I twisted the arm of the bank leasing officer, dropped my house payments to the minimum, and leased the car. Some car purchases require no financial sacrifices at all, but this sure wasn't one of them – I was prepared to literally live on bread and water to have the car of my dreams!

The handling

Obviously how you regard the handling of a car depends very much on what you are used to. And if I'd been pedalling around a pedestrian car, I'm sure I would have thought that the GT-R handled very well. But I hadn't been. My Liberty RS had been shod with super-sticky track-use A008-RS Yokohamas, and the car's standard all-wheel drive made wheelspin impossible – even with a modified 280-odd horsepower. The Liberty had a slight understeering characteristic, but the ability to put power down (even before the apex), and then just haul out of the corner was mind-boggling.

Even while I had been test driving the GT-R, I had been surprised by the amount of power oversteer that was present, but I had put that down to the odd wheels and tyres that the previous owner had fitted. I assumed that decent tyres would fix the car (I mean, this was a GT-R!) and asked that different tyres and the original rims be refitted before I took delivery. However, even with the new tyres and standard 16x8 rims, the GTR power oversteered – no matter what I did when cornering.

If you went in hard, the car would understeer a smidgin before the rear end came out in a big slide when you applied power after the apex. The understeer felt about right – it happened at much higher limits than the Subaru – but all you could do about the oversteer was to accelerate less quickly! If you wanted to stay on your side of the road, you soon ran out of opposite lock: you could literally have the car sideways in the lane. It looked great (bystanders used to actually point and clap!), but you were going nowhere pretty fast. And if you made a mistake, it was diabolical.

It seemed to me that the four-wheel-drive system was simply too slow to come into action. You see, unlike most four-wheel-drive high-performance cars, the GT-R is rear-wheel drive most of the time. It is only when rear wheelspin occurs, you use a very large throttle opening, or (so I then thought) the G-sensor detects enough lateral acceleration, that the front wheels are powered.

So, I figured, maybe you needed to keep the power on harder and later to make the front wheels start to work? I drove the car harder and harder, holding the oversteer slides with armfuls of opposite lock and getting used to viewing the road literally through the side glass. About this time most passengers started developing a strong urge to get out ...

So perhaps it was still the tyres? I couldn't afford to buy another set of expensive tyres (with the car repayments, we could barely afford the bread and water), but I managed to source some slightly worn replacement tyres of a different type. The car handled just the same.

I talked to many experts and other GT-R owners after that, concerned that there was something very seriously wrong with my car. Much to my amazement, most of the other GT-R owners had never driven their cars hard enough to slide them anyway! They were strictly 'pose-

The all-wheel drive R32 Skyline GT-R had massive power oversteer in standard form. Put your foot down and it was as if the car was rear-wheel drive only, not all-wheel drive.
(Courtesy David Bryant)

6. MODIFYING OTHER ELECTRONIC CONTROL SYSTEMS

and-straight-line-go' merchants. Other owners were totally shocked at what I told them. Mostly, it appeared that they were shocked because someone dared to criticise their favourite toy. The experts just told me that I obviously couldn't drive, and to fit bigger rear tyres.

I took one GT-R driver for a ride (he refused to drive the car himself because, he said, that it would be impossible to safely slide it in urban conditions) and the car oversteered around every corner as usual. Flamboyant, slow and dangerous.

"It's handling exactly as I would expect a GT-R to handle," the other GT-R owner said from the passenger seat. "You just need to feed the power in gradually as you are exiting corners. There is a lot of power there, you know."

After the modified Subaru this was simply rubbish – my old modified Subaru went just as hard as the standard GT-R, and you could tromp it out of every corner without understeer or oversteer!

You can imagine my utter disappointment. The twin turbo engine was superb, the seats, steering and brakes wonderful – but the car handling! I was so disappointed that I seriously considering selling it. Remember, this was the car that truth be known I really couldn't afford anyway. Perhaps I would have sold it, but the next development in the story put the car off the road for a while. What happened? I bumped it into a tree – purely driver error, nothing to do with the handling. And only the second time I had ever left the road in many years of very hard driving. During the time that the car was being repaired, I made a decision – I was going to modify the four-wheel-drive control system to bring on four-wheel-drive much earlier and stronger. After that – well I'd see.

Where to start?

The Japanese manufacturers Blitz, Field and HKS produced torque split controllers for the GT-R. I managed to source a Japanese magazine review of one of the controllers and get the article translated. The translation showed that the device plugged into the centrally-mounted G-sensor and that it certainly altered the handling characteristics!

I figured that it must change the output of the G-sensor, perhaps increasing it so that the four-wheel drive ECU thought that the car was cornering harder than it really was. Wouldn't this direct more torque to the front wheels?

Armed with a multimeter and an assistant, I measured the output of the G-sensor in all sorts of driving conditions. I soon found that there isn't just one G-sensor – there were three. Two measure longitudinal acceleration (ie acceleration and braking), and the other measures lateral acceleration (ie cornering). All three sensors had a 0-5V output signal. When there was no acceleration, the sensors all had about 2.5V output. The harder the car was accelerating, the higher the voltage output from the longitudinal G-sensors. When decelerating, the sensor voltage dropped to below 2.5V. With the lateral G-sensor, the voltage decreased below 2.5V on right turns and increased above 2.5V on left turns.

I called on some experts to design an electronic interceptor. This device would boost the output swings of the lateral accelerometer, while at the same time leaving the 2.5V 'stationary' output untouched. The interceptor was designed and fitted to the lateral sensor output, but sadly there was absolutely no difference in the car's behaviour. The standard dash-mounted torque split gauge also behaved as standard.

Another interceptor was developed, and weeks of experimentation followed. The interceptor was working – when it was connected to the longitudinal sensors I could boost straight-line acceleration torque going to the front wheels, but with only limited benefits when cornering. But when it was connected to the lateral accelerometer, little changed. Did the lateral accelerometer even work? I wondered.

Stunning

The first breakthrough came when I took the unlikely step of rotating the G-sensor package through 90°. This meant that the lateral G-sensor became the longitudinal, and the longitudinal became the lateral. Sounds terrible, doesn't it! This was done because I realised that the longitudinal sensors had a very powerful influence on the torque split – something that the lateral sensor didn't appear to have.

Turning the G-sensor package through 90° made a stunning difference to the handling. Instead of being a car where every slow corner exit required a gentle foot and/or opposite lock, now full throttle could be used with near-impunity. For the first time it simply felt like a proper four-wheel-drive car.

But there proved to be some negatives. Firstly, in tight corners taken with a high entrance speed, the car could now understeer excessively. Also, the confused torque split computer allowed some wheelspin high in the rev range in first gear. Still, the handling was so much better than standard that I stuck with this modification for some time.

Then it rained. The predictability and stability that the car had in the dry was immediately gone. Not that it was as bad as I'd found it in standard form, but it just didn't feel quite right, and straight-line wheelspin was even more pronounced.

By this stage, two more important things had happened: I'd had a chat to a bloke who had been involved in the Bathurst race GT-Rs, and my tame electronic guru had developed the third version of the G-sensor adjustable interceptor. This was being used to slightly reduce the

WORKSHOP PRO: MODIFYING THE ELECTRONICS OF MODERN CLASSIC CARS

The torque split controller made an extraordinary difference to the car's on-power handling. Here, it is just gripping and going, exiting the corner at full throttle in second gear.
(Courtesy David Bryant)

output of what (with the G-sensor rotated) was now the lateral sensor.

The ex-Bathurst race team man said something that immediately caught my attention. "The computer reduces the front torque split the harder you corner. So if you remove the influence of the lateral G-sensor, the car will go into four-wheel drive earlier." Gulp! I'd always figured that the computer would increase the front torque split as you cornered harder. That's why I'd initially tried to amplify the signal – I should have been trying to reduce it all along!

"One way to do this is to feed a constant 2.5V signal to the computer input for the lateral G-sensor," he continued. "Course, it understeers like a pig then," he added.

Even better

I raced out to the car and effectively returned it to standard, but for feeding a constant 2.5V input lateral signal to the ECU. He was right – with this modification, the car did understeer excessively. But that was in the dry. In wet conditions, this modification made the car just ballistic – unbelievably good, with so much traction and cornering prowess that it was uncanny. The difference was so immense that in the wet conditions I found myself 'ABS'ing' up to roundabouts, so good was the car's grip in every other situation.

Okay – if holding the lateral G-sensor input fixed at 2.5V gave too much front-wheel drive in the dry but was perfect in the wet, why not use the interceptor to vary the lateral G-sensor signal from 1:1 (ie signal unchanged from the standard voltage swings) right through to 0:1 (ie signal fixed at 2.5V input)? In other words, be able to change the influence of the lateral G-sensor all the way from factory standard to none! That way I'd be able to dial up any cornering torque split from huge oversteer right through to heaps of front-wheel-drive for wet conditions.

And that's just what the final configuration was like.

A knob on the dash allowed variable selection across the whole range. The knob was calibrated from 0-10, with 0 being standard and 10 being for full wet weather. Generally, I was around '7' in dry conditions, with '6' being used in really tight low speed corners when I wanted the tail to come out a bit to turn the car in. When the road got wet, '8' or '9' was selected, and when it was streaming with water I was a '10' man!

Set up in this way, the handling was absolutely fantastic. It's hard to believe, but the electronic mods made more difference to the handling of the GT-R than any other suspension modifications I have ever made to a car. And that includes the changing on previous cars of wheels, tyres, sway bars, springs, dampers, bushes – the lot. The difference simply cannot be overstated.

> Footnote: Writing this some 20 years later, two things strike me. The first is that I think I could probably have just used a simple pot to reduce the voltage swing from the lateral sensor, the technique shown earlier in this chapter in the coverage of the Prius power steering. Second, I think if anything, I understated the difference it made to the car's on-power handling. It still remains the most revolutionary change to car handling I have ever experienced.

IMPROVING AUTO TRANSMISSION SHIFTING

While I am not going to delve into major automatic transmission control modification, there are two techniques that are simple and can make a real difference to how well the automatic transmission in your car works on the road. And the first isn't even an electronic modification!

Cable adjustment

In some cars, engine throttle position information is conveyed to the transmission controller by a Bowden cable. This transmission throttle cable can be adjusted by simply undoing a couple of nuts on the throttle bracket fitting. By making such an adjustment, the transmission can then be made to think that more throttle is being applied than is actually the case.

I did this modification – or, perhaps expressed more accurately, this adjustment – on a 1991 Lexus LS400. The improvement in transmission behaviour was major. The transmission changed more crisply, it better used the (sweet) rev range of the 4-litre V8, and when climbing hills or when the driver wanted better in-gear acceleration, the transmission dropped back a ratio more readily. In fact, for zero cost and five minutes, the improvement was amazing!

Auto triggering the power/economy switch

If the automatic transmission has a power/economy button, an interesting modification can be made. What it does is

6. MODIFYING OTHER ELECTRONIC CONTROL SYSTEMS

By adjusting the Bowden cable that transmits throttle position to the automatic transmission controller, much better transmission control was achieved at zero cost on this Lexus LS400.

automatically select the transmission mode, depending on how hard you are driving. It brings another step of intelligence to transmission control, and on the road the driving difference is significant.

The electronics module that does this is a kit called 'Quick Brake'. (It's called this because the primary application of the module is to quickly turn on the brake lights when the throttle is abruptly lifted, as you do before standing on the brakes in an emergency.) I came up with the idea about 15 years ago, and an electronics magazine called *Silicon Chip* developed the project. The project has passed through a few iterations since, but the current version works in much the same way as the original module. (Search for 'Quick Brake electronic kit' – and note it is a kit, so you'll need electronic kit building skills. At the time of writing, it is available from www.altronics.com.au.)

So how does it work? The module constantly watches the voltage coming from the throttle position sensor. When this changes quickly enough, it trips a relay that then stays on for a pre-set time. Because the module can be configured to activate only when the voltage is rising quickly, it's perfect for watching how fast you're putting your foot down.

Need a spurt of power to get out into traffic? Or you're pushing down the throttle fast as you thread the car out of a hairpin corner? In both cases, the module will sense what is going on and pull in the relay, which can be used to automatically set the Power mode of the transmission. If the timer period is set for 10 seconds, Power mode will stay activated for that period. If you're hard on the throttle any time within the timer period, the timer starts counting again.

Because the module is measuring the rate of throttle application (not just how much throttle is being applied) the driver doesn't have to push the throttle to the floor to get the result. Instead, just a quick movement halfway through the throttle range is enough. In practice – since a switch from Economy to Power modes often results in a down-change – this makes the transmission so much more responsive, it's amazing.

To suit your preferences, you can set how long Power mode stays engaged after the last quick throttle movement. In practice, the difference between (say) 8 and 12 seconds is quite noticeable, and so being able to set this time accurately is important. And of course, you can set exactly how quickly the throttle needs to be moved to trip the relay – this will depend quite a lot on your individual driving style.

In use, the module is wired to the TPS, as described in the kit instructions. Power and ground are also connected.

The next step is to access the Power/Economy wiring of the transmission gearlever selector. The connector for the wiring will be at the base of the lever, hidden under the centre console. Remove trim bits until you can get hold of it and then pull the connector apart. Use the multimeter (set to continuity) to work out which terminals of the connector get connected together when the switch is pushed. For example, in the case of the Nissan Maxima Turbo being used here, the connector contained four terminals. Two of the terminals were electrically connected whenever the power/economy switch was in 'Power' mode. Therefore, to cause the transmission to switch to Power mode, these two terminals needed to be connected together. (The other two connections were for the light in the gearlever which comes on in Power mode. These were left untouched.)

With the right wires sorted out, two extra wires were soldered to these cables, and run to a temporary switch. Flicking the switch was therefore the same as pressing the power/economy button and selecting 'Power' mode. In the Maxima, fourth (top) gear is locked-out whenever Power mode is selected, so it was easy enough to drive down the road, operating the new switch and seeing if the transmission mode changed. It did.

Assuming that your car's auto transmission power/economy button joins the two wires to activate Power

If you have an automatic transmission that has a power/economy switch (or similar), a substantial improvement in transmission behaviour can be gained by using a module that automatically triggers the switch when you are driving hard.

141

WORKSHOP PRO: MODIFYING THE ELECTRONICS OF MODERN CLASSIC CARS

The original Quick Brake kit that can detect how fast you are moving the accelerator pedal, and then triggers a timed relay. This kit is no longer available, but a replacement with similar functionality is.

mode, you will need to connect the wires to the module relay's adjacent common and normally open terminals. If the switch opens to put the car into Power mode, then make the connections to the common and normally closed connections of the module's relay.

You should be able to do the initial testing without even starting the car. With the ignition switched on and the module visible, check that when you move the throttle fast, the module's LED comes on, and then stays on for about 10 seconds. In most cars, there will also be an indicator light that shows when the car is in Power mode. Leave the button in the Economy mode position, and then check that the Power light comes on when you push the throttle fast. It should – of course – stay on for as long as you have the timer set.

If all is working as it should, go for a road test. You will almost certainly find that you need to reduce the sensitivity of the module – in real driving, you tend to push down on the throttle more slowly than when you're just playing in the garage! Adjust the sensitivity until 'slightly more spirited' driving selects Power mode. With this set correctly, trigger the Power mode with a decisive throttle movement and then immediately go back to driving gently. Does the transmission stay in Power mode too long? If so, adjust the timer to suit. Then do the opposite: drive hard for (say) a minute and make sure that the transmission stays in power mode for the whole time. If not, again tweak the timer control. One of the beauties of the system is that it can be set to exactly suit your preferences.

In normal driving, leave the selector in Economy and let the system automatically select Power as required. However, you can still use the Power/Economy button manually if you want to keep the car in Power for an extended period (eg if towing).

The modification is cheap and brilliantly effective. If you have a car where Power mode of the automatic transmission gives you a much-improved shift schedule – but Economy is far better when you're driving gently – then this modification works incredibly well. You soon get used to driving the car's auto transmission mode with your right foot – much as drivers in current cars with very smart automatic transmissions do. In fact, after having the prototype in my Maxima for a few months, I had to take the module out for another application. The change in driving performance was so dramatic that I had it back in the car within days – it felt like I'd lost so much midrange power. It won't improve your full-throttle acceleration times, but in real-world driving, the improvement in performance can be fantastic.

> **Kickdown?**
> I haven't tried it, but on transmissions that don't have power/economy modes, you could probably use the module to trigger the kickdown switch. The sensitivity would probably be set at a lower value, and the time would be set to a shorter period.

FULLY PROGRAMMABLE CAR CONTROLLERS

So far in this chapter I have covered modifying non-engine management car systems largely with simple intercepting circuits using passive components. And, for the near-zero cost, the results can be extraordinary. However, for many car systems, such an approach will not give the outcome that is desired. For example, if you want to map the revised behaviour of the system so that it changes more in some situations than others, or if you want to change the operating logic of the ECU itself, simple passive component interceptors won't work. Instead a full new controller will be needed.

The same need for a new controller arises if you are adding a complex system to a car that previously did not have it (eg air suspension). One approach to achieving these outcomes is to use small and cheap micro controller boards, for example PIC-based boards. Let's first look at doing this, and then I'll look at a much more expensive, complex and capable programmable controller. Air suspension is being used here as the example, but the same ideas could be applied to controlling a variety of car systems, including active aerodynamics, dynamic damper control and so on.

AIR SUSPENSION CONTROL

The car that has featured quite often through this book – my Honda Insight – has been fitted with air suspension.

6. MODIFYING OTHER ELECTRONIC CONTROL SYSTEMS

Air suspension has considerable advantages over steel springs in both ride and handling. (See *Custom Air Suspension*, also published by Veloce, for more on this.) The movement of air into, and out of, the air springs is controlled by electrical solenoid valves, with the high-pressure air provided by an onboard 12V electric compressor and a storage tank. So what functions does an air suspension electronic control system need to perform?

Functions

At its simplest, an air suspension control system needs to monitor ride height and then trigger air solenoid valves appropriately to maintain this height at a constant level. That sounds easy, so now let me add some real-world complexities.

Firstly, how do we measure ride height? Typically, potentiometer-based sensors are used, mounted in parallel with the springs. As load increases, the springs are compressed, and so the ride height decreases. But the springs also compress and extend with bumps – that's their purpose! So how can we differentiate the measured ride height variations being caused by bumps from the variation being caused by 'true' ride height changes? An averaging function is needed, so that the fast movements of bumps is ignored but the slower movements of ride height changes are measured.

But then it gets more complex. How long should this averaging period be? Movement of the springs with bumps are fast, so if we were to average over – say – 10 minutes, we'd get a good indication of the actual ride height. But what happens then, if the car is stopped – and five people get into it? The suspension will compress, and we can't wait 10 minutes before correcting the ride height – the suspension will be on its bump stops!

Therefore, we really need two suspension height measuring modes – one that reacts quickly (people getting in and out) and one that reacts slowly (change of ride height caused by use of fuel, variation in load of mud or snow, or the temperature of the air within the springs changing). The 'fast reaction' averaging could be over (eg) 10 seconds, and the 'slow reaction' averaging over (eg) 10 minutes.

So why not just make the averaging time 10 seconds in all conditions? Imagine a long corner, where the car rolls and the outer suspension compresses. In that situation, a 10-second averaging technique will cause the system to try to inflate these outer springs. Thus, two different averaging modes really are needed.

But how will the system know which mode to select? In the case of the controller covered in a moment, I decided to select the correct averaging mode by detecting how often the ride height sensor was changing in output. This is because when the car is being driven on the road, the suspension height outputs change often per second over bumps, while when the car is stopped, the ride height outputs change much less frequently.

Now what about a driver control? A potentiometer input is also fitted for the driver, allowing the manual selection of ride height over a pre-fixed range. Finally, in the case of the Honda, I decided to control the rear air springs as a pair and the front springs separately.

So let's see how this is all coming together. The controller requirements:

Analog 0-5V inputs:
- Rear suspension height
- Front-left suspension height
- Front-right suspension height
- Driver height control pot

Digital outputs:
- Rear suspension up and down solenoids
- Front-left spring up and down solenoids
- Front-right spring up and down solenoids
- Mode LED (this shows whether the slow or fast averaging mode has been selected by the software)

PIC-based controller

Working with electronics company eLabtronics, I developed a suitable controller based on the company's STEMSEL module. This is a prebuilt (but unboxed) module that uses a PIC18F14K50 microcontroller. The module has 12 digital

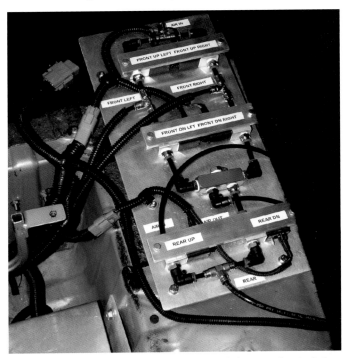

The air solenoids that control the flow of air into, and out of, the air springs.

WORKSHOP PRO: MODIFYING THE ELECTRONICS OF MODERN CLASSIC CARS

The two eLabtronics PIC-based controllers, working in conjunction with two quad relay boards. (Two of the relays are spares.)

In initial use, the cheap control system worked very well. Ride height was maintained at what I thought was about ±10mm (±0.4in), and the system responded well to changes in car load. So all seemed to going along fine – until two things changed.

First, I connected the suspension height measuring sensors to my MoTeC ADL3 digital dashboard, so that I could monitor actual suspension height real-time, to the millimetre. This showed that in fact ride height (averaged over 3 seconds) was varying more than I had thought – the system was holding ride height to more like ±25mm (±1in) versus the setpoint. In most cars with air suspension, that would not be a problem, but the Honda has as standard a low ride height – and at times I wanted to run it even lower than standard. If the ride height was varying by this much, the situation could easily develop where it was low enough that a bump could cause the suspension to hit the bump stops.

The second problem that developed was one of my own making. While working on another book, *Modifying the Aerodynamics of Your Road Car*, I added a full length undertray with rear diffuser to the Honda. Subsequently, I added a small rear wing. The result of these two aerodynamic modifications was substantial downforce,

or analog input/output pins, four low-current driver outputs that can drive relays (and so operate the solenoids), and an onboard USB connector, that allows reprogramming from a PC or laptop, and can be powered from the car's 12V supply. In addition, it has a regulated 5V output that was used to power the suspension height sensors and the dashboard height selection pot.

The presence of only four driver outputs means that two STEMSEL modules were required – one for the front suspension, and one for the rear. These worked with two commercially available quad relay boards, with six of the eight relays used.

Now what about programming the module? While I devised all the logic of the control system, writing a program to achieve this is beyond my capabilities. However, eLabtronics at the time had University of Adelaide intern engineer Daniel Calandro working with them, and Daniel eagerly took up the challenge. Using the proprietary eLabtronics visual CoreChart coding approach, Daniel developed the air suspension controller software.

Part of my software design request was that key aspects of the software could be user-tuned to suit the application. These adjustable parameters were:
- Hysteresis (the required difference between the driver height request and the actual suspension height before the system reacted)
- Modes 1 and 2 averaging periods
- Sensitivity of the mode selection subroutine

The CoreChart program could be brought up on a laptop screen, the parameters in specific boxes altered, the laptop cable plugged into the module, and the revised program loaded into the module.

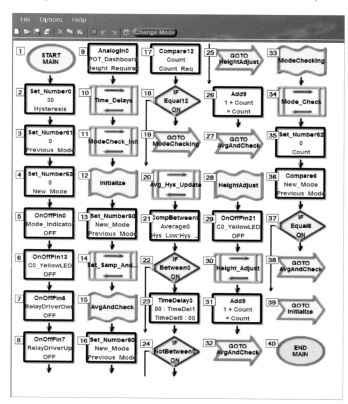

The CoreChart programming developed by Daniel Calandro and eLabtronics to control the Insight's air suspension.

6. MODIFYING OTHER ELECTRONIC CONTROL SYSTEMS

even at normal road-going speeds. (Incidentally, the downforce could be measured quite accurately with the combination of the air suspension height sensors and the MoTeC dash.) The logic of the air suspension control was never designed with the idea that a car could effectively get heavier as it went faster, and even more so that this change in 'weight' could occur in less than 10 seconds as I accelerated from a standstill to 100km/h (~60mph) or more!

For its cost, and in most car air suspension applications, the eLabtronics STEMSEL-based control system was – and remains – an outstanding controller. However, for my application, I decided that I needed something much more sophisticated. In short, my additional requirements were that the controller have:

- 2D and 3D maps
- Real-time laptop tuning
- Ability to easily alter controller logic
- Laptop gauges showing actual suspension height
- PID control (see Chapter 4)

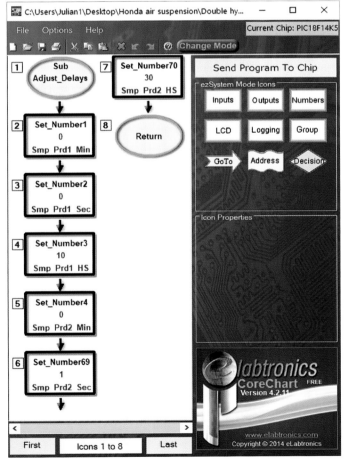

This subroutine allowed the altering of time delays. With the program changed, the revisions could then be uploaded to the module. No 'live' tuning was possible.

- Road speed input

If this was starting to sound like the requirement for a fully-programmable engine management ECU – (but with user-access to the operating logic), that's no coincidence. I therefore approached programmable engine management manufacturer Adaptronic, and asked if they had something suitable. To be honest, what they provided just blew me away, and it's a product that has major implications for anyone wanting to control other car systems.

ADAPTRONIC E1280S FULLY PROGRAMMABLE CONTROLLER

The Adaptronic e1280s is normally sold as a programmable engine management controller. However, its unique configurability using a special flowchart programming approach means that control systems can be developed for literally any system in a car. The best overall description of the controller's capability was written by the company:

"*The e1280s patented, completely configurable tuning system allows the tuner to implement any control function that can be articulated, using a system of basic elements like maps/tables, PID controllers, timers and basic arithmetic and logical functions.*

The ECU introduces a whole new level of configurability. This has been made possible by the use of an FPGA (a type of programmable logic device) rather than a microcontroller, for a majority of the calculations.

The best way to describe this new functionality is as follows. ECUs currently on the market can be viewed as having fixed 'equations' built into their microcontroller, with a firmware upgrade needed to change anything more than some constants and gains. The e1280s, on the other hand, allows the user to change the actual equations and logic at any time (no firmware change needed), so if the default configuration doesn't do everything you need, you can make the necessary changes yourself using the software."

For those of you who have used any computer programs that utilise block diagrams to implement certain functions, that is what you can expect from the e1280s software. You are able to see a block diagram representation of the entire logical configuration inside the FPGA. The block diagram is separated into a number of 'pages' which can be selected from a list.

The blocks (called 'elements') in the diagram could include the following, and more: adders, multipliers, logical operators, comparators, general purpose 2D and 3D tables, general purpose PID controllers, timers, delays, stepper motor blocks, and gauges. There are also 'dedicated elements' for the trigger inputs, analogue inputs, digital inputs, aux outputs, injector outputs and ignition outputs.

WORKSHOP PRO: MODIFYING THE ELECTRONICS OF MODERN CLASSIC CARS

The block diagram represents the equations and logic that are implemented in the FPGA chip, and at any time you can add, remove, or modify any elements on the block diagram. This means that you can have as many 2D and 3D tables as you want (memory permitting) with arbitrary axes and size! And if there is something that we haven't considered or implemented in the default ECU configuration, you can freely add any elements you need to make it work for you. Also, you can do closed loop control, of anything by using as many PID controllers as you need.

Finally, there is a built-in scope function in the software, which has all the same features as a real oscilloscope, and allows you to observe what the FPGA is seeing on its inputs, etc."

The Adaptronic fully programmable controller.

Let's look at this in more detail.
1. If you have the capability, you can develop the control program from scratch, without even knowing how to code. (I don't know how to code, so that's very useful for me.)
2. Developing the control program logic has strong parallels with developing other technical systems – for example, I see analog circuit design cues in the thought processes that you follow. If you've done any type of design work previously, that makes it easier.
3. The software has a simulation mode that allows you to test the operation of the program that you've designed (or modified). In this mode, the inputs are able to be altered, and the action of the logic and outputs then assessed. This can be done working purely on a PC – the ECU doesn't need to be connected.
4. If given sufficient detail on what is required, Adaptronic has the ability to develop custom logic for the customer. The software is then sufficiently transparent that you can make changes as you wish. Those changes can be normal 'mapping' changes (as when you are tuning a system) or can be changes to the logic of how the system actually functions.
5. With 16 analog inputs, 8 digital inputs and 16 outputs, the e1280s is likely to be able to simultaneously control many different car systems, not just one. That is important because the system is not cheap – it's similar in price to a normal engine management ECU. Being able to control multiple systems obviously makes the approach more cost-effective.

Because it can be hard to see how all this works, let's step-by-step develop a controller for a water/air intercooler system on a turbo car. We want the system to control how fast the pump runs based on engine load. Once we've done that, I'll add some further functionality.

WATER/AIR INTERCOOLER PUMP SPEED CONTROL

ECU Wiring — Adaptronic e1280s Connector / Factory Connector

Pin		Pin		Pin		Pin	
101 Inj 1	0	4 Aux 1	Pump	23 Knock 1	0	42 CAS 2	0
102 Inj 2	0	5 Digin 1	0	24 Knock 2	0	43 VSS 2	0
103 Inj 3	0	6 Digin 2	0	25 Aux 8	0	44 Digin 5	0
104 Inj 4	0	7 Aux 2	0	26 Ana 6	0	45 Ana 15	0
105 Inj 5	0	8 Digin 3	0	27 Ana 7	0	46 Digin 7	0
106 Inj 6	0	9 Aux 3	0	28 Ana 12	0	47 SGND	0
107 PGND	0	10 PGND	0	29 Ana 13	0	48 +5V out	0
108 PGND	0	11 Ign 2	0	30 SGND	0	49 Digin 8	0
109 +12ign	0	12 Ign 4	0	31 Ana 3	0	50 PGND	0
110 Inj 7	0	13 Ign 6	0	32 Aux 7	0	51 CAS 3	0
111 Inj 8	0	14 Ign 7	0	33 VSS 3	0	52 CAS 4	0
112 Inj 9	0	15 Ign 8	0	34 Ana 4	0	53 VSS 1	0
113 Inj 10	0	16 Aux 4	0	35 Ana 5	0	54 Ana 11	0
114 Inj 11	0	17 Ana 1	IAT	36 Ana 9	0	55 Ana 14	0
115 Inj 12	0	18 Aux 5	0	37 Ana 10	0	56 Ana 16	0
116 PGND	0	19 Digin 4	0	38 Ana 8	0	57 +5V out	0
1 Ign 1	0	20 PGND	0	39 VSS 4	0	58 +12cnst	0
2 Ign 3	0	21 Aux 6	0	40 CAS 5	0	59 Digin 6	0
3 Ign 5	0	22 Ana 2	MAP	41 CAS 1	0	60 PGND	0

The first step is to nominate the inputs and outputs. Once named in this table, they are available within the logic of the controller. Here I have nominated Auxiliary 1 output as Pump, Analog input 1 as IAT (intake air temperature) and Analog input 2 as MAP (manifold absolute pressure). Incidentally, Analog input 1 can be biased for temperature sensing (ie use a pull-up resistor) or not, as required – if this sensor is also being used by the engine management ECU, then no bias would be applied as the engine ECU is already providing it.

6. MODIFYING OTHER ELECTRONIC CONTROL SYSTEMS

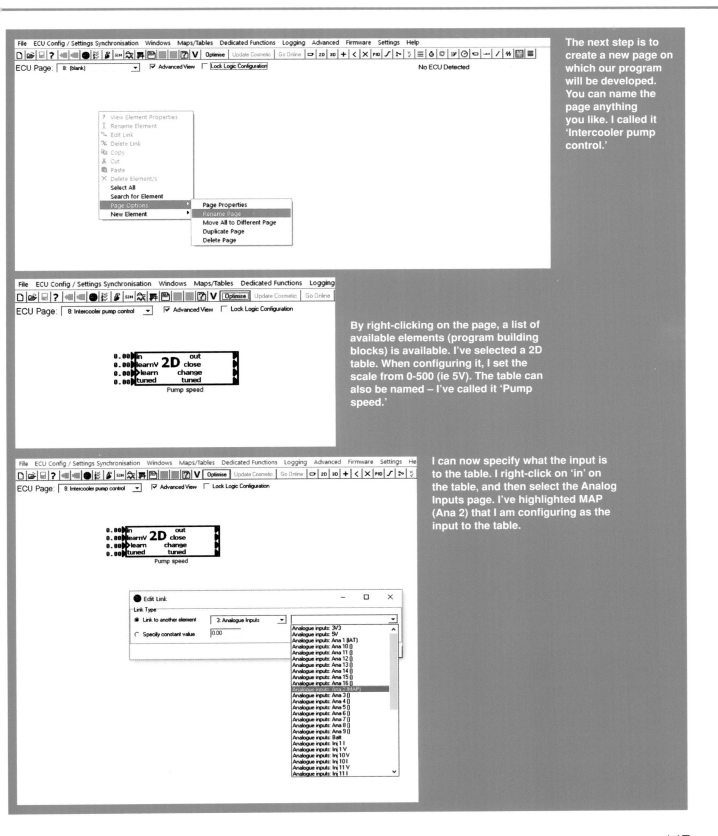

The next step is to create a new page on which our program will be developed. You can name the page anything you like. I called it 'Intercooler pump control.'

By right-clicking on the page, a list of available elements (program building blocks) is available. I've selected a 2D table. When configuring it, I set the scale from 0-500 (ie 5V). The table can also be named – I've called it 'Pump speed.'

I can now specify what the input is to the table. I right-click on 'in' on the table, and then select the Analog Inputs page. I've highlighted MAP (Ana 2) that I am configuring as the input to the table.

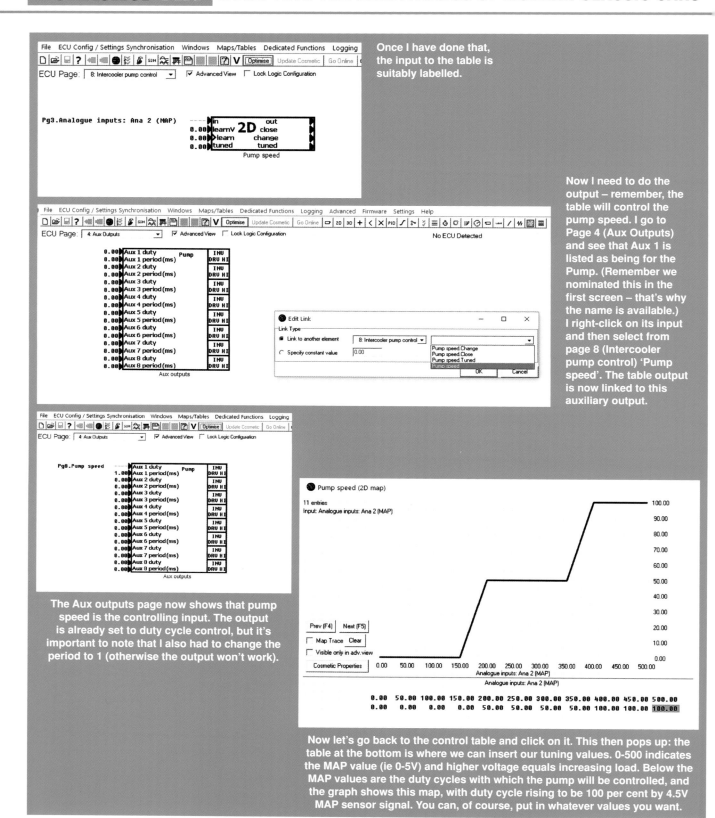

6. MODIFYING OTHER ELECTRONIC CONTROL SYSTEMS

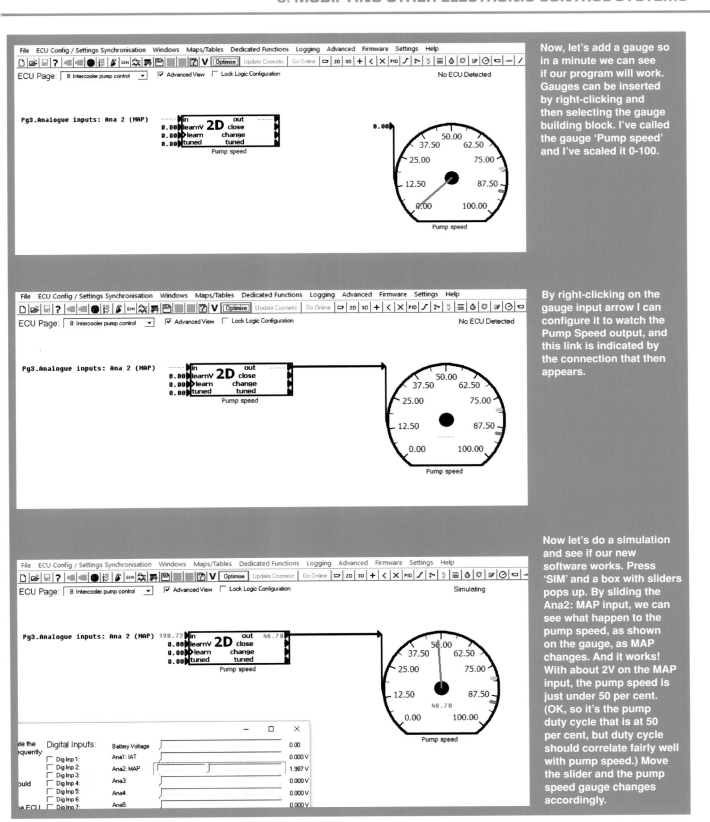

Now, let's add a gauge so in a minute we can see if our program will work. Gauges can be inserted by right-clicking and then selecting the gauge building block. I've called the gauge 'Pump speed' and I've scaled it 0-100.

By right-clicking on the gauge input arrow I can configure it to watch the Pump Speed output, and this link is indicated by the connection that then appears.

Now let's do a simulation and see if our new software works. Press 'SIM' and a box with sliders pops up. By sliding the Ana2: MAP input, we can see what happen to the pump speed, as shown on the gauge, as MAP changes. And it works! With about 2V on the MAP input, the pump speed is just under 50 per cent. (OK, so it's the pump duty cycle that is at 50 per cent, but duty cycle should correlate fairly well with pump speed.) Move the slider and the pump speed gauge changes accordingly.

WORKSHOP PRO: MODIFYING THE ELECTRONICS OF MODERN CLASSIC CARS

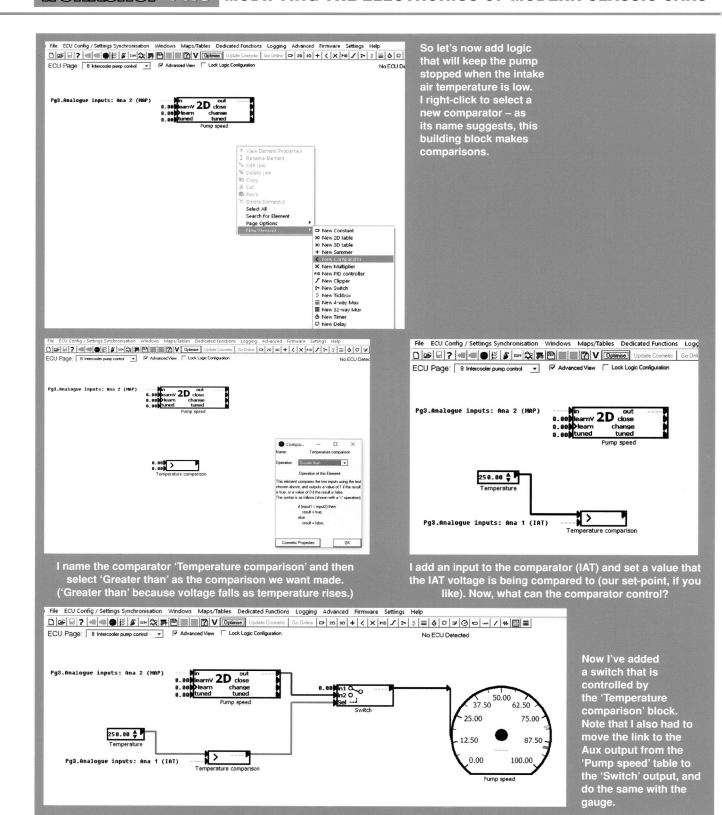

So let's now add logic that will keep the pump stopped when the intake air temperature is low. I right-click to select a new comparator – as its name suggests, this building block makes comparisons.

I name the comparator 'Temperature comparison' and then select 'Greater than' as the comparison we want made. ('Greater than' because voltage falls as temperature rises.)

I add an input to the comparator (IAT) and set a value that the IAT voltage is being compared to (our set-point, if you like). Now, what can the comparator control?

Now I've added a switch that is controlled by the 'Temperature comparison' block. Note that I also had to move the link to the Aux output from the 'Pump speed' table to the 'Switch' output, and do the same with the gauge.

6. MODIFYING OTHER ELECTRONIC CONTROL SYSTEMS

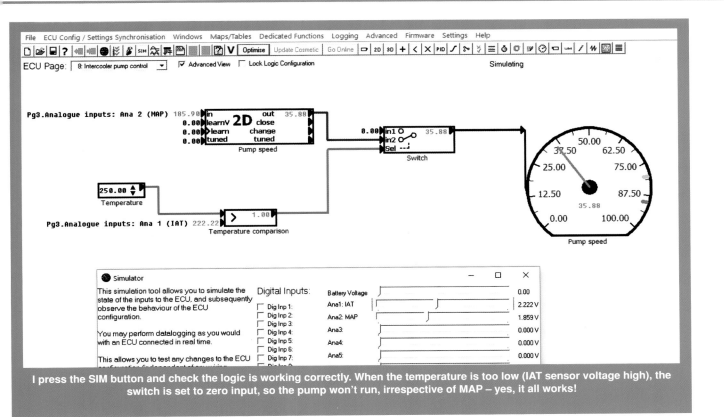

I press the SIM button and check the logic is working correctly. When the temperature is too low (IAT sensor voltage high), the switch is set to zero input, so the pump won't run, irrespective of MAP – yes, it all works!

The most important point about the above description of the intercooler water pump speed controller is not that we were able to develop a controller: it's that we can give that controller any logic we want!

- Like to map pump speed against both intake air temperature and manifold pressure? Use a 3D table instead of a 2D table
- Decided that it would be better to use throttle position and manifold pressure? Just change the inputs to the table
- Want a driver input switch that causes the pump to constantly run at full speed when you're driving on the track? Just add a digital input and another software switch to the logic

And even with this water pump controller using a digital input, two analog inputs and an auxiliary output, there are still enough inputs/outputs to have another eight such controllers! (Or of course, a smaller number of more complex controllers.)

Incidentally, when setting up the intercooler pump speed controller software, I left the sensor inputs as raw volts, but I could easily have added 2D tables that converted these values to engineering units, eg kPa (psi) or °C (°F), and then worked with these figures.

So let's look now at the e1280s software controlling the Insight's air suspension. The initial development of the software was done to my requested requirements by Andy Wyatt of Adaptronic. With the software logic established and working, I was then able to make major changes to it. In other words, with a framework established, it was not hard to add or change functions – but writing the software from scratch would have been beyond my capabilities. I'd suggest that would be the case for many people.

Changes that I have made include:
- Adding gauges so I can watch what is happening 'live' while the car is being driven (absolutely invaluable)
- Creating a mode that changes how quickly the system reacts when road speed is altering fast
- Altering the P factor in the PID controller on the basis of the required system reaction speed
- Adding an auxiliary output that lights a dash pilot light if any solenoid is open
- Adding smart switching of the solenoid that joins the rear air springs, to soften bump and roll
- Adding a function that causes the system to react fast if the driver is turning the suspension height pot
- Adding an input from the MoTeC dash that inhibits height correction when cornering

I think the potential for the Adaptronic controller is massive for modifying and controlling car systems.

WORKSHOP PRO: MODIFYING THE ELECTRONICS OF MODERN CLASSIC CARS

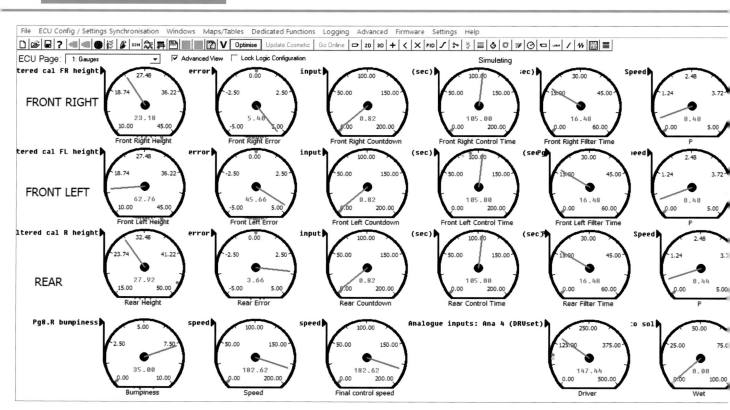

This very complex gauge laptop dashboard shows me a range of critical factors including suspension heights, suspension height error (ie how far the actual suspension height is from the set-point), suspension height sensor filter times (ie how long the sensor output is being averaged for), the Proportional (P) factor being used in the PID controller, and the driver input knob setting. Gauges can be created to watch any number changing anywhere in the software logic flow – not just the inputs and outputs. Having these gauges available when tuning the response of the system has been invaluable.

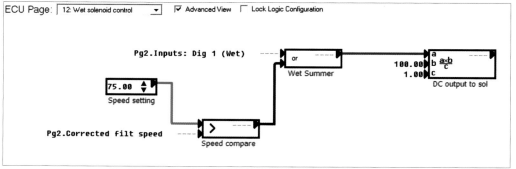

This additional logic is used to switch a solenoid that connects the rear springs. (When they are connected, the ride is softer but the resistance to roll is also less.) This logic opens the solenoid at speeds of less than 75km/h (47mph) or when a driver control switch is activated. (I flick the switch when the road is wet – the lower rear roll stiffness improves cornering grip.)

Examples include torque split control, stability control, climate control, suspension control, electric power steering control, active aerodynamics control, and many others.

The cost of using the Adaptronic controller is much greater than other techniques covered in this chapter, but the capability of the controller and its ability to achieve the required outcome is also far greater than is achievable using simple interceptors. In some cases, as with the air suspension control, it allows a whole new control system to be developed from scratch.

6. MODIFYING OTHER ELECTRONIC CONTROL SYSTEMS

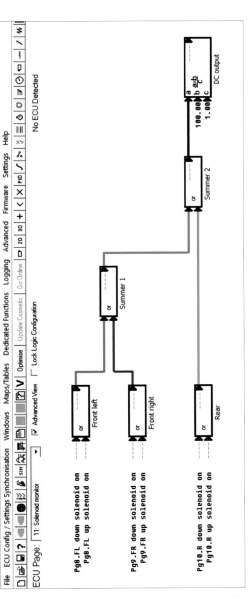

This additional logic is used to light a dashboard monitor light whenever any solenoid is open. Four 'OR' logic blocks are used to trigger an auxiliary output that runs the pilot light.

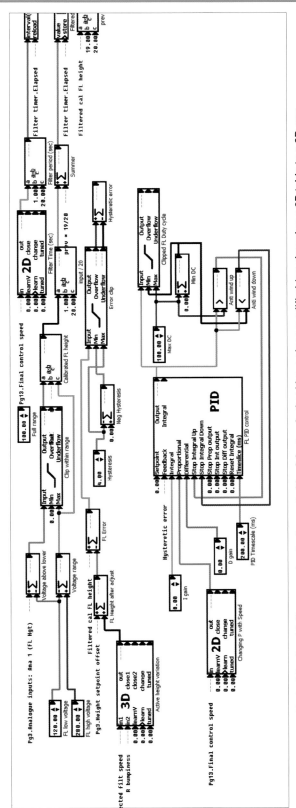

Part of the front-left suspension logic created by Adaptronic that I have since modified (eg changing a 2D table to a 3D one with an additional input, and changing a fixed variable into a 2D table).

WORKSHOP PRO MODIFYING THE ELECTRONICS OF MODERN CLASSIC CARS

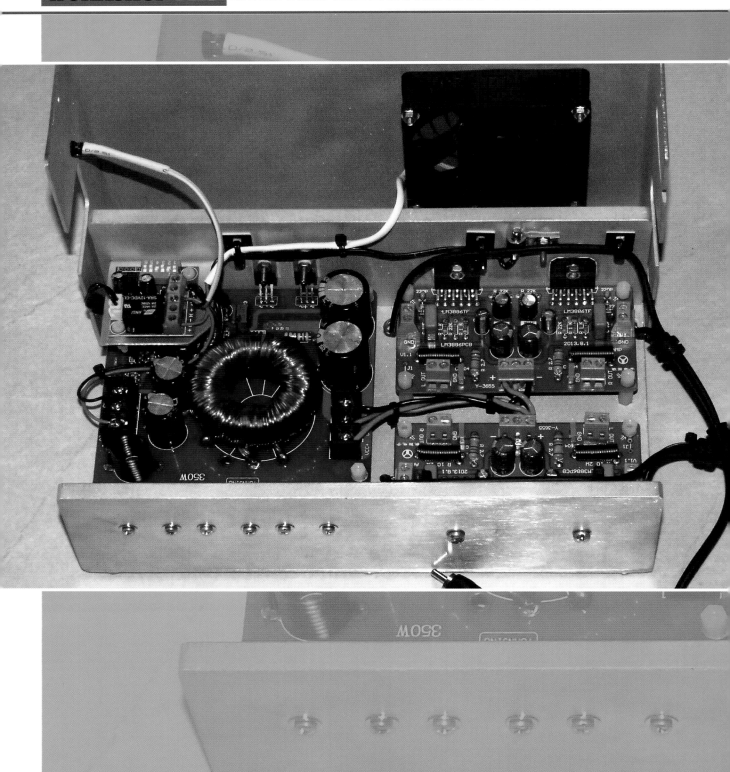

Chapter 7

Modifying sound systems

- **Components**
- **Wiring**
- **Suggested upgrade approaches**
- **A budget upgrade using secondhand parts**
- **Technical tips**
- **Upgrading amplifier cooling**
- **Installing component speakers in the front doors**
- **Software subwoofer design**
- **Installing a free-air subwoofer**
- **Building a spherical subwoofer**
- **Building a four-channel amplifier**

WORKSHOP PRO: MODIFYING THE ELECTRONICS OF MODERN CLASSIC CARS

So you'd like to upgrade the sound system in your car and you don't know where to start? Well, with your car, you have an immediate advantage. Sound systems in cars of the 1990s and 2000s usually have a standard-sized opening in the dash (either one-DIN or two-DIN), and the sound system is not normally interconnected with other car systems. These two factors make upgrading relatively straightforward.

However, there are so many ways of upgrading a sound system, it's easy to get lost. To avoid that, let's start by looking at the components, and then cover some common wiring and modification approaches. We'll then look at some examples of systems that I have upgraded, usually quite cheaply. Finally, I'll look at building a unique subwoofer, and then your own 4-channel amplifier.

COMPONENTS
Head units

It's the head unit in the dash from which all sound signals originate. Therefore, if this is of a poor quality, no matter how good your amplifier, speakers and cables are, the system will still sound poor. Better quality head units have line-level (RCA) outputs that allow you to easily connect an amplifier. Usually there will be four of these outputs (ie front-left, front-right, rear-left and rear-right), but some head units also have a dedicated subwoofer output. Note that if the head unit doesn't have line level outputs, it's possible to use a converter that changes the speaker level signals into line-level outputs suitable for an amplifier – but the sound quality will not be as good.

Many cars of the era covered by this book have cassette-radios, and unless you're after a retro look, you probably won't want a cassette function. Later cars have CD radios – and again, it's not likely that you'll be playing many CDs. If both media have now fallen by the wayside in

An upgrade head unit that incorporates video screen playback but still fits in a one-DIN opening. The space below the head unit was not used to allow room for a new panel with added switches.

Even if there is room for just a single DIN unit, a major upgrade over standard is possible. (Courtesy Adam Trites)

your music listening, you'll probably want to upgrade to a new head unit that has a front-face input (to take the feed from a smartphone or similar), Bluetooth connectivity or can take a plug-in memory stick. If the dash has a two-DIN aperture, you'll be able to fit in a double size unit with a large screen. If you buy a suitable unit, you can then display navigation, a reversing camera, and even play-back video.

Amplifiers

If the head unit has a good four-channel amplifier, and you do not wish music played loudly, you may have no need for an additional amplifier. However, if you like your music loud, and/or wish to add a subwoofer, you will need to budget for an external amplifier.

When selecting an amplifier, select one with good specifications – a high power output and low Total Harmonic Distortion (THD). Any THD over about 0.1 per cent *at maximum power* is likely to be able to be heard; many commercial car sound amplifiers have far greater

If your car's dash has a double-DIN aperture, you can use a new large-screen head unit. Installing a modern head unit allows you to integrate the system with your phone, in addition to using the screen to display navigation. This unit is also quite shallow in required mounting depth. (Courtesy Alpine)

7. MODIFYING SOUND SYSTEMS

distortion than this. In amplifiers, you largely get what you pay for; spend peanuts and you'll typically get something that's not very good! Brand-name amplifiers are worthy – if they're too expensive, look for secondhand.

The number of channels an amplifier has refers to its number of speaker outputs. For example, a two-channel has two speaker outputs, a four-channel has four speaker outputs, and so on. But in some amplifier designs it is possible to 'bridge' the output. When this is done, the number of amplifier output channels halves but the power of the remaining channel(s) rises accordingly. Note that in bridged mode, the minimum speaker impedance usually doubles (eg from 2Ω to 4Ω).

An equaliser is very useful in fine-tuning a system. Usually it is a set-and-forget item, so it doesn't need to be mounted in plain view. I use this unit in one of my cars – it's highly effective and can also provide a low-pass, adjustable level filter for adding a subwoofer. (Courtesy Clarion)

This four-channel amplifier integrates two variable crossovers that can be configured as high- or low-pass. The low-pass channels can drive woofers (eg in the back shelf) and the high-pass channels can drive door speakers. (Courtesy Alpine)

Three amplifiers being assembled in stacked form on an aluminium frame. Space was tight and this was the only way to fit them in. (Courtesy Adam Trites)

Equalisers

Equalisers are used to boost or cut frequencies. This is just like the action of the normal bass and treble controls, but instead of working with only two centre frequencies, the equaliser can work with eight or ten frequencies. This gives you greater precision, allowing you to change just the frequency you want. Having this functionality is very useful, and the equaliser doesn't have to take up any space on the dash – it can be hidden away after the system is set up.

Speakers and crossovers

In many standard systems, the speakers are the weak link. They're also the area where careful selection and good quality installation can make an enormous difference to the final sound. First, let's look at speaker specifications.

A speaker's sensitivity describes how loud a speaker is for a given input power. Sensitivity is measured in sound pressure level (SPL), normally with a 1W input and at a distance of 1m. If you have the choice of two speakers, and one is 3dB more sensitive than the other, you will need a lot less power with the more sensitive speaker to get the same music loudness. Amplifier power is expensive, so if all else is equal, better to get the more sensitive speaker and save on amplifier cost. It's for this reason that most original equipment speakers are very sensitive.

The frequency response of a speaker describes the lowest and highest sound pitches it can produce. Frequency response though, is seldom accurately quoted for car speakers. The actual response depends on how they are mounted, and a frequency response specification without relative loudness level (eg ±3dB) is even more worthless. (That is, the stated spec might say the speaker has a frequency response of 50-10,000Hz – but without a ±dB figure, it might be inaudible from 8000Hz upwards!)

Speaker power ratings are also often rather moot. The speaker might be rated at 100W, but at that power figure is it distorting badly? It could well be, with the '100W' figure merely indicating that the voice coil hasn't melted with this input power!

Speakers also have a rated impedance. Car speakers were once all typically 4Ω, and most aftermarket head units are designed to work with this impedance. However,

WORKSHOP PRO — MODIFYING THE ELECTRONICS OF MODERN CLASSIC CARS

These component speakers (sometimes called splits) use separate tweeters and bass/midrange speakers. Crossovers are also provided that incorporate adjustable tweeter levels.
(Courtesy Kicker)

manufacturers have been moving to lower impedance speakers, and so many are now 3 or even 2Ω. The biggest danger in mismatched impedances (eg the speaker impedance too low for the amplifier) is that the amplifier will be over-driven and die. Ensure that you are using speaker impedances that match those the manufacturer of the amplifier says can be driven.

No speaker can produce the full audible range of frequencies, so speakers are built with specialist functions – woofers for low frequencies, mid-range speakers for (yes!) mid-range frequencies, and tweeters for high frequencies. To split the signal into the different frequency ranges suitable for each speaker, crossovers are used. A 'low-pass' crossover allows low frequencies to pass and blocks high frequencies, while a 'high-pass' crossover allows high frequencies to pass and blocks low frequencies. The simplest crossover, a high-pass design that prevents bass frequencies getting to a tweeter, is a non-polarised capacitor. Complex crossovers are fully electronic and often have adjustable frequencies.

Many amplifiers have a low-pass, adjustable crossover built in to allow the direct driving of a subwoofer.

Subwoofers
Subwoofers produce the low frequencies. This is the hardest frequency range to produce in a car, and so

Midrange and treble is more directional than low frequencies, and so mounting these speakers high and forward gives good imaging. This midrange speaker is shown part way through the custom mounting process on an A-pillar trim.
(Courtesy Adam Trites)

A prebuilt subwoofer. This one uses a sealed enclosure and is rated at 200W – however sensitivity is only 87dB.
(Courtesy Rockford Fosgate)

7. MODIFYING SOUND SYSTEMS

subwoofers need their own dedicated enclosures. Subwoofers are normally driven in mono from summed audio channels – that is, you do not need a pair. Because low frequencies are not directional, the subwoofer can be placed anywhere in the car. More on subwoofers in a moment.

Cabling

Normal heavy-gauge hook-up wire can be used for all speaker and power connections except in a couple of instances. Firstly, a dedicated amplifier draws a lot of current and so should be fed directly from the battery; this cable should be much thicker and incorporate a suitable fuse close to the battery. The cable and fuse ratings can be determined from the value of the onboard amplifier fuse – or you can simply use the thickest cable that the amplifier's terminals with accept. Don't forget that the amplifier ground wire needs to be of similar thickness to the +12V feed. Long runs of heavy power supply cable may be cheaper if the cable normally used for welding is used. Do not use CCA (copper-clad aluminium) cable.

The other cables needing to be thicker than usual are those connecting the amplifier to the subwoofer. Commercial kits for both types of wiring are widely available.

Interconnections between sound system components

Neat power and signal wiring, and an ancillary fuse holder. (Courtesy Adam Trites)

(like the amplifier and head unit) should be made with shielded, good quality cables, normally equipped with RCA connectors. Some of these cables incorporate a power wire to turn on the amplifier.

WIRING

One of the aspects that often confuses people considering sound system upgrades and changes is the cabling – what goes to what? So in this section I want to show the wiring of some common upgrade paths.

An amplifier wiring kit that includes the power, ground, turn-on, and signal wiring for amplifiers up to a fuse rating of 60A. There's sufficient length of signal wiring to allow the amplifier to be mounted at the rear of the car. (Courtesy Crutchfield)

WORKSHOP PRO: MODIFYING THE ELECTRONICS OF MODERN CLASSIC CARS

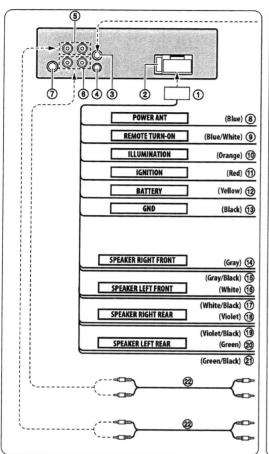

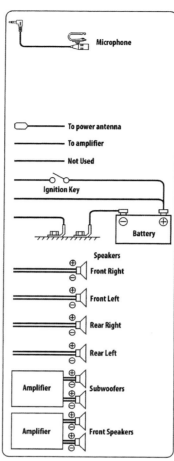

A typical wiring set-up when installing a new head unit. A provided wiring loom (1) plugs into the head unit. Four speakers are driven from the in-built amplifier – front-left, front-right, rear-left and rear-right. Note the polarity is shown of the speakers. There are six other connections – (8) to the power antenna (if fitted), (9) remote turn on (to an amplifier, if fitted), (10) to the dashboard lights, (11) to ignition-switched 12V, (12) battery 12V, and (13) to ground. RCA outputs are provided for two 2-channel amplifiers, which here are shown driving front speakers and subwoofers. The microphone is for Bluetooth phone functionality. (Courtesy Alpine)

Below: Driving four full-range speakers from an added amplifier. The head unit (20) is connected via four RCA cables (24) to the inputs of the amplifier. These RCA connections are for front-left, front-right, rear-left and rear-right. The amplifier drives the front speakers (18) and rear speakers (19). Note the power and ground connections for the amplifier – a fused 12V connection to the battery (7), ground connection (8) and 'turn-on' power feed from the head unit to the Remote input of the amplifier (9). With this amplifier type, an input switch needs to be set to select 3 or 4 inputs. (Courtesy Alpine)

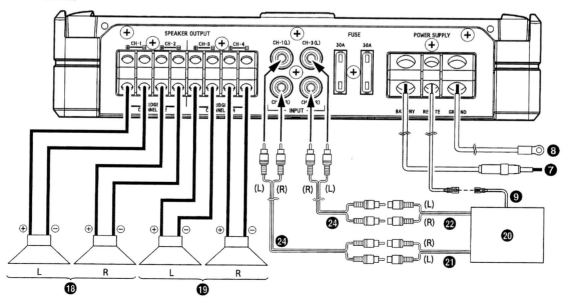

7. MODIFYING SOUND SYSTEMS

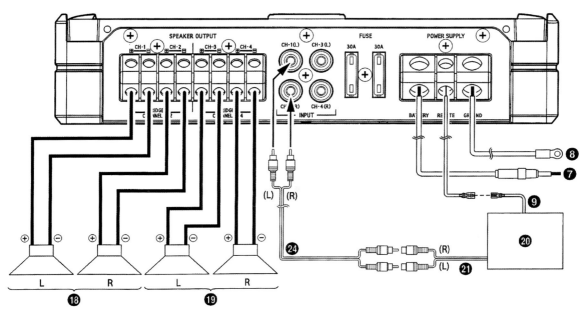

If the head unit has only two RCA outputs, the added amplifier is wired like this. Only two RCA leads (24) from the head unit (20) to the amplifier are used. Note that this system does not provide fader (front/rear balance) – the more efficient speakers will be louder. (Courtesy Alpine)

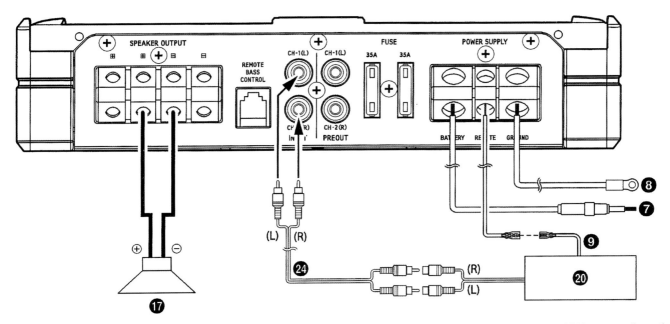

This shows an amplifier (top) added to a head unit (20) to drive a single subwoofer. Left and right line level (RCA) outputs from the head unit connect to the amplifier inputs (24). The amplifier is driving the subwoofer (17) in bridged mode – before connecting an amplifier like this, ensure it is able to operate bridged. (Courtesy Alpine)

WORKSHOP PRO: MODIFYING THE ELECTRONICS OF MODERN CLASSIC CARS

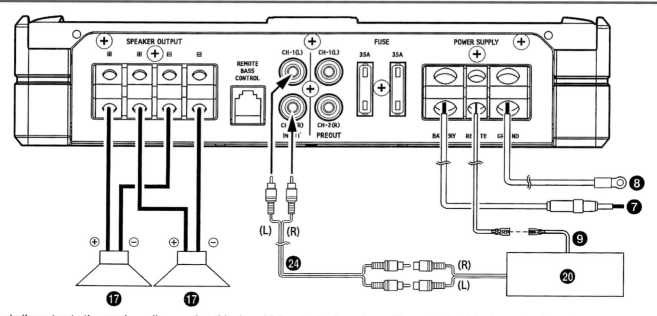

A similar setup to the previous diagram, but this time driving dual subwoofers (17), one for the left channel and one for the right. This type of approach can be used if you use two smaller subwoofers – for example, 8-inch speakers located in the rear deck of a sedan. (Courtesy Alpine)

SUGGESTED UPGRADE APPROACHES

Original head unit, new speakers

A simple approach is to retain the existing head unit and replace the speakers. This works well when the original speakers have fallen apart due to time and use – a quite common occurrence.

The advantage of this approach is that if the original head unit was a good one, you can use its inbuilt amplifier to drive the better quality speakers. However, ensure that you select sensitive speakers (high dB) or you may well find that the speakers are no longer loud enough. (Or you have to drive the head unit amplifier so hard that it distorts, so you end up with worse sound than before.)

In addition to replacing the front speakers, adding separate high-mounted tweeters is a good approach. High frequencies are very directional, so the tweeters improve the imaging. Putting these tweeters on the 'sail' panels (the triangular panels inboard of the mirrors) works well, and the crossover (often just a simple capacitor) can be hidden in the door.

In terms of wiring, replacing speakers is easy. You can also normally find speakers that will slot straight in without physical changes to metalwork being required.

Original head unit, new four-channel amplifier, new speakers, optional subwoofer

By adding a four-channel amplifier but retaining the original head unit, you can use new aftermarket speakers that are less sensitive, and then drive them harder. However, unless the original head unit had line-level outputs, you will need to use two adaptors to turn the speaker level outputs into line level inputs for the amplifier. In addition, you will probably need to run new cables from the amplifier to the speakers, and a new power feed from the battery to the amplifier. Finally, a 'switch on' signal (typically, the ignition-switched power feed to the original head unit) will need to be connected to the amplifier.

If you're happy with the existing head unit, or want to keep the appearance of the car original (just hide away the amplifier), this approach works well.

You can also add a subwoofer to the system. If you select the amplifier carefully, you can bridge two of the four channels to give you a subwoofer drive. But doesn't that leave you with only a three-channel amplifier – and five speakers? One approach is to use the head unit to drive the front two speakers – these typically need less power. Drive the two rear speakers with two channels of the amplifier, and the subwoofer with the third. (Of course, if you want to, you can instead add an extra single channel amplifier specifically for the subwoofer.)

Full new system

Or you can replace everything – using a new head unit, new speakers, new amplifier(s) and new subwoofer. This is certainly the route to go if you decide to upgrade to a more modern head unit (eg one with navigation, etc). Never run a new head unit with the original speakers – too often they weren't very good in the first place or the cone suspension has perished.

If you're unfamiliar with sound system upgrades, you

7. MODIFYING SOUND SYSTEMS

might think that cost is the big inhibiting factor – and indeed the parts can cost a lot. However, don't underestimate the installation task. Sound system upgrading is one area of electronic car modification where, inevitably, it all takes a lot longer than you expect.

I'd like now to cover some of the sound system upgrade approaches I've taken. I typically spend my money carefully, upgrading only the parts of the system I think need it, and using secondhand equipment on occasion.

A BUDGET SOUND UPGRADE USING SECONDHAND COMPONENTS

If you're after a budget sound upgrade, one good approach is to buy a secondhand original equipment head unit and speakers from a more expensive (or more recent) car. Taking this approach can be very cheap for the quality of outcome that you gain. But what do you buy – and how do you make sense of all those wires? In the following section I install a new (secondhand) system in a Nissan Maxima.

The head unit

A single CD AM/FM head unit was bought. Manufactured by Eurovox, it's installed in lower line Mercedes-Benz vehicles sold locally, which explains the 'Mercedes-Benz' and star on the faceplate. I liked the relatively plain appearance, and the quality long-distance radio reception (in my experience, a characteristic of Mercedes radios). Not so good was a lack of line-level outputs (only direct speaker feeds were provided), and a similar lack of explanation of what the 13 wires connected to the provided loom actually did. At least the security code was available – an absolute necessity on any secondhand gear.

The first step was to get it home and do a web search. 'Eurovox 4880MB' found only one entry on Google – and that was no help in sorting out the wiring. Dropping the

No information was available on the function of each wire, but some detective work sorted that out.

'MB' (for Mercedes-Benz, I assume) also revealed nothing, and a search of Eurovox sites found a complete dearth of technical support. But my lack of success shouldn't be off-putting: a web search is the first thing to undertake if buying a head unit of unknown specs.

The next step was to closely examine the provided wiring harness. The harness was a short one with two plugs. One end plugged into the head unit, and the other was obviously designed to connect to the Mercedes body wiring loom. That loom would have the connections for the four speakers (ie eight wires), constant power, ignition-switched power, earth, and probably an electric aerial connection.

But how to work out which was which? The eight speaker wires were likely to be grouped in pairs, and a quick examination indeed showed four pairs of colour: two yellow, two green, two orange and two white. One of each pair was marked with a trace, showing the positive. Hmmm, looking good – that's the speaker wires out of the way.

The power and ground wires are critical to get right – you don't want to get them reversed! The ground wire is most often black or green, and – yes – here was a thick wire that was black and hadn't been assigned a job. Paired with the black wire was a thick pink, and also in there was a red. Given that the pink was paired with the black it seemed most likely that these were the constant 12V and ground wires, while the red was most likely the ignition-switched 12V. (On some units there are fuses in the 12V and ignition-switched 12V leads – a dead giveaway.) That left another yellow (separate from the 13-pin plug), and, given that it had a bullet connector on the end, that one was most likely for the electric aerial. Hmmm, and the remaining grey and orange wires? – no idea! (Reading the handbook later indicated that one of these was probably

This ex-Mercedes head unit was bought secondhand. It came with its security code – a must-have. If you want a more sober-looking unit to match an older car, secondhand original equipment is a good way to go.

an input for a telephone mute control – ie the radio mutes when a call is received. And the other is probably an instrument panel lighting input, allowing the display, to auto-dim.)

I gingerly applied power to the black and yellow wires, and to my relief watched the red 'security' LED come up flashing. (If you have a variable voltage power supply, limit the current and start the voltage low and bring it up slowly.) Then, when I connected the red wire (ignition-switched 12V), the security code request came up. I plugged that in and the radio appeared to work. Of course, there was no sound at this stage – no speakers were connected.

Connecting a single speaker to each pair of 'like' colour speaker wires and then twiddling the fader and balance controls soon allowed me to sort out which speaker wire pair was which. Make sure that you label the wires as soon as you know their function!

So my new unit was working. Now to get it into the car.

The old one out

The first step was to remove the (previous) standard system and do a reverse sorting-out of which wires were which. The sound system in the Maxima was the original system, comprising an AM/FM radio and attached equaliser/amplifier. Removing the standard radio required, firstly, undoing screws holding an escutcheon in place, and then undoing more recessed screws that held the radio brackets in place. After that the whole unit could be pulled out forwards.

Before the plugs are disconnected, it pays to take a long, hard look at them. The plugs included connections for:
- separate up/down electric aerial pushbuttons (to be retained with the new head unit)
- ground
- ignition-switched power
- the four speakers
- and, in this case, the cable connections between the two units

Using the same rules of thumb that were used to work out the power and ground supplies on the new head unit, I guessed which plug was the one that had the power feeds. Then, using a multimeter (set to volts) and probing the plug terminals, I found the constant 12V, ignition-switched 12V and ground connections.

Sorting out the speaker wires is always fairly easy. First, you'll need just a normal 1.5V battery with a couple of wires connected to it. What you do is apply the voltage from this battery across the terminals that you suspect of being connected to a speaker (having measured them first and found no 12V feeds on them!). When you have found a pair of speaker leads, that speaker will make a scratchy pop as you connect and then disconnect the battery. Furthermore, when the positive lead of the battery is connected to the positive lead of the speaker, the speaker cone will move forward. (If you can't see the speaker, a sheet of paper over the grille will usually indicate the direction of cone movement.)

With all the connections sorted, it's then just a case of soldering all the right wires together. Note that if the power and ground supplies for the radio are small in diameter, there may be an unacceptable voltage drop when you put in a higher-powered head unit. The way around it is to run heavier gauge wires, eg straight to the battery.

The new one in

How you mechanically fix the new head unit in place will vary with each installation. In this case, I used a mixture of glue (yes, glue) and two aluminium brackets that I made to suit. The head unit came supplied with a slide-in cradle that allows the unit to be removed by the insertion of two long-pronged tools. But I don't like the cradle system and so decided to install the unit permanently in place.

The head unit sat up against one of the original

The old unit being removed. It was so old, it used cassettes!

The new head unit was held in place with a mix of glue and two new brackets made from scrap aluminium sheet.

7. MODIFYING SOUND SYSTEMS

This is what the job looks like finished. As can be seen I chose to use a single DIN unit so that I could add a gauge panel in the centre console. At this stage I have placed only a digital inlet air temperature gauge in the new panel (made from grained ABS sheet) – other controls and indicators will follow. The new head unit is very rigidly held in place, and the simple controls allow easy operation.

brackets. With a large area touching, I used contact adhesive to bond the two surfaces. This was never going to be enough to completely hold the radio, but it prevents lateral and twisting movement. The two aluminium brackets were also used. One locates on the rubber-encased bolt protruding from the back of the unit, and the other supports the weight of the head unit, pushing the two top glued surfaces together. The brackets were folded-up from scrap sheet aluminium, using a bench vice.

And how is the sound? Well, those four old speakers needed an upgrade, too – and here's how I did that.

THE SPEAKERS

The first step was to listen to the speakers that I already had. In the case of the Maxima, use of the fader and balance controls soon demonstrated that the front right door speaker had a terrible 'pop' in it, the front left speaker was very faint, while the two rear speakers sounded okay – but there was a real lack of treble all-round. A visual inspection (off with the door trims) showed the reason for the popping – the rubber roll suspension of the speaker was completely missing. And the other (faint) front speaker? Most of the cone was gone!

At the back, the reason for the lack of treble could also be quickly seen – new rear seatbelts had been fitted. They went where originally the tweeters sat in the two-way rear speakers, so out had come the tweeters, with the rear grilles chopped in half with a saw. Hmmmm.

Obviously, the front speakers would need to be completely replaced, and the rear speakers had to be neatened in appearance in addition to being improved in sound.

Buying

As with the head-unit, there are some real advantages in going secondhand original equipment. The first is that the speakers are cheap – really cheap. At a car dismantler, very little money can get you a pair of 150mm (6in) speakers from a late model car, while if you're prepared to go smaller in size, you'll probably be able to pick up two pairs for the same amount. The second major advantage is that, as mentioned, standard speakers tend to be very efficient. That is, they are relatively loud with the input of little power. Because amplifier power costs a lot, the car manufacturer saves by specifying efficient speakers and keeping the amplifier power down. For the same reason, these speakers also suit our needs! The downside is that you're not going to find any cheap, high-power speakers of huge aftermarket-style performance.

Measuring one of the Maxima's original front door speakers showed that a 150mm (6in) speaker just wouldn't fit. That's a pity, because many commonly available speakers are of this size. What was needed in this particular application was a 125mm (5in) speaker, no more than 50mm (2in) deep. (Be careful about measuring the available depth: door speakers normally have a very small clearance to the window and its winding mechanism.) For the Maxima, the new rear speakers could be either quite large – or alternatively, no more than 100-125mm (4-5in) in diameter. The difference depended on whether I wanted to go to the trouble of cutting rear deck metalwork and making a new shelf. In this budget, quick and simple upgrade, I was leaning towards the smaller speakers – but it depended on what I could find.

Traditionally, when picking speakers that you aren't going to be able to listen to before buying, you select the heaviest speakers – those with the biggest magnets, widest voice coil diameter, longest-travel suspension and highest power handling. And these rules still apply, but not quite to the same extent as when buying new. In this application, better to:
1. Sort the selection of available speakers down to the sizes that suit your application.
2. Move the cones up and down manually (spread your fingers and apply pressure across the face of the cone) to make sure that there's no binding.
3. Inspect for water damage, perished rubber suspensions, etc.
4. See if they're branded.
5. Then pick on magnet weight, power rating, etc.

Don't get too fixated on the marked wattages. Unlike really cheap aftermarket speakers, original equipment speakers tend to have at least semi-realistic power ratings. A rating of 10W, for example, might sound really low, but for normal in-car listening (as opposed to blasting), 10W of power handling *by an efficient speaker* is quite loud.

Some speakers will have a roll surround (front suspension) that's reversed. Rather than the curve of the suspension projecting outwards, it curves inwards. This allows the speaker to be mounted either with the face of the flange against the panel (or grille) or the back of the flange against the panel. Curiously, these speakers look less attractive, and so may be cheaper! Dual cones (where there is a much smaller secondary cone nestling around the voice coil dust cap) stretch the high frequency (treble) response, and so if given the option, select dual cone rather than single cone designs. Also keep in mind whether you need grilles and any other mounting hardware in your particular application. Unless you are using an amplifier that specifically specifies otherwise, always buy 4Ω speakers.

I bought two pairs of speakers. The first pair had a diameter of 125mm (5in). They were mounted in a weird plastic arrangement, but easily came out of these mouldings when rotated. These speakers used the reverse suspension mentioned above, were dual cone and had quite a small magnet. The other pair looked much higher in quality, although at only 90mm (3.5in) in diameter, they were probably from the dashboard. They had a built-in dust-protecting grille, and were also dual cone.

Testing

The last thing that you want to do is spend time doing a custom installation – only to find that the speakers are worse than the ones you are replacing! The first step should therefore be to bench test the drivers. You can use the car sound head unit, but I chose to use my domestic amplifier (which has a power meter). At low power levels there is no problem with using 4Ω car sound speakers with a domestic amplifier.

Tested bare on the bench, the speakers will sound terrible, but there shouldn't be any pops or buzzes. Furthermore, you should be able to get a feel for the power handling. I compared the cheap and nasty looking 125mm (5in) speakers with the better looking (but smaller) 90mm (3.5in) speakers – and found that the bigger speakers were better! Then it occurred to me that I should test the speakers from the Maxima in the same comparative way. The front door speakers were no good, but what about the rears? Removal of a speaker showed it to be a 100mm (4in) single cone design (it was originally matched with a tweeter, remember). And on the bench, I thought it sounded about midway between the two speakers I had bought.

Decisions, decisions …

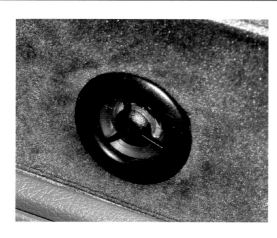

The new tweeters flush-mounted in the door panels.

Not mentioned so far is that I'd also bought two brand new dome tweeters. These were also ex-original equipment, being a special purchase made by a wholesaler. They came complete with their crossover capacitors. The tweeters were going to be mounted on the front doors, which meant that the main mid-range speaker didn't need to be a dual-cone type. Hmmmm. So what I decided to do was to use the old rear speakers in the front doors, together with the new tweeters. The back deck could then get the 150mm (6in) speakers. And the surplus 90mm (3.5in) speakers? Put into storage for now.

Front speaker installation

The front speakers in the Maxima are located behind grilles in the door trim. Unlike some more recent cars, the grille is just a grille – rather than being acoustically coupled to the speaker. That meant that nothing fancy was needed in the speaker mounting, other than to make sure that the speaker didn't project too far outwards from its mounting surface.

The new speakers (the ones previously used in the rear deck with the missing tweeters) were smaller in diameter than the originals. However, an adaptor ring was easily cut from some thin surplus aluminium sheet with an electric jigsaw. (Don't use particle board or similar to make adaptor rings for doors. There's plenty of water inside a door and the particle board will, over time, disintegrate.)

That was easy but a bit harder was the tweeter installation. Firstly, the crossover capacitor needed to be unsoldered from the tweeter (allowing it to be remotely mounted and providing more clearance), and then the sticky gasket on the front surface had to be peeled off. The tweeter could then be mounted in the door trim. However, to provide rear clearance, a little panel work was first undertaken with a hammer and dolly.

When mounting the tweeter, the following steps were undertaken:
- The centre of its location was marked on the outside of the velour trim.

7. MODIFYING SOUND SYSTEMS

- A small diameter hole was drilled through at the mark.
- A holesaw was then used from the rear, with the cut going only through the backing hardboard, and not through the velour.
- A knife was then used to cut through the velour, forming four flaps that could be then folded back through the hole.
- The tweeter was pushed into place from the front.

Taking this approach means that no bare backing can peer through the hole – the edge of the trim remains neat and the result looks good.

The tweeter was wired to the main door speaker, with the crossover (preventing bass frequencies getting to the tweeter) provided by the capacitor that came with the tweeter. To make sure that the aluminium adaptor plate sealed neatly against the metalwork (and to prevent any rattles) some silicone sealant was applied to the back of the plate before it was screwed into place.

Rear speaker installation

The rear speakers were even easier. A rectangular adaptor was cut from 10mm (0.4in) particle board (it's okay to use particle board here because it's in the boot – trunk – not the door) and painted black with a spray can before the speaker was attached with nuts and bolts. Black grille cloth was then stretched over the front of the assembly, being tacked into place on the back. The use of a baffle larger than the speaker allowed the easy covering-up of the holes created by the previous speakers, providing a visual lift as well as preventing air from the back of the cone cancelling-out sound waves from the front. (Note that any cloth that is see-through when held up to the light can be used as grille cloth.) The baffles were secured in place with self-tapping MDF screws inserted from below. If the shelf is uneven, a foam rubber gasket may be needed to stop air leakage around the edges of the baffle.

And the result? By using original equipment (and so sensitive) speakers, added high-mount tweeters and a good quality head unit, the sound quality was highly competent. Without an added amplifier and subwoofer, it was never going to shake the car to pieces, but, by the same token, the sound was clear and easy to listen to. For so little cost and relatively little work (apart from the tweeters, no re-wiring, remember), the upgrade was superb.

The new rear speakers in place. The use of a large particle board baffle wrapped in black grille cloth made for an easier installation.

The view from underneath. The size of the particle board baffle can be seen better here.

TECHNICAL TIPS FOR IN-CAR SOUND

When replacing the standard head unit with an aftermarket one, consider wiring-in the new unit so that the original plug remains intact. This can be easily done if, instead of cutting off the old plug, you make connections to the wires upstream of the original plug. The advantage? When you sell your car, you can easily remove the new head unit and then simply plug in the old one – returning to standard is then just a 5-minute job. Remember, you seldom get extra for your car, even when it's fitted with a better than standard sound system. (Of course, to be able to do this, you'd better keep the old head unit!)

However, there's one area where instead of making the connections to the original head unit's wiring, it's best to put in some new wiring. And that area is the power supply. Most standard head units have very small gauge power supply and ground wiring, and if you're not to suffer a voltage drop when using the internal amp of a high-powered head unit, you'd better run new wires to supply it with the needed juice.

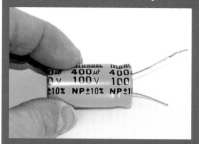

A non-polarised capacitor like this can be used to roll-off the bass being fed to smaller speakers.

If you are using an added amplifier for the rear speakers and/or the subwoofer, and driving the front speakers from the head unit, here's a trick – make the front speakers operate just as mid/treble units. The easiest way to do this is to install crossover capacitors to stop the deepest bass reaching the small speakers. In a 4-ohm system, a non-polarised 400uF capacitor wired in series with the speaker feed will roll-off frequencies below 100Hz. You'll need two – one for each front channel – and you can buy them very cheaply online.

The GPS antenna for an in-built nav system must be mounted so that it has as wide a viewing angle of the sky as possible. In a modern sedan with a very shallow angled rear window, placing the antenna in the middle of the rear deck works well. But what about in a hatchback? If the parcel tray is never to be removed, again putting it directly under the middle of the rear glass is fine. But in some cases, you'll need to look at sticking it to the underside of rear glass or the windscreen, or mounting it in the middle of the dash. Get the location wrong and the number of satellites accessed will plummet, affecting accuracy.

Getting the correct speaker polarity (as described earlier) is important. You won't blow up a speaker if it is connected backwards – and in fact, if you have all the speakers connected backwards, there will be no problem at all. But if you have some connected one way, and others connected the other way, there will be a lack of bass and the imaging will be odd. The lack of bass is because as one speaker cone is pushing forwards, the other will be pulling backwards. In other words, one speaker will cancel the other.

An easy way to quickly see if this is the case is to move the balance/fader control to all the extremes. If the system has more bass when only one speaker (or one pair of speakers) is selected, something is wrong in the

When building a subwoofer, it can be easier to start with a pre-built, carpeted enclosure. However, you don't have to stick with the design – here I am making a hole for a port to convert this normally sealed enclosure into a ported enclosure.

connections of the positives and negatives (normally called phasing). Reverse the connections, *one speaker at a time*, until it all sounds right.

UPGRADING AMPLIFIER COOLING

Amplifiers live and die on the basis of heat. If they get too hot, the output power drops and the likelihood of distortion rises. Get an amplifier really, really hot and it will stop working forever. That's especially the case with subwoofer amplifiers that work longer and harder than any other amplifiers in the car. It's also harder to hear subwoofer distortion, so it's easy to have the subwoofer turned up a bit too high for long-term amplifier health. However, improving amplifier cooling is easy – and you don't need to be an electronics expert. The cost can also be near zero – all you'll need is a small fan, some surplus heatsinks and a few hours of time. And, as a bonus, you can also improve the amplifier's appearance.

7. MODIFYING SOUND SYSTEMS

The amplifier that I modified was a cheap secondhand unit that, while nothing fantastic, when run in bridged mode produced adequate power for a high-efficiency subwoofer.

Turning it over reveals a pressed sheet steel cover …

… that is easily undone by removing four screws. Nearly all amplifiers use similar construction.

This is what the amplifier looks like inside. As you can see, the printed circuit board normally hangs upside down when the amplifier is mounted horizontally. A thick aluminium extrusion forms the main body of the amplifier and the electronic devices that need heatsinking are clamped against its inner surface.

Here are the power transistors clamped against the inside of the extrusion – the whole housing then acts as the heatsink.

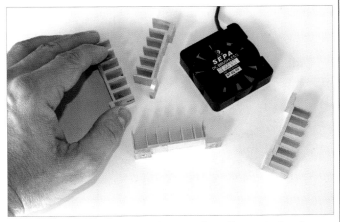

Improving the heat transfer away from these electronic components required some heatsinks and a fan. These heatsinks were salvaged from other equipment, as was the fan. You'll find heatsinks in nearly any piece of electronic gear that's thrown away. For their new role, the heatsinks were shortened with a hacksaw and then smoothed with a file.

169

WORKSHOP PRO — MODIFYING THE ELECTRONICS OF MODERN CLASSIC CARS

Heatsink compound (available online) should be applied to the top of the components before attaching the heatsinks. The compound thermally connects the heatsink and the component, providing much better heat transfer.

Both ends of the amplifier used heatsinked components, so new heatsinks were added at the other end as well.

The heatsink compound was applied to each component and then ...

Here are the new heatsinks in place at one end ...

... the new heatsinks were screwed into place. While these heatsinks look puny compared with the heatsinking abilities of the whole case, the fact that these will have airflow passing over their fins makes them much more effective than their size would suggest.

... and at the other.

7. MODIFYING SOUND SYSTEMS

I decided to replace the original bottom cover with a new one on which the fan could be located. To improve appearance, the new cover was made from 5mm (³⁄₁₆in) thick clear acrylic. Here, the acrylic is shown with its protective paper in place – leave it there until the job is finished.

The new panel in place.

The hole for the 57mm (2¼in) fan was cut with a holesaw. In this case the fan could be located inside the amplifier – there was just enough clearance. If there isn't room, it can be mounted on the outside. The fan draws air from inside the amplifier. So that this air flows over the new heatsinks, intake air holes (shown by the green arrows) were positioned directly over the new heatsinks.

The wiring for the fan was brought out through a new hole drilled in one of the end plates. It was then connected across the remote and earth inputs, so that the fan operates only when the amplifier is turned on.

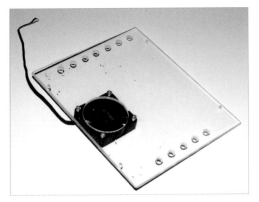

The new bottom panel, yet to be cleaned but with all the holes in it. The upper section of each hole was chamfered slightly with a pointed grinding stone rotated in an electric drill. The panel was attached to the amplifier using the original screw holes.

The fan used is a 24V design. But it works fine on 12V, is compact, and was available for nothing (like the heatsinks, it had been salvaged from other equipment, and, in fact, came from a discarded photocopier).

WORKSHOP PRO: MODIFYING THE ELECTRONICS OF MODERN CLASSIC CARS

The fan was a tight fit inside the amp – the leads of this capacitor needed to be bent slightly to create room.

A Velometer Jnr instrument was used to measure the airflow through just one of the intake holes. As the instrument shows, there is more than 200 feet per minute airflow through each hole.

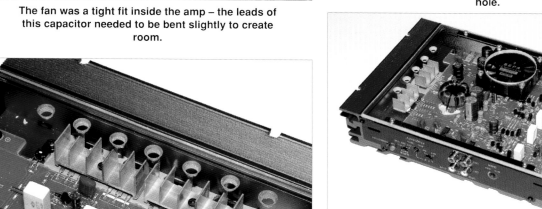

Here are the chamfered intake holes positioned directly above the heatsinks at one end of the amplifier.

No before/after testing was done of the temperature build-up in the amplifier but you can be certain that with the added heatsinks and the fan-forced air movement, cooling has been massively improved. The amplifier can now be either mounted 'bottom-up' to show off this view, or alternatively, mounted conventionally but spaced a little above the mounting surface to allow for air movement.

INSTALLING COMPONENT SPEAKERS IN THE FRONT DOORS

Upgrading the speakers fitted in car doors is one of the most common ways of improving car sound. However, there are plenty of traps for the unwary.

Selecting drivers

First, is the door going to be acting as a sealed or ported enclosure? While it is technically possible to build a ported enclosure in a door (and many premium factory systems do just that), in the real world of retrofitting, it's hard to do this. As I cover later, doors in wet weather have a constant stream of water running through them, so traditional speaker enclosure approaches (like building a ported enclosure from MDF) are doomed to long-term failure – the wet MDF turns to mush. That leaves the use of plastic (including fibreglass) to build the enclosure – but by the time you make an enclosure that will clear the moving window, fit into the available space, and be waterproof, it's all getting pretty tough.

The vast majority of upgraded speakers in doors therefore use the sealed enclosure approach: the volume of the door cavity becomes the volume of the speaker enclosure. The use of a sealed enclosure of 40+ litres/1.4+ cubic feet (even small doors have quite a lot of internal volume) has major implications for driver selection. This is the case because normally with speaker designing, the enclosure is sized to suit the speaker. But in this case, *the speaker needs to be specified to suit the enclosure*. The easiest way to do that is to pick an aftermarket speaker

7. MODIFYING SOUND SYSTEMS

or original equipment speaker *specifically designed for in-door use*. In the example shown here, I chose to use the high-quality front door speakers from a Lexus IS300 – and I was extremely happy with the results.

HEIGHT?
Most speakers in car doors are mounted low down in the doors – more at ankle level than ear level. Therefore, as is done with many current original equipment systems, it makes sense to use 'splits' in car doors – the mid-bass driver down low and a separate tweeter mounted up high. Using co-axial two-way drivers in the doors is therefore *not* the best approach.

Mounting door speakers
As mentioned earlier, doors have a lot of water within them. The rubber window seal, especially when the window is moved up and down, doesn't stop water entry into the door. This isn't such an issue from a panel corrosion point of view, as drain holes in the base of the door allow the water to escape and air to circulate. (Although you certainly don't want those drain holes blocked!) But from a speaker installation perspective, that water can damage cones, rust steel speaker frames, and cause composite boards (like MDF) to very quickly fall apart.

To avoid water damage, take note of the following:
- When installing speaker adaptors and mounts, use only materials that can tolerate being wet. Plastic – for example, HDPE (high density polyethylene) – is a good material to use in these applications. (HDPE is also used to make kitchen chopping boards.)
- Standard speaker water shields should be retained, or equivalent shields made from new material.
- Plastic waterproofing membranes (and similar) used inside door linings should be retained, and any holes in them resealed.
- It is preferable not to use foam seals on the 'wet' side of the speaker – they will retain moisture.

After you've ensured that water won't degrade your speakers, the next point is to develop solid mounts. As standard, door speakers are variably mounted on either the inner metalwork of the door, or on the door lining. Some speakers use through-bolts that connect to both.

When mounting new speakers, aim to create the most solid mount possible. This is quite difficult – door linings are generally flimsy, and the inner door metalwork not much better. You may choose to add stiffening plates (eg aluminium sheet, polycarbonate sheet or HDPE sheet) to the inner door metalwork, better connect the door elements (eg bolting the door lining to the metalwork), or add a lot of damping/soundproofing material to the door and/or the lining. (Or, of course, do all of these!)

Door speaker mounting ring turned from HDPE.

Speaker mounted in the ring. (Both photos courtesy Adam Trites)

By far the best (and most unforgiving) way to see how your door installation is going is to connect the speaker to an amplifier and frequency generator (suitable apps are available for smartphones) and drive the speaker over the range from about 30-150Hz. Doing this will allow you to find buzzes, air leaks and whistles – almost certainly, there will be plenty! Your task then is to track them down, one at a time, and eliminate them.

INSTALLING NEW DOOR SPEAKERS

These speakers, originally used in the doors of the Lexus IS200 and IS300 models, were to be installed in a Honda's doors. Lexus (and other premium car makers) often have quite good sound systems, so if these speakers can be sourced cheaply, they're worth using. Note the generous water shield positioned at the top of the driver.

WORKSHOP PRO: MODIFYING THE ELECTRONICS OF MODERN CLASSIC CARS

The Honda uses large door grilles behind which hide pedestrian speakers.

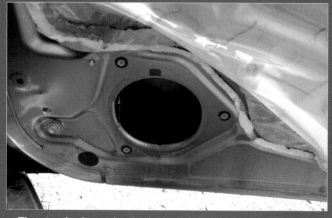

The standard opening in the metalwork is too small for the Lexus speakers.

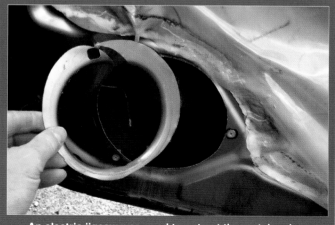

An electric jigsaw was used to cut out the metalwork …

… and then a large diameter grinding stone in an electric drill was used to bring the hole precisely up to size. In the case of the Honda Insight, this material is made from aluminium and so the metalworking was easy. If working with steel, ensure all the filings are gone before the speaker arrives!

Rivnuts were inserted, with washers used either side to prevent them pulling through the thin metal.

The speakers could then be screwed into place. A spacer was needed on one screw, and a foam strip was placed around the rear of the plastic speaker housing before it was attached. This foam is in contact only with the plastic surround, not the metal of the speaker frame.

7. MODIFYING SOUND SYSTEMS

The new driver in place – together with a crossover that will drive a high-mounted dome tweeter. The next step was to rework the door lining so it would again fit into place.

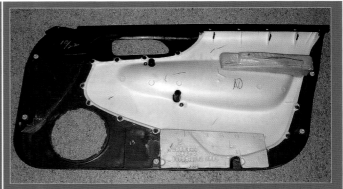

The next step is to take the standard door lining …

The rear of the standard door trim. There was not enough clearance for the new speaker so …

… the plastic was cut away and then smoothed.

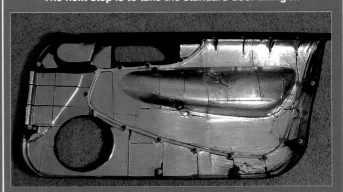

… and add extensive sound deadening material. The material better seals the enclosure and prevents the trim from vibrating. (Last two pics courtesy Adam Trites)

The door trim now fits over the new speaker perfectly – and the grille clips into place as it did with the standard driver. Note how the plastic housing of the Lexus speaker has the correct forward extension to match the revised door lining. Taking this approach avoids the need to make a spacer ring.

The installation prior to the door trim going back on. A Clarion dome tweeter (arrowed) has been mounted in the 'sail' area (the triangular panel at the top inner corner of the door) and an Alpine crossover next to the lower driver. The crossover has multiple link positions to allow the setting of the tweeter level. The final sound is very good – even outstanding, considering the modest cost of the components.

WORKSHOP PRO: MODIFYING THE ELECTRONICS OF MODERN CLASSIC CARS

SUBWOOFER DESIGN

If there's one thing that everyone wants in a car sound system, it's deep bass. The amount of bass is the major criterion which many people use to judge the quality of sound – can it really shake you, or do you just hear the notes? Building and installing subwoofers is also one area where the careful modifier can make a huge difference to the quality of the in-car system.

How speakers work

Before we get into subwoofer enclosure design, you need to know something more about how a speaker works.

A speaker operates by moving its cone back and forth. When the cone pushes forward, it creates a high-pressure wave which travels through the air, and when it moves backwards, it causes a low-pressure wave – sometimes called a 'rarefaction.' The ear picks up these air pressure waves and turns them into sounds.

A woofer will sound poor if it is not mounted correctly. Foremost is the need to separate the front and rear of the speaker. To see why this is needed, imagine a bare woofer sitting on a bench. The speaker is being driven by a powerful amplifier, and is running music with a lot of bass. The cone will visibly move forward and backwards with the music, but the bass will be poor. The reason for this is that when the cone moves forward, a 'proper' pressure wave in the air isn't created. Instead, the air simply moves around the edges of the speaker frame, filling up the low-pressure area created behind the cone. When the cone moves back again, the air flows the other way. Instead of pumping bass into the cabin, all the cone is doing is pumping air around its edges.

Separating the front and rear of the speaker doesn't mean that the driver must be in a sealed box – although that is one approach that works well. The separation of the pressure waves needs to occur *acoustically*, and in fact the enclosure design may well allow air to flow from the back of the speaker out the front through a port. But – and this is a key point – the enclosure needs to be designed so that the rear pressure waves *add* to the sound, not *cancel* it out.

Different box designs

Four major types of speaker boxes are used. (Note that in each case it doesn't matter how many drivers are put into the box, whether they're facing in or out, etc.).

The designs are:
- sealed
- ported
- passive radiator
- bandpass

Sealed

In this type of box, the sound waves coming from the back of the cone are effectively wasted. Instead of contributing

A sealed enclosure is a simple box with the driver mounted in it. The design is more forgiving of sizing errors, but typically does not go as deep (or is as efficient) as ported or passive radiator designs. (Courtesy Crutchfield)

to the sound that moves you, they're dissipating in the acrylic fluffy lining inside the sealed box. However, while this type of box produces less SPL (ie they are less efficient on a watts per dB basis) they are easy to make, have only a gradual bass drop-off as the notes get lower (and that drop-off can be counter-balanced by the rise in bass response that occurs within the closed confines of a car), and can potentially handle more power than a ported enclosure.

The effects of mismatching the speaker and its enclosure is also less severe when a sealed design is used, so if you have a driver of unknown specs, a sealed box is generally the only way to go. You can recognise a sealed box design easily – there are no openings in the box except for the driver(s).

One form of sealed enclosure is free-air subwoofer installation, typically where the driver makes use of the whole volume of the boot. A free-air installation is covered later in this chapter.

Ported

As briefly mentioned above, a ported enclosure additionally makes use of the energy coming from the back of the cone. It does this by using a connecting port (or vent) that joins the inside of the box with the outside. The port diameter and length are carefully sized so that the plug

A ported enclosure. In this case, the port comprises a slot at the bottom of the front face. Ported enclosures are more efficient than sealed enclosures and typically go deeper in bass. However, they require careful design. (Courtesy Crutchfield)

7. MODIFYING SOUND SYSTEMS

of air contained within the vent is excited into back and forth motion, but its movements are delayed just enough that when the cone of the driver is moving forward, so is the plug of air inside the port. In this way, the two air movements complement each other.

The advantages are twofold: firstly, the efficiency of the system is greater (ie more SPL per watt of amplifier power), and secondly the bottom-end bass response of the system can be improved over the use of a sealed box.

The downsides are that if the port isn't just right for the driver and box, boomy one-note bass can be the result, and even with well-designed ported boxes, at ultra-low frequencies the cone of the woofer becomes unloaded – which can cause it to be destroyed if you're not careful in how you set up the system. And, while the bass response holds on to lower notes, once it does start to fall away, it does so more quickly than with a sealed enclosure.

Ported designs can be easily recognised by the presence of one or more openings that connect the rear side of the driver to the atmosphere.

Passive radiator

A passive radiator design is relatively rare. This type of design uses a passive radiator (a driver without a magnet) to act as a port – the cone of the passive radiator moves back and forth like the plug of air within a port, but at no times can the main driver become completely unloaded as is the case in a ported enclosure. The disadvantage of making your own is that passive radiators are not as easily bought as conventional drivers – and finding the detailed specs on the passive radiator (needed to do good designs) even rarer!

A passive radiator design looks like it has two drivers in a sealed box (sometimes by just examining the system you won't be able to tell that one of the 'drivers' is in fact a passive radiator) although the passive radiator is usually bigger than the woofer.

Bandpass

A bandpass design is a very tricky thing. Rather than producing frequencies from as high as the woofer can go – and then trailing them off at the other end as is determined by the enclosure design – a bandpass lets just a narrow spread of frequencies be emitted from the box. Because it is producing only frequencies from (say) 30-90Hz, it can be more efficient than the other box designs – it's concentrating all of its energies just in this narrow field of frequencies. All the sound comes from the ports, with the driver itself buried inside.

There is a variety of bandpass designs – some mount the driver on the division between two boxes and vent just one volume, while others vent both boxes. Still others connect one box to the other by means of a vent and then have a further vent in the first box. Bandpass boxes need very tricky design indeed, and if you're not careful, either the frequency spread (the range of bass notes produced by the sub) or the efficiency (how loud it is for a given input power) can drop right down. In addition, the bandpass boxes are much more complex to make than other designs – there tends to be lots of pieces inside that need to be airtight, and fitting the ports inside the box can be difficult. However, get it right and you can have a loud and strong subwoofer which is small and effective.

In a bandpass design the driver won't be visible (unless there's a plastic window fitted) – the speaker's cone does not connect directly with the air, except through a port.

A bandpass enclosure. In this type of design, the drivers are coupled to the cabin only through the port or ports – here they're mounted behind clear plastic. A bandpass enclosure produces sound only over a narrow band of frequencies.
(Courtesy Crutchfield)

Different designs

You can buy many different pre-built subwoofers, or you can make your own. I've been building speakers for many years and find it to be a fascinating and rewarding pursuit. Designing and building your own subwoofer also

A passive radiator enclosure combines some aspects of sealed and ported enclosures. Technically, very good results are possible with these designs.
(Courtesy Crutchfield)

177

allows you to make it exactly as you want – in terms of size, cost and sound. For a subwoofer, key design criteria include:
- How loud do you want it?
- How much room do you have for it?
- What is the performance of the rest of the system? (Are there other drivers that will take the mid-bass load or does the sub have to do that as well?)
- Do you have a preferred driver? (For example, it's on special at a bargain price this week.)

The design of subwoofers has been revolutionised by design software. Using it can help shape answers to a lot of the above questions – it's easy enough to find out, for example, that your chosen driver will need an enclosure that is too big for the available space, or that if you use a bandpass design, you might need to beef-up the rest of the speakers to carry the response down to 100Hz.

Using subwoofer design packages

Above I covered the fundamentals of different sub box designs: sealed, ported, passive radiator and bandpass. Each has different characteristics in terms of the frequency response, power handling, sensitivity, and how large the enclosure needs to be. In fact, there are so many variables that have an impact on these outcomes that back when dinosaurs were strutting the earth, most box design consisted of informed trial and error. Like, what's this box design sound like? Hopeless: try again … Is that better? Er, well not much …

The breakthrough came when two Australians – A N Thiele and Richard H Small – realised that the behaviour of a speaker could be modelled in different enclosures if certain electrical and mechanical characteristics of the driver were known. These Thiele-Small (TS) characteristics, as they came to be known, are now available for all decent subwoofer drivers. Here are some of the parameters:

Resonant frequency (Fs)
In general terms, this is the lowest frequency at which the speaker's cone can effectively couple itself with the air – there's only so much that fancy box designs can do. Subwoofer drivers should have a resonant frequency of less than 50Hz.

Vas
The Vas figure is measured in either litres or cubic feet, and is an indirect measurement of how stiff the cone suspension is. The number actually indicates the volume of a closed box which would give the speaker the same stiffness as its suspension does. A low Vas number means that the cone suspension is fairly stiff, while a high number shows that the cone suspension is more floppy.

Q – Qms, Qes, Qts
In the same way that a car suspension needs to have its springs damped, so the moving mass of a speaker cone needs to be damped. 'Q' refers in fact to the *opposite* of damping – it's how much the resonance of the speaker is magnified. Qms refers to the control exerted by the speaker's mechanical suspension – the spider and roll surround. Qes refers to the control exerted by the driver's electrical system – the voice coil and magnet. Qts is the total Q and is derived from both the Qms and Qes.

Xmax
Xmax refers to the available travel of the cone. In other words, it's how far the cone can move before it hits the stops or the voice coil comes out of the magnet core. A high Xmax figure means that the same size driver can move more air. Xmax can be measured in inches, cm or mm.

Others
In addition to the above specs, you'll also find other variables like Re, Mms, BL, Sd and so on. Basically, the more data like this that is available from the manufacturer of the driver, the better. You don't have to have an in-depth knowledge of what each of these parameters means – but you do have to have enough of them to plug into the program.

Designing
So you've found a driver that has available lots of Thiele-Small parameters, you've got a good idea of the space available for the sub, and you want to start doing some virtual designs. Now what?

The next step is to find a subwoofer design software package. If you do a web search, you'll find a variety of free programs, or you can buy one. However, as you'd expect, the ones that you pay money for are a lot better than the freebies – better in terms of the data that can be generated, the different types of box designs that can be modelled, the 'help' support, and the accuracy of the results. I use BassBox, developed by US company Harris Technologies. In the 'Lite' version it's relatively cheap but can still do an excellent job.

For beginners, one of the real virtues of BassBox is that the provided help files represent a complete course in subwoofer design. The notes are to the point and very clear. In fact, this material alone is worth probably half of the cost of the software package.

By using this (or other) design packages, you can easily trial different enclosure designs and sizes before you start any woodworking. You can also model commercially-available, pre-built enclosures that use drivers for which specs are available. If you do this, you'll find many of them turn out to be appalling, typically giving boomy, one-note bass.

7. MODIFYING SOUND SYSTEMS

I've used BassBox Lite for years to design car subwoofers – it's a good package.

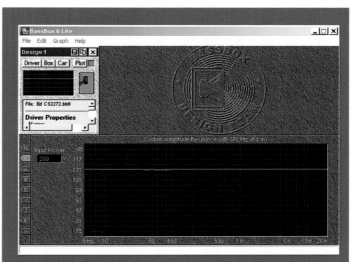

One button press later, the recommended sealed box volume is shown – just 31 litres when only small amounts of acrylic filling are used within the box. (And even smaller when the box is packed with acrylic filling, which has the effect of increasing the apparent box size.) Another button press, and the predicted in-car response is shown – it's ruler flat. This means that the predicted SPL stays the same at all frequencies – there's no major drop-off in bass.

AN EXAMPLE SUBWOOFER DESIGN

Let's do a subwoofer enclosure design with BassBox for a 12-inch driver that has these specs:
- Impedance: 4 ohms
- Power handling: 200 watts RMS
- Fs: 25.6Hz
- Qts: 0.409
- Vas: 93.4 litres
- Cone area: 490cm^2
- Xmax: 10.5mm

(Apologies if you prefer working in Imperial units – using conversions all through this section would make things overly complicated.)

With that data entered in the software, the first indication of the box type appears – a marker which is placed a little closer to 'closed box' than 'vented box' – indicating that this driver 'prefers' a sealed box design. However, this is not a hard and fast rule, and the fact that it's not right at one end of the continuum shows that there's some flexibility available.

Next, the program can be asked for its suggestions, under each box type. (But first the 'in-car' response needs to be turned on – this takes into account the dramatically increased bass that a sub can generate in a car compared with the open air.)

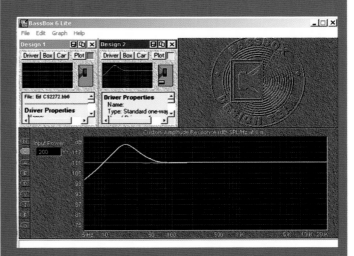

Next, what about a suggested ported box? This time the program suggests that for maximum bass, a massive box – 124 litres in size – should be used with a port 105mm in internal diameter and 370mm long. And the response? The yellow line shows that at just 20Hz there is a major increase in SPL – it's up to 122dB, compared with the 111dB of the sealed design (both with an input power of 200W). This is probably overkill, but at this stage we're just working with the program suggestions.

WORKSHOP PRO: MODIFYING THE ELECTRONICS OF MODERN CLASSIC CARS

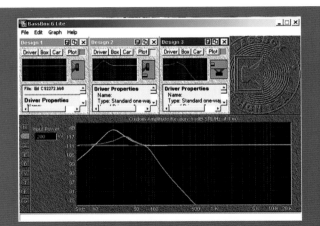

So what about a single-tuned bandpass design? Press the buttons and it suggests chamber volumes of 30.6 litres and 26.4 litres, but as can be seen by the green line, the result is not that great – there's a distinct peak at 31Hz. The suggested port size is also a whopper – 162mm diameter and 850mm long ...

There are other box designs that we could trial, but let's concentrate on these three. While it first it looks like the decision is already made – after all, the sealed box was both smallest and gave the flattest response – but there's something else to keep in mind: how efficient the different enclosures are. The frequency plot above has been made against SPL with an amplifier power of 200W – this gives the clearest picture of what is going on. An increase in loudness by 3dB means you can halve the amplifier power needed – so if we could get the response of the ported and/or bandpass enclosures smoother but still staying loud, we'd have a winner. If the box is compact as well, anyway ...

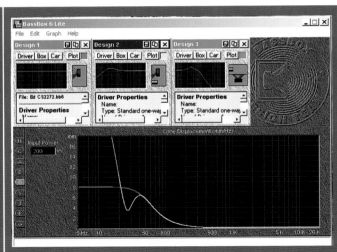

Another thing to check with this design is that the cone won't reach its Xmax – that it won't be bouncing off the bump stops. This graph shows both the sealed enclosure (red) and the revised ported enclosure (yellow). The change in density of the yellow line indicates that with an input power of 200W, the driver's Xmax will be reached at 20Hz, a very low frequency. That's fine – if the figure had been 35Hz, there would have been real problems!

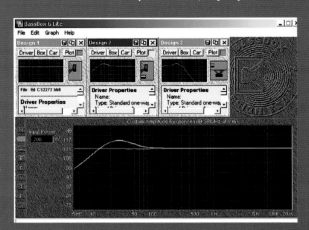

By reducing the ported box volume to 55 litres and using a port which is 75mm in diameter and 270mm long, we can get the response shown by the yellow line. The red line (the sealed 31 litre box) has been kept on-screen as a comparison. The bass response of this modified ported design starts to rise smoothly from 100Hz and then peaks at 28Hz before smoothly dropping away again. That will give smooth but strong bass.

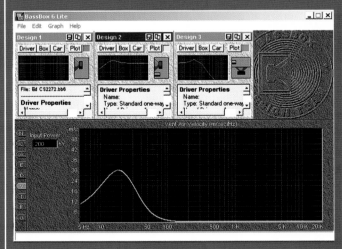

A final check is that the speed of the air movement inside the port won't be so high that whistles or buzzes are generated. This graph shows that the maximum port speed will be 30 metres/sec – which again is fine.

7. MODIFYING SOUND SYSTEMS

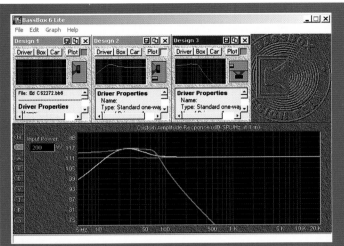

So that all looks great – but can we do better with a single tuned bandpass box? With one chamber at 50 litres and the other at 35 litres, and with the sound coming out through a single 110mm diameter, 171mm long port, the result is as shown here by the green line. As can be seen, there's effectively no bass above about 75Hz, but below that it's strong and loud. However, without other good quality speakers to carry the load down to 75Hz, there's going to be a mid-bass hole.

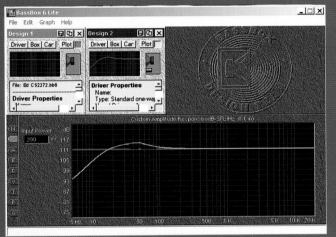

Okay, so the 'winner' is the ported enclosure – so let's see if we can now get it down in size. Pre-built boxes of 35 litres are available for 12-inch drivers – so is it possible to use one of these? Well, yes and no. You can see here (yellow line) that when the vented box is dropped down in size to 35 litres, the efficiency also goes down – it's now only a very small amount better than the sealed, 31-litre box (red line).

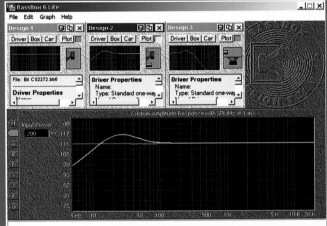

So, to summarise, there are two major options – a sealed 31-litre box (red response) or a ported 55 litre box with a 270mm long, 75mm vent (yellow response). Note that if you go for the ported box you could realistically get away with a lot less amplifier power – perhaps half as much. And of course, there are plenty of other options – probably an infinite number of combinations, in fact. However, this brief overview shows the basics of the design approach that can be followed. As you can see, the opportunities to tailor the result to your preferences, budget and build skills are almost limitless.

SELECTING AND INSTALLING A FREE-AIR SUBWOOFER

Nearly all manufacturers of good in-car sound systems use them – including Lexus, who, arguably, makes the best quality original equipment car systems of all. What am I talking about? Free-air subwoofers, where the enclosure comprises only the volume of the boot (trunk), rather than a separate tuned box. Installing a free-air sub is much easier than building and fitting in a boxed subwoofer – there's less woodworking (a *lot* less!) and the design process is simpler. The result also takes up much less boot space and is lighter.

And the downsides? You must select the correct driver for the application, the bass probably won't be as 'tight,' the approach best suits a three-box sedan, and of course the quality of bass will vary with how much you've got in the back. So let's put the approach into action.

WORKSHOP PRO: MODIFYING THE ELECTRONICS OF MODERN CLASSIC CARS

The free-air subwoofer from a Toyota Soarer. Impedance is 2Ω, so if using this driver, ensure the amplifier is happy driving such a low impedance.

Selecting the driver

As described earlier, a speaker designed for a ported enclosure has certain characteristics, and a speaker designed for a sealed enclosure has other, different characteristics. The same applies for a free-air application.

There are two ways of obtaining such a speaker. The first is to buy new – look for a manufacturer that lists the speaker as appropriate for a free-air or infinite baffle application. The other way is to buy secondhand. You can source a driver from a car that ran a standard free-air sub, including Lexus and Mercedes, among others. Be sure when selecting these speakers that you have an amplifier to suit – some of these speakers are only 2Ω and not all amplifiers will be happy running a speaker with such low impedance.

Finally, you can come from left field and use a speaker that was never designed for car use but which has the right mix of attributes. For example, home theatre 4 or 6Ω speakers designed to mount in the ceiling or walls are perfect for free-air car use. In fact, that's the approach taken here, where a fire-damaged Sonance three-way 8-inch speaker was purchased for near-nothing, and the high-performance woofer removed from the assembly for car use.

In all cases, the free-air speaker should *not* have a super-floppy suspension. In other words, there should be some stiffness when you carefully push the cone back and forth with your spread fingers. You still want a long travel but in a free-air application, a super floppy suspension will easily allow the cone to overshoot and bottom-out. (This type of speaker is best mounted in a relatively small sealed box, where the trapped air adds to the springiness of the cone suspension.)

INSTALLATION

As with a number of cars produced over the last 15 years, the guinea pig car already had a cut-out in the rear deck designed to take a subwoofer. In this case, the hole was designed for a 200mm (8in) driver. This means that installing the new sub was as easy as …

Removing the two-way rear speakers …

… and lots of trim pieces …

… and more trim pieces …

… before the rear deck cover could be pulled forward …

7. MODIFYING SOUND SYSTEMS

... to reveal the standard subwoofer speaker cut-out.

The new speaker just dropped straight into the hole, although the mounting holes in the plastic speaker frame needed to be elongated a little with a file to match the holes in the metalwork. There's something else important to note in this photo. Unlike cars of old, the centre seat position has a lap/sash seatbelt, with the inertia reel mounted on the rear deck. This means that the structural integrity of the rear deck must be maintained – that is, be very wary of cutting big holes in it to mount a speaker!

A foam rubber gasket was cut out and sandwiched between the driver and the metalwork. This prevents airflow occurring around the edge of the frame.

THE SUBWOOFER AMPLIFIER

The new subwoofer of course needed an amplifier, and I used the one we saw earlier in this chapter, when I gave it better cooling. When selecting a subwoofer amplifier looks for these characteristics:

- The amplifier should be able to be bridged.
- The amplifier should feature an in-built crossover. Preferably, the filter frequency should be able to be altered to tune the point at which the subwoofer operates.
- Variable gain is important, as it allows you to tune the loudness of the subwoofer within the system.
- If the amplifier has the ability to take both left and right signal inputs and then sum them (ie L+R), you'll be able to easily develop a dual-channel bass output. If the two channels of the amp can be bridged, this summing will occur automatically.
- Another control that's useful is an amplifier bass boost switch.

The amplifier was mounted on a square of black-painted particle board which was then bolted to the underside of the rear deck between the new subwoofer and one of the original rear speakers. Note that the amplifier was spaced away from the board so that the fan could still operate effectively.

Heavy power supply cabling was then run from the amplifier all the way forward to the battery in the engine bay. Most cars need to have a hole made in the firewall through which the power supply cable can pass – here it was drilled in an inset panel that also contains the steering column. Taking this approach avoids having to drill through a firewall that may be double-skinned.

WORKSHOP PRO: MODIFYING THE ELECTRONICS OF MODERN CLASSIC CARS

The cable must be protected where it runs through the firewall – either by a grommet (as shown here) or by having a short length of thick-walled hose slipped over it.

At the battery, the negative cable connected straight to the negative terminal. But the positive cable uses an interposed fuse – you must always install a fuse close to the battery when running power cabling to an amplifier.

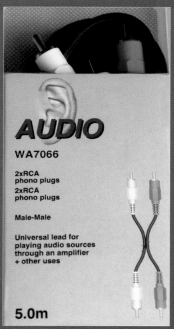

The audio feed for the sub amp was via twin RCA line level cables. These were run down the other side of the car to keep them away from the power feed (a possible source of electrical noise). A 5m (16ft) cable is the right length in most cars.

This head unit came with line-level outs. If there aren't any RCA sockets on the back, you'll need to use a speaker-to-line-level converter – you can pick up the audio signal from the rear speakers.

The rear deck cover was moulded in shape, so making a replacement wasn't an easy option. Instead, two pieces of foam rubber were placed either side of the subwoofer so that the cover would easily clear it. (Note also the circled foam strip placed under the brakelight to stop it rattling.)

Then the cover was re-installed with a new speaker grille placed over a hole cut in the cover. The cover has enough flexibility to conform to the new required shape and looks good.

And with everything tucked up out of the way, the loss of boot space is very small. One of the hardest aspects of this sort of installation is stopping bits of the car rattling when you crank up the level. As shown above, a foam strip was inserted under the brakelight, but even so, when the system is loud, the cover can rattle a little, as can the centre inertia reel seatbelt! However, with that qualification, the sound is fine and the bass improvement massive.

7. MODIFYING SOUND SYSTEMS

A SPHERICAL SUBWOOFER!

Would you like to have a compact, light and easily built subwoofer? How about this spherical subwoofer, then? But what on earth is a subwoofer doing looking like a sphere? And how is an enclosure of this shape easy to build?

This subwoofer enclosure design is easy to build because it's based on two commercially available bowls that normally have plants put in them – that's right: two flower pots! The pots are made from a new material – fibre-reinforced clay. This material is stiff, relatively light and acoustically dead. The pots are available from garden suppliers.

So how is the enclosure constructed? In short (more detail on the process in a moment) the flattened surfaces on which each bowl normally sits are ground smooth using an angle grinder. In one of these flats is cut the hole for the driver; in the other is cut the hole for a speaker port. The driver and port are installed in their respective bowls, then a layer of polyester wadding is placed inside the bowls. The bowls are then glued together using industrial adhesive.

The result is a subwoofer enclosure that is very stiff (the curved walls flex little) and is shaped so that standing waves (internal reflections that normally occur off the flat walls of a subwoofer enclosure) are not present. The enclosure is also very quick to build (under an hour, easily) and it does not require woodworking tools or machinery.

The downsides? These designs are based on smaller drivers that need to have a low Q – the one shown here uses a 130mm (5in) driver that had a Qts of 0.55. Also, depending on the interior arrangement of your car, the spherical shape may make fitting the subwoofer an easy task (eg mounted largely inside the dish of a spare wheel) or alternatively, it may be a more difficult task than fitting in a conventionally-shaped square enclosure.

Finally, to develop a really good compact subwoofer using a small driver, you'll need either the Thiele-Small specs of the driver, or have a way of measuring those for yourself (eg by using a hardware/software combination like Woofer Tester 2 – a very good speaker test equipment package).

If you use a driver salvaged from a discarded household speaker (as was done here), the total cost can be very low – especially considering the quality results.

The spherical subwoofer is easy to make and, for its size, sounds very good. It uses a rear port.

MAKING THE SPHERICAL ENCLOSURE

Note: The following step-by-step sequence shows a slightly larger spherical speaker enclosure being built. However, the process is identical to developing the compact subwoofer.

The first step is to buy a pair of the bowls. At the time of writing, these bowls were available in 280mm, 340mm, 400mm and 510mm diameters, and in black or white finishes. The one used in the subwoofer design is 280mm in diameter with a black finish. The internal volume of the bowl is best measured by filling it with sand using a graduated jug (eg 1 litre). Fill one bowl and then double it for the internal volume of the subwoofer, minus port and driver volumes. The described subwoofer design has a volume of just 8 litres. (The larger 340mm bowls shown in this step-by-step sequence give a 16-litre final design.)

Here is the as-bought bowl turned upside-down. Although not clear in this photo, the bowl rests on moulded 'feet,' which need to be ground off to obtain a flat surface.

A normal metal-grinding disc mounted in an angle grinder does a good job of removing the 'feet.' However, it's dusty and so it's best done outside with the operator wearing goggles, hearing protection, and a dust mask.

Here's the view with the base smoothed. To get it as flat as possible, follow-up the grinder with the use of a belt sander, or moderately coarse sandpaper and a sanding block.

WORKSHOP PRO — MODIFYING THE ELECTRONICS OF MODERN CLASSIC CARS

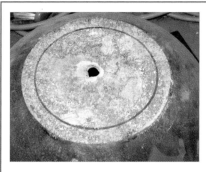

The next step is to mark the cut-out required for the driver. Be careful in sizing of this hole – too small and there will be the need for lots of further grinding; too large and you will not get the driver to seal in the hole. Note the original pot drainage hole.

The hole for the driver can be cut out with an electric jigsaw – this is the piece that's been removed. Use a coarse wood-cutting blade – and expect to get only a few holes out of the blade before it is blunt. Don't go like a bull in a china shop – you can crack the bowl if you push too hard.

Here is the hole immediately after cutting out. It's difficult in the clay-fibre mix to follow the line accurately (there are small pebbles in the material too), so it's better to err on the side of undersize rather than oversize. The hole can then be ground back to the line by carefully using the angle grinder.

It's hard to see, but the arrow points to one of the speaker mounting holes that has been drilled. If you have a masonry drill bit, you could use that (not with a hammer drill!), but these holes were drilled with a normal high-speed steel bit. Drill the holes sufficiently large that machine screws can be inserted – nyloc nuts and washers will go on the inside of the bowl.

Use silicone sealant under the edge of the speaker frame. This is important because despite smoothing with the grinder and sandpaper, the mounting surface is likely to not be perfectly flat. You don't want air leaks around the edge of the driver.

The driver installed in the hole. Note the use of washers and nyloc nuts on the speaker mounting screws. Tighten these when the sealant is yet to set – but be careful not to over-tighten or you could crack the bowl. If you want to mount a grille to protect the driver, and it needs to be bolted into place, do this step next. Any subwoofer mounting brackets should also be installed at this point.

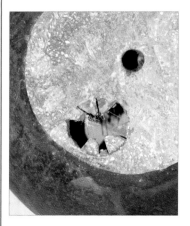

With the driver installed, it's time to tackle the other bowl. The other bowl, its 'feet' already ground off, needs a hole made for the port tube. I found the easiest way to do this was to use the jigsaw to cut a series of radial slots like this, with the pieces then broken off by the careful use of pliers. The hole was then filed to final shape. The hole doesn't have to be perfect – the industrial glue used to secure the port in place will fill any small gaps. (Don't use an expensive holesaw to make this opening – the teeth will soon be blunt and the holesaw ruined.)

The port tube needs to be the correct diameter and length to match the enclosure design. PVC pipe is cheap and easy to use.

7. MODIFYING SOUND SYSTEMS

The port glued into place. For this task, and also gluing together the bowl halves, I used water clean-up 'Liquid Nails.' Note that this glue has also been used to seal the original drainage hole in the pot.

Now is also the time to feed the cable through a hole drilled in the bowl, and seal it with glue. Either solder the wires to the woofer's terminals, or equip the cable with push-on terminals to make this connection.

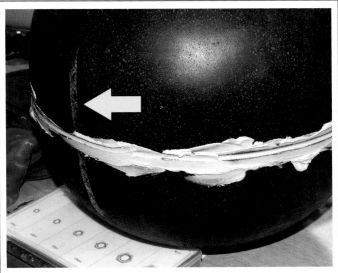

The bowls are pushed together and then a wet finger used to smear the glue around the join, ensuring that the gap is completely filled. A damp cloth is then used to carefully remove the surplus glue. (The glue must be of the water clean-up type!) Note the chalk witness line (arrowed) that shows the correct alignment of the bowls as they are joined. This line is made earlier when the best rotational fit is found – because their lips aren't dead-flat, the bowls fit together better in some orientations than others.

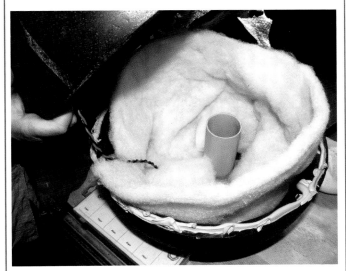

This photo shows two steps. The first is that polyester quilt wadding (available from dressmaking supply shops) has been placed around the inside of the lower bowl. (These reduce internal reflections.) Ensure that there is enough wadding projecting upwards from the lower bowl so that when the two bowls are glued together, the wadding covers the inside surfaces of the upper bowl as well. Also ensure that the rear of the driver and the port are not blocked by the wadding. The next step is to glue together the two halves. Use a generous amount of glue around the rim of the lower bowl. The upper bowl is then carefully lowered over the lower bowl, ensuring the wadding goes inside the upper bowl as it is lowered. (This job is best done by two people.)

The finished subwoofer installed in rear of a small wagon. The grille has been removed for this pic.

187

WORKSHOP PRO: MODIFYING THE ELECTRONICS OF MODERN CLASSIC CARS

Design – and results

The spherical subwoofer shown above used a 130mm (5in) speaker salvaged from a JVC mini sound system. Using Woofer Tester 2 to test the driver, the specs were:
- Fs (resonant frequency): 58Hz
- Qts (total Q): 0.55
- Vas (compliance): 5.35 litres

The enclosure was modelled in the Woofer Tester software as an 8-litre volume, tuned to 51Hz with a port 37mm in diameter and 130mm long.

When tested with the frequency generator built into Woofer Tester 2, the completed subwoofer shows smooth in-car response down to about 40Hz. Note that *this is not a 'one note bass' subwoofer* – lots of subs have a very peaky response around resonance, and so all bass notes sound the same.

In the real world, most door and dash speakers have little or no response below about 100Hz, so getting down to 50Hz potentially adds an extra octave of response. This makes this sub an ideal add-on to a standard system. It's not the sort of design that will shake your car to bits – its output is more in line with the standard subwoofers fitted to

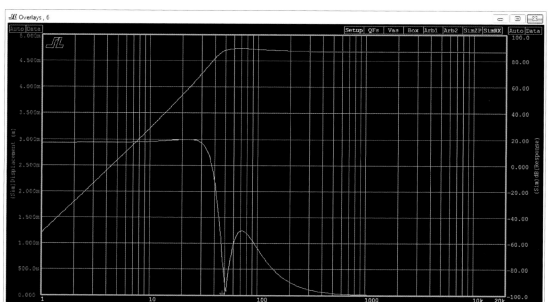

Here is the predicted frequency response and phase curves of the design. The system has a resonant frequency of about 50Hz – very good, considering the size of the driver and enclosure.

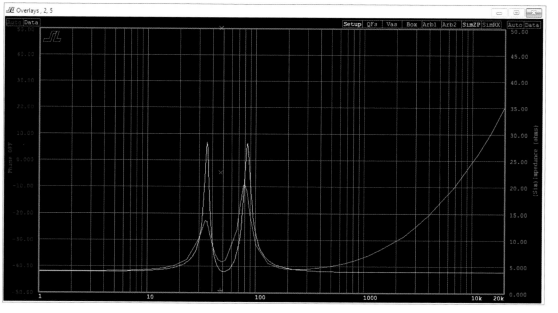

Once the subwoofer was built, Woofer Tester 2 was used to measure the actual impedance curve and compare it with the modelled prediction. As can be seen here, the two curves match well (especially in frequencies) in the area where the subwoofer is working, meaning that the modelling was accurate.

7. MODIFYING SOUND SYSTEMS

current premium wagons and hatches – but it will make a distinct improvement to many car sound systems.

BUILDING A FOUR-CHANNEL AMPLIFIER

Here is a four-channel car sound amplifier with a maximum output of 68 watts per channel. That's more than enough, even with relatively inefficient car speakers, to give you plenty of volume and punchy bass. (But it's not enough to run a subwoofer – this amplifier is for the door and/or rear deck speakers.) And if you're happy to do some of your own metalwork and you already have some hardware like fasteners and spacers, the cost is very low. Quality? Far better than the vast majority of similar power car amplifiers!

One of the requirements was that the 270-watt amplifier be reasonably compact and light. The final item has a mass of 1.75kg (~4lb) and dimensions of 250 x 140 x 75mm (10 x 5.5 x 3in).

Starting points

The heart of the amplifier comprises four LM3886 ICs. This audio amplifier IC has been around for a while – it's an oldie but a goody. Each is capable of 68 watts into 4 ohms at a maximum distortion of 0.1 per cent.

However, rather than start with the bare ICs, I used two prebuilt, two-channel modules available on eBay. Note that the selected modules require a plus/minus 28V DC supply, rather than the AC transformer supply that most of these modules are configured for. Therefore, when sourcing these modules, ensure they look exactly as pictured. To find them, search for 'Assembled LM3886TF Dual channel Stereo Audio Amplifier Board 68W+68W 4Ω 50W*2 8Ω.'

Next up, you'll need a power supply capable of driving these modules. Previously, developing such a supply would have been expensive and time-consuming – but now one is available off the shelf. It's called '1PC Switching boost Power Supply board 350W DC12V to Dual ±20-32V for auto' (or similar).

While the output of the power supply is ±32V as it arrives (the onboard pot allows adjustment), the LM3886 is happy with up to ±42V, so that's fine.

I also chose to use a fan for heatsink cooling and triggered it via another eBay module. This module is called '20-90° DC 12V Thermostat Digital Temperature Control Switch Temp Controller New.'

The completed four-channel, 12V amplifier. The power supply module is on the left, and the two 2-channel amplifier modules on the right. When the lid is placed on top, it locates the fan above the amplifier modules. (A similar sized vent hole to the fan is located under the power supply module.)

This prebuilt stereo amplifier module is available on eBay, with two of the modules needed for this amplifier. The module requires at least ±28V DC to run and, in use, plenty of heatsinking is also required.

WORKSHOP PRO — MODIFYING THE ELECTRONICS OF MODERN CLASSIC CARS

This power supply module generates ±32V DC from 12V DC and can be directly connected to the amplifier modules.

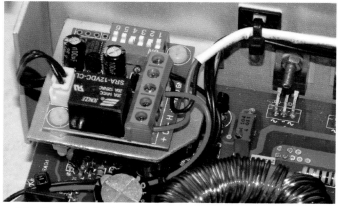

This tiny module triggers the cooling fan on the basis of the temperature selected with the DIP switches. In this case, the temperature for fan activation was set at 40°C (104°F). The remote temperature probe is located between two of the LM3886 amplifier ICs on one of the heatsinks.

Wiring

The electronics aspect of building the amp is straightforward. The power supply board input GND, K and 12V terminals are connected as indicated – GND to chassis ground, and the 12V terminal directly to the positive of the car battery. Use a high-current fuse in this battery supply – eg 20A. The K terminal requires 12V to switch on the power supply – normally it's connected to the 'power aerial' output of the head unit. (Or if you don't have this, you could connect it to any ignition-switched 12V supply.)

The power supply board's output VCC+, GND and VCC- terminals are, respectively, connected to the (+), GND and (-) terminals of the two amplifier modules.

The line level inputs from the head unit are connected to the IN amplifier module terminal blocks (observe correct polarity), and the speakers connected to the OUT terminal blocks (again observe correct polarity).

And that's it for wiring!

Building

You need to provide plenty of heatsinking capacity – either by the use of substantial heatsinks, or by using smaller heatsinks but adding a fan. In use, most of the heat is generated by the four LM3886 modules – despite appearances, the power supply module heatsinking requirements are more modest.

I wanted a compact box, so made one from aluminium sheet specifically to suit the required dimensions. The overall dimensions of the box were about 250 x 140 x 75mm (10 x 5.5 x 3in). The heatsinks were formed by using 8mm (5/16in) thick aluminium plate for two walls of the box. A salvaged 12V fan was placed in the top panel (and there is a matching size hole in the bottom panel) and the fan is triggered by the temperature sensing module. The fan is set to turn on at 40°C (104°F).

The power supply module comes with the required insulating washers and collars for mounting the transistors to the heatsink, while the amplifier modules uses plastic encapsulated ICs and so do not require any extra insulation.

To mount the boards, you'll need to provide the

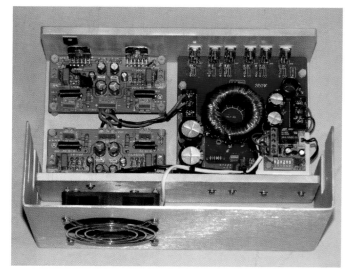

This view shows the six switching transistors used by the power supply module (top right). This module is supplied complete with the insulating washers and pads for heatsink mounting. However, you need to supply your own board mounting stand-offs.

7. MODIFYING SOUND SYSTEMS

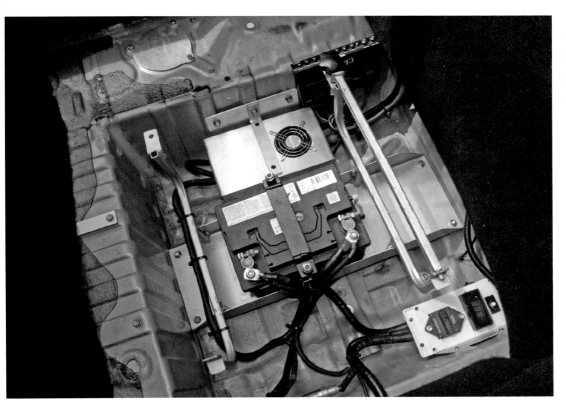

The amplifier (top) in the car. It is mounted next to the (relocated) battery in the spare wheel well, under the inverted spare wheel that's been removed for this photo.

insulated standoffs and screws, washers and nuts.

Rather than place connectors for the inputs and speakers on the box, I chose to directly wire these connections to the boards. These leads were run through rubber grommets that slide up appropriate channels when the lid is screwed into place.

Obviously the type of housing you place the components in is up to you – you could even use a discarded car sound amplifier enclosure that incorporates its own heatsink. But remember, whatever approach you take, you'll need either quite substantial heatsinking or need to add a fan.

Results
The only financial outlay I had was for the amplifier and power supply modules. And for that outlay, this is an unbeatable amplifier. The sound is excellent – better than commercial car sound amplifiers costing two or three times much, and so much better than the typical four-channel amplifier built into a head unit that it's not funny!

WORKSHOP PRO MODIFYING THE ELECTRONICS OF MODERN CLASSIC CARS

Chapter 8

Upgrading lighting

- **Fitting more efficient or higher power bulbs**
- **Fitting driving lights**
- **Stopping driving light vibration**
- **Fitting LED bulbs**
- **Upgrading interior lighting**
- **Fitting additional interior lights**

WORKSHOP PRO — **MODIFYING THE ELECTRONICS OF MODERN CLASSIC CARS**

Many 1990s and 2000s cars have poor lighting. Headlights are weak, filament brake lights relatively slow to respond, and interior lighting dim and poorly distributed. That's the bad news – the good news is that all of these deficiencies can be addressed!

IMPROVING FORWARD LIGHTING

If you live in a country with empty roads (especially if those roads are unfenced), having good forward lighting is an absolute must if you're to travel fast with safety.

More efficient bulbs

The first, easiest and cheapest step you can take to improve forward lighting is to fit more efficient headlight bulbs in place of the standard items.

Incandescent bulbs are available that make claims like '60 per cent more light,' all without drawing any more than standard power. So a 50W high beam remains a 50W high beam – but more light is produced. The fact that the same power is drawn is important for a number of reasons. First, the standard wiring doesn't have to supply any more than standard current, so the voltage drop is no higher than previously. Second, the upgraded bulbs are legal. And third, the heat produced by the bulb is likely to be the same as the standard bulb, so not cause degradation of the reflector or the (possibly plastic) lens.

However, it's hard to get something for nothing with much the same technology, and the greater light output these bulbs produce is often nowhere near the claims. In fact, having tried many of these bulbs over the years, I'd suggest two things.

First, buy upgrade bulbs only from reputable manufacturers like Narva or Philips. Second, don't go for 'blue' or other colour-shifting bulbs: just get ones that have more light output. These will generally be whiter in light than the original bulbs, anyway. (The reason you don't want colour-shift bulbs is that their output is usually lower than non colour-shifting bulbs.)

I've previously fitted Narva 'Plus 60' bulbs. The blurb claims 60% more light and a 20 metre longer beam. However, the difference was subjectively closer to perhaps 10 or 15 per cent. So it's worth doing – but try to get the bulbs when they're being sold on discount, and expect only a minor improvement in light output.

Higher power bulbs

Instead of going for more efficient bulbs, you can instead go for higher power units. This approach can be taken if: (1) you don't care about legality (and you are very unlikely to ever get pinged unless you have the headlights aimed much too high), (2) the headlights are made of glass and steel, and (3) the standard wiring is heavy-duty. All of

'Plus' type bulbs will improve forward lighting, but subjectively not by as much as claimed. I've found the difference to appear to be more like +10 or 15 per cent – still worthwhile doing, but not a radical improvement. (Courtesy Narva)

Driving behind good lights transforms open-road night driving. Here two HID Narva lights, 225mm in diameter, are being used. Both cornering and pencil beam versions are fitted. On a straight road like this, roadside reflectors are visible 2km away.

8. UPGRADING LIGHTING

If your car uses metal and glass headlights, and has heavy-duty standard wiring, more powerful bulbs than standard can be fitted. I upgraded this Mercedes 230 from 55/60W to 90/100W bulbs to gain a subjective 50 per cent improvement.

these aspects apply to my old Mercedes 230. So rather than buying replacement 60/55W H4 bulbs, I upgraded to 90/100W H4 bulbs, again from Narva.

The heat output and current draw of the new bulbs are substantially greater than the standard bulbs, so this swap is not a great idea if you have plastic headlights. Also ensure that you increase the current rating of the headlight fuse or fuses. In the case of the Mercedes, the upgrade is very good. I'd say that, subjectively, the light output is about 50 per cent greater than before.

HID replacement headlight bulbs

High Intensity Discharge (HID) kits are available for standard headlights. That is, you can buy – say – an H4 replacement kit to replace the standard H4 bulb with a more powerful HID bulb, complete with remote ballast. However, invariably this gives a poor result – if not for you, then for approaching drivers. The problem is in beam control. The reflector was not designed for the HID bulb geometry, and so low beam cut-off and edge control are often very poor, blinding drivers coming towards you. If your car has separate reflectors for high and low beam, you may be able to fit an HID bulb upgrade kit to just the high beam, where accurate control is not so important.

LED replacement headlight bulbs

LED replacement bulb upgrades are a different story to HID replacements – so long as you buy quality LEDs from a reputable manufacturer. The difference is that, because the individual LEDs can be very small, they can be arranged to replicate the layout of the original incandescent filament for which the reflector and/or the lens was designed. That is, unlike HID replacements, the original beam pattern can be retained. (Some producers of LED replacement bulbs don't worry too much about this, which is why I said to buy reputable brands.)

I have fitted Philips Ultinon Essential LED HL 6000K replacement bulbs to H4 headlights. (The bulb kit is part number 11342UEX2.) The bulbs have the required high and low beams, and with 6000K colour temperature, they are very white. They also have a measured output much higher than incandescent bulbs. Furthermore, they use about one-third the current of conventional incandescents. And, to top it all off, at the time of writing, they were only about twice the price of the 'Plus 60' style incandescent bulbs that, in my experience, give only a marginal improvement!

Philips claim '+200 per cent' improvement in light output over conventional bulbs for some of their LED replacements – but how that is calculated or measured isn't explained. However, using a light meter to measure the intensity at a distance of about 3 metres (10ft), I recorded a value *three times greater* than the original incandescent bulbs. The incandescent bulbs were not new, but there was no darkening of their glass or any other hints that they'd lost performance.

However, despite the kit being sold as a 'plug and play,' I did run into a small problem in fitting the LED bulbs. So let's take a step backwards and follow the installation process.

The kit is supplied with two bulbs with H4 bases. Because the LEDs generate heat that has to be drawn away, the bulbs come with copper braided heatsink strips that protrude from the back of the bulb. These are flexible and so they can be bent around obstacles – contrasting with some LED bulbs that use solid heatsinks. However, even with that flexibility, the LED bulbs are much more difficult to fit than conventional bulbs. If you have little room behind the headlights (for example, you can only *just* get your hands in to replace conventional bulbs) you will probably need to take the headlights out to fit these bulbs. In some cars, that involves removal of the grille, bumper cover, etc, so it may not be a five-minute job.

I removed the bumper of my Honda (it's been off many times, so not a big deal), and then removed a headlight. The small power supply box for the LEDs connects to the bulb via a multi-pin plug, and the other end of the supplied loom plugs straight into the H4 connector on the car's original loom. I checked that the headlight worked, and then did the same on the other side. But I then realised that low beam was not working – irrespective of the position of the high/low beam stalk, high beam was all that was available.

What was puzzling was that I could not even hear the standard high/low beam relay clicking. So, was it a blown fuse? I checked – no. I mulled over the issue and measured various things with my multimeter – to no avail. I unplugged the new bulbs and plugged in the old – and they worked fine! I then connected one of the LED bulbs on one side, and one of the old bulbs on the other side – and then both the old and the new worked on high beam! It was all very odd – almost as though the system needed the presence of an incandescent bulb to make it work. (But this car has no 'intelligent' sensing of headlight current, or anything like that.)

The wiring diagram for the headlights was of no help (but then I do not have the manual for Australia – the country in which the car was sold – and headlight wiring does vary, depending on market), so I was starting to make guesses. Perhaps the system needed a greater current flow? Perhaps there was a relay whose coil was in series with the high beam? (I know that sounds odd, but the manual for Canada shows a relay wired like that as part of the high beam/low beam switching.)

I then experimented with a 33Ω, 5W resistor, connecting it between the various terminals. When the resistor was connected between the high beam connection and ground, both LED low and high beams worked. (I later I found that even the much smaller resistance of an automotive relay coil across the high beam/ground connections was sufficient to make the low beam LEDs work.) I must confess that I don't understand why a pull-down resistor is needed on the high beam, but in my application it was.

And on the road? Firstly, the not-so-good news. High beam was improved, but nowhere near to the extent achievable by fitting good ancillary lights. The light was

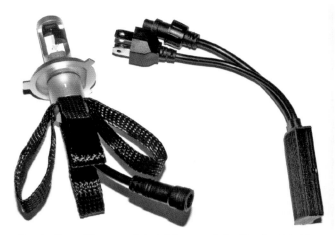

One bulb and its associated driver module. The flexible blue braided strips coming from the bulb's base are copper heatsinks Note the H4 plug on the loom – this kit is sold as plug-and-play.

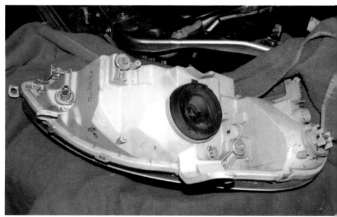

The headlight removed for bulb fitment. If there's plenty of room behind the headlight, you should be able to fit the bulbs without removing the headlights, but if space is tight, the headlights will need to come out. Be careful not to scratch the lens, especially if it is plastic. Here the headlight is resting on an old towel.

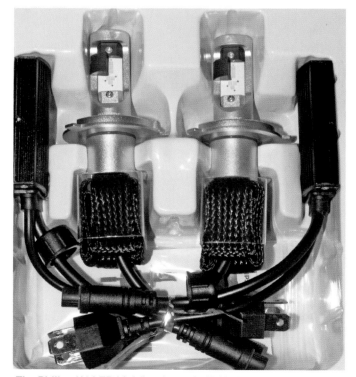

The Philips H4 LED high/low beam bulb kit. Note in the bulbs the vertical array of LEDs, and the shields placed to give the correct beam pattern when working with H4 reflectors.

better than standard, but the sheer distance of vision achievable with extra lights was not there (see below for more on fitting extra lights). Now the good news – the low beam was transformed. The beam pattern was in fact better than standard, with a beautifully even spread of light that was whiter and much brighter than with the incandescent bulbs. Recommended!

Additional lights

The only way to get a massive upgrade in high beam light output is to fit additional lights. Ancillary forward-facing lights come in three distinct types:

- Fog lights – these have a very low beam cut-off to avoid shining the light into the fog and dazzling the

8. UPGRADING LIGHTING

The new LED bulb in place. The rubber cover needed some strategic slitting to fit over the slightly larger bulb base.

normally available in identical bodies, the pair of lights will look much the same, with perhaps just slightly different lens designs. (Some driving lights are available in all three configurations – fog, cornering and pencil. So be sure to know what you're buying!)

When comparing like with like, the larger the reflector, the more light that you will get out of it. Driving lights come with various bulb wattages, but going from a 50W to a 100W bulb, for example, won't make anywhere near the difference of upsizing from a tiny 50W driving light to a big 50W design.

There are plenty of unbranded driving lights around, often at very low prices. However, for durability and functionality, I highly recommend that you go only with a major, longstanding brand like Hella, Narva, Cibie, Bosch – or something similar. Have a look at the brands of lights used by those who really need them – long-distance trucks and buses are a good starting point.

If you buy a driving light of adequate size and from a good maker, it's unlikely that you'll need to go higher than standard in wattage. That is, the recommended bulb and wattage for that lamp should be sufficient.

At the time of writing, most ancillary lights use traditional filament (incandescent) bulbs. However,

driver with the reflected light. Their reach is therefore limited, typically being about 150 metres. Some fog lights are coloured, which further helps reduce dazzling. For driving fast at night in non-foggy conditions, fog lights (whether original equipment or aftermarket) are useless. That makes sense – they're not designed for that purpose.

Driving lights (cornering lights, sometimes called 'spread beam'). Cornering lamps aren't just for going around corners; the broad beam that they provide can also be useful in seeing any wildlife or stock that is about to step onto the road. Cornering lights have a typical range of about 250 metres.

Driving lights (pencil beam). As the name suggests, these lights have a long, narrow beam. A good pencil driving light will have a useful range of up to 2000m – that's 2km. What 'useful range' means in this context is that you'll be able to pick up roadside reflectors at that sort of distance – on dead-straight roads, anyway.

A serious night driving set-up might use one cornering beam and one pencil beam. Since these variations are

To gain a major increase in high beam lighting, you will need to add extra lights. For a given quality brand, the bigger they are, the better the light output.

The difference in colour temperature between the LED (left) and incandescent halogen (right) is obvious in this photo taken during daylight – the LED bulb colour is similar to daylight. With the LED bulbs fitted, low beam was substantially improved.

197

WORKSHOP PRO: MODIFYING THE ELECTRONICS OF MODERN CLASSIC CARS

some are available with HID bulbs, often with the ballasts built into the light body. Unlike aftermarket headlight replacement HID kits, ancillary lights designed for HID bulb use are typically excellent in performance and beam control, and the HID light source is extremely bright. I use HID-equipped Narva ancillary lights on my cars and I cannot speak highly enough of them.

In recent times, LED light bars have become popular. These use multiple LEDs with small reflectors. These light bars do not have anywhere near the range of pencil beam traditional (or HID) driving lights; instead they have a wide, very white and even beam. In this way, they are similar to the beam pattern of traditional cornering lights. If you drive mostly at night, at relatively low speed, on winding, narrow roads, these lights may suit your needs. Certainly, their shape makes them easier to integrate with the front of the car.

Buying

Before you venture to a retailer, work out the general type of light that you want. Driving lights are available in round and rectangular (and some oval) designs, but in addition to this, they also vary in depth and mounting technique. Will the light be hung from its mount (pendant style) or will the light be standing upright? Most driving lights can be easily adapted to either configuration, but some can't – so you need to know before you go buying. Is a very thin body required, or is mounting depth unlikely to be a problem? What's the maximum size that you feel comfortable with from an aesthetic point of view? The more recent the car, the more difficult it will be to visually integrate the lights.

And of course, money comes into it, too. The sky's the limit when it comes to spending money on ancillary lights. If the price of good quality lights is too great, look for secondhand driving lights (although remember to make sure that you're not getting fog lights when you're thinking of driving lights!).

Mounting positions

Because of the angle that the beam of light makes with the road, the higher the lights are mounted, the better. However, in practice, mounting the lights above or below the bumper will make very little difference to the end result. More important to the positioning are two factors – visual preference and available mounting points.

Many years ago, when all cars had chrome-plated steel bumpers, it was easy – you drilled a hole either in the top or the bottom of the bumper and bolted the lights in place – a 15-minute job. But with plastic-covered bumpers, integrated brake ducts and undertrays, finding decent mounting points can be very difficult. Invariably, a pair of custom brackets will need to be made. Another major reason that very firm mountings are needed is that it's

Aluminium bracing straps added to large driving lights to reduce movement. These straps were folded by hand, using a bench vice and some careful, step-by-step bending. A new hole was drilled in the light body for the mounting screw.

important that the driving light is as rigid as possible. A light that vibrates – or worse, moves an even greater amount under aerodynamic and vertical loads – will have a shorter bulb life, in addition to potentially flashing the beam all over the place. More on these aspects in a moment.

Wiring

While if you keep a few things in mind, the mechanical fitting of extra lights is straightforward enough, some people then drop the ball when they get to the wiring. However, despite the need to use a relay, if you make one vital decision, the wiring can in fact all be pretty simple. The vital decision? That's the one where you decide that whenever the high beam is on, so are the driving lights. If you configure the system like this, all the wiring can be done under the bonnet (hood) without any need to run cables back into the cabin – in most cars, that makes things much easier.

With this approach, power is picked up directly from the positive terminal of the battery, and the relay is triggered

Always fuse the battery high current feed to the lights. The fuse should be mounted as close to the battery as possible. Tape a spare fuse nearby – just in case.

8. UPGRADING LIGHTING

Finding the high beam wire – it will have battery voltage on it when the high beam light is on. In this car, that has a separate high beam light, that was an easy process.

as close to the battery as practicable. Second, use wire of adequate gauge for the wiring discussed so far. The wire from the battery to the fuse holder, and then to the relay, needs to carry all the current that the two lights will draw. That could be as high as 15A – make sure that the wire is rated to suit.

You'll need to make a connection to the high-beam wiring to provide the switch-on signal for the relay. Use a multimeter to probe the back of the headlight until you find a wire that has 12V on it when the high beam is on – and 0V when it is off. If the car uses a separate high beam light within the headlight assembly, finding this wire will be even easier. Tap into the high beam power supply wire and then run the new wire to terminal 85 of the relay. Terminal 86 (the other side of the relay's coil) then gets connected to ground.

(In some cars, power is always fed to the high beam, with the factory switching working by earthing the light to turn it on. In cars where the wiring is like this – you'll soon know because you won't be able to find a 12V feed at the light that turns on and off – just wire the relay coil in parallel with the high beam. The other connections remain the same.)

The way the new electrical system works is this. When you turn on high beam, power also flows through the relay's coil, creating a magnetic field which pulls the relay

straight from one of the high beam power feeds. Note you can always have a switch under the bonnet to disable the new lights, should a situation develop where when you are running high beam, you don't want the ancillary lights on. (In the real world I have come across that requirement only once – when the car had a failing alternator and the extra current draw of the driving lights was gradually flattening the battery.)

The first step is to wire the two grounds of the driving lights to the car body. Some older design lights, that use metal bodies, ground themselves through their mounting bolts. However, with the plethora of plastic in cars, it makes sense even in these designs to run an additional ground from the bodies to the metalwork of the car. The other wires from each of the lights (ie their power wires) run back to the relay, connecting to terminal 87. Terminal 30 (the other side of the relay 'switch') connects to the positive of the battery via an in-line fuse.

There are a few things to pay attention to when doing this work. First, always make sure that you use a fuse. Without a fuse, any short-circuit that might develop could lead to a car fire – and that's the case even with the car locked and you having walked away. Always place the fuse

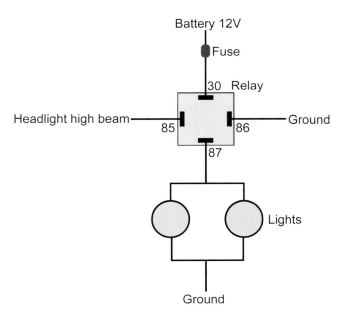

Power is fed directly from the battery to the relay via a fuse. The lights are wired in parallel and are fed from the relay. The relay is triggered by the headlight high beam, allowing all wiring connections to be performed under the bonnet. If you want to be able to manually disable the extra lights, fit a switch in the headlight high beam feed to the relay.

contact over. This allows power to flow from the battery, to the fuse, to the driving lights, turning them on.

If you want to put a switch in the circuit, place it in between the headlight high beam power pick-up and the relay, ie break the wire heading to terminal 85 and put it in there. Note that, unlike the power feed for the driving lights, these wires can be quite thin in gauge without any problems being caused.

> **WARNING!**
> These instructions refer to traditional car lighting. If your car has standard high intensity discharge (HID) lighting, you should not probe into the wiring of the system. Dangerously high voltages are present in these systems.

Aiming

So you've bolted on the new lights and wired them up. Think you're now finished? Wrong! Aiming of the lights is one of the most important steps in their fitment.

What you need is a dark, empty and straight road, preferably with roadside reflectors. You'll also need two people, one to adjust the lights and the other to sit in the driver's seat and call the shots. The actual process is simple. With the car idling and in a normal position on the road, adjust each light until the beam reach and illumination is best. Typically, that will mean starting with the lights too high then gradually dropping them down in aim until distant reflectors start to be highlighted – then down just a fraction more. The driver's side light should be aimed right down the middle of the road while the passenger side light should be angled a little more towards the edge of the road. Incorrect aim – especially of pencil beams – can make the difference between having fantastic additions or having lights that you may as well not have bothered fitting. The difference can be that great.

STOPPING DRIVING LIGHT VIBRATION

The longer the range of ancillary driving lights, the better needs to be the mounting system that locates them. I ran across this problem when I installed a pair of large Narva driving lights on a 2006 Honda Legend. Solving the problem also involved some tricky engineering, so it's worth recounting here.

Starting points

As with nearly all cars of the era, the Legend uses a plastic bumper cover that has no structural strength at all. This means that attaching lights to this cover, or to the number plate mounts (commonly done via a small, right-angled plate), is a recipe for disaster.

Behind the bumper cover you'll find an intrusion beam, sometimes covered with a collapsible plastic. The beam behind the plastic makes a stiff mounting point for driving lights. However, on some cars getting the bumper cover off can take several hours – and is especially difficult if you don't know where all the often-hidden screws and bolts are. It's also very easy to scratch the side panels or the cover itself.

An alternative approach, one that I've now taken on three cars, is to make a frame that fits behind the grille.

The frame I made to mount the big driving lights. This is the view with the grille removed and the frame installed ...

... and the frame out of the car. The long verticals were added to prevent the horizontal beam from twisting. Even with a frame this extensive, I was unhappy with driving light movement and so a further fix was needed.

8. UPGRADING LIGHTING

This frame attaches to solid mounting points on the car. The driving lights in turn mount on projections from this frame that pass through original openings in the grille. This approach has several advantages. If the frame is made correctly and the body pick-up points are strong, the driving light mounts will be rigid. The lights end up mounted fairly high (good for lighting the road) and the lights are removable without leaving any mark. This is the approach that I took with the Honda Legend.

Another important point to note is that most large driving lights are insufficiently stiff in their factory mounts … irrespective of how strong the metalwork is to which they're bolted. To overcome this, some manufacturers sell adjustable length stays designed to attach the top of each light to the bodywork. However, these straps often don't match convenient spots on the car where it's possible to bolt them. An alternative is to make your own straps from flat steel or aluminium, bent to an appropriate shape. Again, this is the approach I took on the Legend.

Building the frame

The first step on the Honda legend was to remove the top plastic cover over the radiator and then unclip the grille. This could then be removed without taking off the bumper. Directly behind the grille was a heavy gauge metal box – presumably, in models with radar cruise control, it holds the radar. This box was attached to the bonnet locking platform by three 6mm bolts. It looked a good place to start mounting a driving light frame.

I unbolted the box and then welded to it a piece of 40mm square, 2.5mm wall thickness RHS mild steel square tube. To provide additional stiffness, I then added two steel lugs (32 x 5mm) that further attached the RHS back to the locking platform. Two more steel lugs (again 32 x 5mm) were then welded to the tube, projecting forwards and providing the mounting spots for the lights. These were angled and shaped so that the grill could be slid up over them, allowing the grille to be replaced with the frame in position.

The lights were then mounted on the frame and the stiffness of the system assessed. This is easiest done by 'bumping' down on the lights with the end of a closed fist – so inputting a 'step' vibration.

Doing this showed that the lights had plenty of movement in them – a surprising amount, considering the stiffness of the frame, and its five pick-up points on the car's body.

I then added some aluminium straps that connected the top of each light to the bodywork. These straps, made from 32 x 5mm aluminium, and bent by hand in a vice, were shaped to fit through slots in the grille, and so could be attached with the grille in place. The big Narvas, previously fitted to a car I owned, already had holes drilled through the upper surface of their housings – I'd used similar straps on the previous car. Note that before you fit these straps, the lights must be properly aimed!

With the extra straps fitted, the 'bump test' showed much less light movement, so I waited for that evening and then went for a test drive. But I was unhappy with the result – with such a bright beam and long reach, even a tiny movement is easily seen in the beam flickering.

The next step was to add long verticals that tied the mounting frame to the base of the car. These verticals, made from 20 x 1.5mm square tube, prevented the bonnet locking platform from twisting (so causing the beam aim to move up and down). The new parts of the frame needed to be made as bolt-ons (otherwise the frame could no longer be inserted and removed), and these new parts attached to a tubular crossmember via four existing 6mm bolts.

It was time for another night test drive, and this time the movement of the lights was much reduced. I think most people would have been pretty happy at this stage, but I wondered if I could get the lights even steadier. Especially on dirt and rough bitumen, the lights could still be seen to be moving a little.

I'd made the frame as stiff and strong as I thought was pretty well possible, and without pulling half the car to bits, there were no stiffer parts of the body to which to bolt it. It was time to take a different approach.

Measuring and ideas

When the lights were manually bumped, they could be seen to vibrate, with the vibration then dying away. (Of course, with the car running over rough roads, this vibration was constantly being excited.) So how fast was the vibration? A cheap app, called 'Vibration,' exists for iPhones. This app turns the smartphone into a sophisticated, data-logging vibration measurement device. I use the app for measuring suspension natural frequencies, but it can also be used for higher frequency vibrations like those I could see occurring on the lights.

Securely taping an iPhone to one of the lights and then bumping the light repeatedly showed that the system was resonating at about 21Hz. In other words, it was naturally vibrating back and forth about 21 times a second. After a bump input, the logged data showed it took about half a second for this vibration to die away. These are

NO WELDER?

The frame shown opposite was welded together with a MIG. If you don't have such a welder, the easiest approach is to buy a very cheap stick welder (quite cheap secondhand) and then tack together the frame (tacking is easy). You can then take the frame to a professional welder for final welding.

characteristics of a system that is stiff (but not stiff enough to keep the lights absolutely still!) and has little damping.

The first approach could have been to mount the lights on rubber. However, for rubber mounts to be effective, they need to be carefully designed. Designed how then? We know the 'forcing frequency' is 21Hz (that's the frequency of the vibration trying to shake the lights). To stop this vibration being transferred to the lights, we would need a rubber mount that has a natural frequency of about one-third this frequency – so around 7Hz. A mount this soft will need to compress about 10mm with the weight of the light on it. This means the mount would be so soft in fact that accurately locating the light would be pretty well impossible, even with the continued use of the upper strap (that would in turn also need very soft mounts). So in this case rubber mounts wouldn't be effective.

So what other approaches could be taken?

Tuned mass dampers

One approach to vibration reduction is to use a tuned mass damper. With this technique, a mass (a weight) is placed on a springy mount. The mass of the weight and the stiffness of the spring are matched so the system has a resonant frequency the same as the system in which the vibration is trying to be reduced (so in this case, 21Hz). The tuned mass damper is heavily damped, so that vibrations die away quickly after a 'step' input. (An example of a 'step' input is the acceleration caused by one bump.)

So how does it all work? The original system (in my case the lights and frame) vibrate at their natural frequency – 21Hz. This excites the tuned mass damper that is designed to vibrate at this same frequency. But as soon as the input vibration stops, the vibration in the tuned mass damper dies away very quickly, because it is heavily damped. The two systems are rigidly attached to one another, so this causes the main system vibration to die away quickly too. Furthermore, because of the increase in mass, the system as a whole needs more energy to get it vibrating in the first place – or for the same amount of energy, will vibrate less. A tuned mass damper seemed to show the most promise – so I made a pair.

A tuned mass damper needs to use a minimum mass about 10 per cent of the total mass of the main system, and as already said, it needs to have a resonant frequency about the same as the main system. Both values can be accurately calculated, but it is often easier in this sort of challenge to use theory to just get into the ballpark – and then subsequently use a trial and error approach. A system with a resonant frequency of 21Hz (ie 1260 cycles per minute) needs to have only about 0.5mm droop – that is, the tuned mass damper spring deflects by only half a millimetre with the weight on it. That gives you a feel for the required stiffness of the spring.

UNDERSTANDING THE TERMS AND IDEAS

A lot of the ideas covered here sound very complex ... but they aren't.

Get a springy plastic ruler and hold one end firmly on a desk while you let the other end hang out into space. Push down the free end and let it go: the end of the ruler will then bounce up and down. Note how the up and down movement gets smaller each time, but the speed of movement remains the same until the ruler has stopped moving.

From this we can say that the ruler has a resonant frequency (the number of times it moves up and down per second, measured in Hertz (Hz)) and a decreasing amplitude of movement (less distance up and down as time progresses).

Now tape a weight to the free end of the ruler. Push down the free end of the weighted ruler and let it go. You will see that the speed of up/down movement has now reduced – the resonant frequency of the system has been lowered. You should also be able to see that the amount the spring (ie the ruler) droops (its static deflection) increases with the size of the weight placed on it. To put this in a different way: as static deflection increases, resonant frequency decreases.

The length of time the ruler continues to bounce once excited is indicative of the system damping. A system with higher damping will stop bouncing sooner.

The 'ruler experiment' can therefore easily show:
- Resonant frequency of the system
- Changes in resonant frequency with differing mass/spring relationships
- Reduction in resonant frequency with increase in spring static deflection
- System damping

If you have any confusion with the ideas covered here, just a minute or two with the ruler will make it much clearer!

One of the tuned mass dampers. It uses a steel weight supported on a rubber mount that is located horizontally. The assembly is bolted to the light bracket via a right-angled mount.

8. UPGRADING LIGHTING

The two tuned mass dampers positioned under the large driving lights. Their wobbling cancels the driving light wobble!

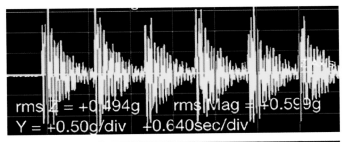
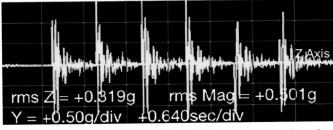

This pair of graphs, reproduced from iPhone screens, show the 'bumped' behaviour of a driving light before and after the tuned mass dampers were fitted. Looking at the yellow trace, you can see on the bottom screen (damper fitted) that after each bump input, the vibration dies away far more quickly.

Tuning of the tuned mass damper can then proceed from that starting point. If, for example, the excited mass visually bounces up and down on the spring too slowly (say at only 5Hz), then the frequency needs to be to raised – so the mass needs to be smaller, or the spring stiffer. (The smaller the deflection of the spring when the weight is on it, the higher the tuned frequency.) If you don't have access to an iPhone, these changes in behaviour can be seen quite easily with the naked eye.

After trialling a number of different approaches, I settled on the following:
- Spring – rubber mount 21mm in diameter and 34mm long, with two 6mm threaded studs, placed horizontally
- Mass – stepped disc of steel, 450 grams, mounted at end of the rubber mount
- Location – on angle bracket directly attached to the mounting bolt of the lights

Note that the rubber mounts of the tuned mass dampers provide a degree of self-damping – not achieved, for example, by a steel leaf spring that I also trialled.

So can I now say that the lights remain vibration-free under absolutely all conditions? It would be nice to say that – but it wouldn't be true. I can say, however, that on all but really rough bitumen and rough or heavily corrugated dirt, the movement of the lights is not detectable, and even on those rough surfaces, it is only just detectable. Success!

LED BRAKE LIGHTS, REVERSING LIGHTS AND INDICATORS

In addition to LED headlight replacement bulbs, LED replacements are also available for brake lights, reversing lights and indicators.

In each case, you need to obtain a replacement LED bulb that has identical physical and electrical connections, so that the new bulb will plug straight into the original socket. That might seem an obvious thing to say, but there are numerous different bulb bases used in car lights – in fact, a chart I have shows over 20 different bulbs for tail-lights alone! Therefore, never buy one from just a specification list – actually take the original bulb out and confirm that the replacement has an identical base, before you buy.

Upgrading to LED brake lights has a significant technical advantage. LEDs are much faster to light than incandescent bulbs, and with following traffic, every tiny time increment impacts stopping distances. In addition to the fact that the LED reaches full brightness more quickly, the suddenness with which the light off/light on transition occurs attracts the attention of the following driver.

Because most brake and reversing lights use a simpler optical construction than headlights, the issues of maintaining beam patterns with LED replacement bulbs is not as important. For example, you can get LED bulbs that incorporate a lens to better focus the output longitudinally – and this will normally result in brighter rear lights.

Indicators are a bit different. Because the electro-mechanical flasher in most cars is designed to get faster

WORKSHOP PRO: MODIFYING THE ELECTRONICS OF MODERN CLASSIC CARS

LED replacement bulbs do not use smart energy management – instead they just have resistors to decrease current flow. If your running car has a voltage much higher than 12V, and you blow lots of LED bulbs, add a series resistor.

LED replacement bulbs for different types of interior lights. The upgrade is as easy as removing the original filament lamp and replacing it with one of these. (Courtesy Auto Lighting UK)

A replacement W21/5 tail-light/brake light LED bulb. As with LED headlight bulbs, I suggest you buy quality branded bulbs. (Courtesy Philips)

IMPROVING INTERIOR LIGHTING
Installing LED bulbs

LED replacement bulbs are available for all interior car lighting. Compared with filament bulbs, they are:
- much brighter
- use less power
- last longer
- have a whiter light

That's a pretty significant list of advantages – sufficiently so, that replacing the interior light bulbs should be on your 'must-do' list. Depending on the light fitting and the bulb type it uses, you might decide to replace the incandescent bulb with a drop-in replacement, or change the whole fitting. Replacing the original filament bulb with an LED replacement is quite straightforward – you just swap the bulbs. However, replacing the whole fitting is a little more involved. Let's take a look.

in its flashing rate when an indicator bulb is blown, the lower current draw that occurs when you replace incandescent bulbs with LEDs will change the flasher speed. To overcome that, you can fit high-power (eg 6W, 50W) resistors in parallel with the new LED replacement bulbs, with one resistor needed for each LED. Alternatively, you can replace the electro-mechanical flasher with a fully electronic one that will drive LED indicators without changing the flasher rate.

Replacement LED bulbs of the sort being described here do not use any form of smart power management. Instead, they just use conventional resistors to limit the current flow through the LED. Therefore, any change in running-car voltage will result in a change in LED current. If the LED is designed for 12V nominal voltage, and your car has available 14.4V with the engine running, then current through the LED is 20 per cent higher than the design current. If you find yourself having to frequently replace blown LED bulbs, you may need to fit additional series resistors to decrease the current through the bulbs.

FITTING A NEW LED LIGHT ASSEMBLY

This LED light provides an even and fairly wide beam. It's a 12V drop-in replacement for household halogen MR14 bulbs, but it is also ideal for in-car use.

8. UPGRADING LIGHTING

I chose to upgrade the hatchback lighting in a Honda. The new LED light, complete with its pins, is 45mm deep, so keep in mind the required rear clearance. (The other major dimension is its diameter, at 50mm.)

A small screwdriver was used to pop out the lens ...

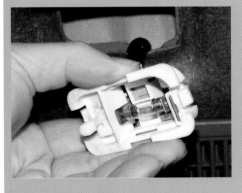

... and then the body of the light itself could be removed.

A piece of grained ABS plastic was cut out, and, to provide better rearwards clearance, a short section of rubber hose was glued to it. The LED light was glued into the tube.

The wiring was then connected. I had available a suitable connection block, but if you don't, solder straight to the lamp's pins or use crimp-on terminals. Insulate the connections.

The panel was held in place with a little more glue. Job finished!

The results are stunning – so much more light comes from the LED assembly that it makes the original look like a feeble, yellow glow-worm!

WORKSHOP PRO: MODIFYING THE ELECTRONICS OF MODERN CLASSIC CARS

INSTALLING ADDITIONAL INTERIOR LIGHTS

Adding extra courtesy lighting is easy and cheap – and it's something you'll appreciate every time you use your car at night.

The best place to source new lights is at a car dismantler – especially one that will let you wander around the yard at will. This is a better approach than buying aftermarket lights because (a) aftermarket accessory lights are often poor quality compared with original equipment, and (b) at a yard you'll be able to pick from a variety of different styles and trim colours. The lights you get depend on how you intend using them – you can pick from map lights, door lights, boot lights, new interior lights, and so on.

FITTING AN EXTRA BOOT LIGHT

Most sedans run a very small light sited under the rear deck. They typically provide poor illumination inside the boot – and very little behind the car.

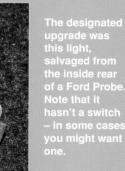

The designated upgrade was this light, salvaged from the inside rear of a Ford Probe. Note that it hasn't a switch – in some cases you might want one.

The first step was to remove some of the fasteners holding the lining in place ...

... and then mark the required cut-out. Make sure that you check that there's adequate clearance behind the trim panel for the light before picking the spot.

The hole in this trim material was easily cut with a pair of sharp scissors.

The light could then be tried for size.

In this car the boot lid opens wide, making it easy to mount the new light where it can do some good (ie lighting up behind the car as well as in the boot/trunk). The interior trim allows easy fitment.

206

8. UPGRADING LIGHTING

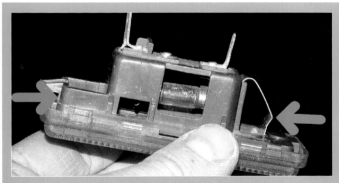

Most lights of this type clip into place, so the cut-out should be sized accurately to take advantage of the spacing between the clips.

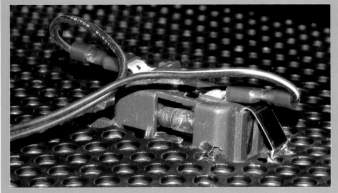

The next step was the wiring. This light required two female spade terminals, which were easily crimped to the new cable.

The cable was then inserted in some spare convoluted tube …

… before being taped to the original loom. The new cable runs inside the factory loom, down the hinge and into the boot.

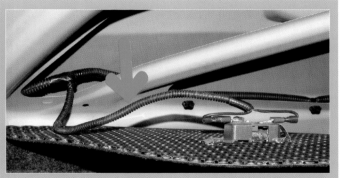

The new loom in place.

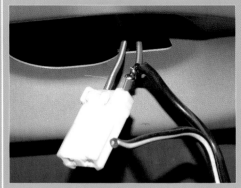

The final step was to make the connections to the original boot (trunk) light wiring. This was done at the light, with the connections soldered …

… then taped, and covered in convoluted tube.

Looks like it was always there! The next step is to swap the filament lamp for an LED bulb replacement.

WORKSHOP PRO: MODIFYING THE ELECTRONICS OF MODERN CLASSIC CARS

FITTING MAP LIGHTS

Map lights are seldom used for reading maps ... but the name has remained (like 'glove' boxes!) Get used to having map lights and you really notice if they're not there. The converse is also true – fit them and you soon get used to having them.

The first step is to have a good look at the headlining where you intend to mount the maplight assembly. In this car, there is sufficient depth in the headlining to allow a recessed map light assembly – in other cars there isn't. Make your light selection after you've done some careful measurement of your car.

The selected light is from a 2000s model Toyota Corolla. As can be seen, it incorporates two maplights. Other designs include the main dome light, and even sunglasses holders.

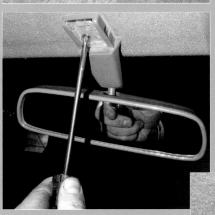

In most cases you won't need to take the headlining right out. Instead, removing the rear vision mirror, sunvisors, and their fittings will allow the front of the headlining to droop down far enough for access.

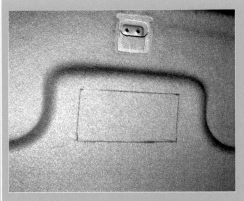

The next step is to mark the required opening. Be very careful when doing this – if you make the hole too large, it's impossible to put back bits of rooflining ...

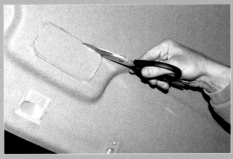

The hole can be cut out with a pair of sharp scissors. Place an old magazine between the headlining and the roof panel so you can't mark the inside of the metal.

It's impossible to make a perfectly neat hole, but concentrate on keeping inside the marked lines – it's easy enough to later enlarge it if the hole proves a little tight.

This light is held in place by spring metal clips – two on this side and two on the other side. However, they're designed for a thicker rooflining, so ...

8. UPGRADING LIGHTING

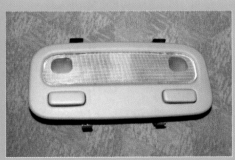

... they were bent outwards and downwards so that they would still hold the light firmly in place from behind.

Power for the new lights is picked up from the centre roof light. This needs to be unscrewed so that its wiring plug is accessible.

The cable that will take power from the dome light to the maplights can then be fed into the space between the rooflining and the roof, until ...

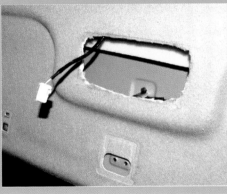

... it appears where the maplights are going to be mounted. In this case enough of the original maplight cable was bought that it didn't need to be extended.

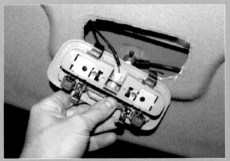

The cable was plugged into the maplight, and then ...

... the assembly was clicked into place in the rooflining. Now to make the lights work!

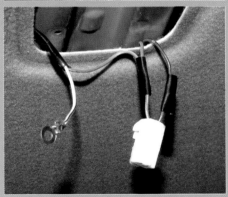

Most interior lights have two wires leading to them. One has power on all the time, and the other only when a door is open. Connect one of the wires from the maplight to the powered wire, and connect the other maplight wire to ground (here, the eye terminal).

Then put everything back together and check that the new lights work! Again, they look like they've always been there ...

WORKSHOP PRO MODIFYING THE ELECTRONICS OF MODERN CLASSIC CARS

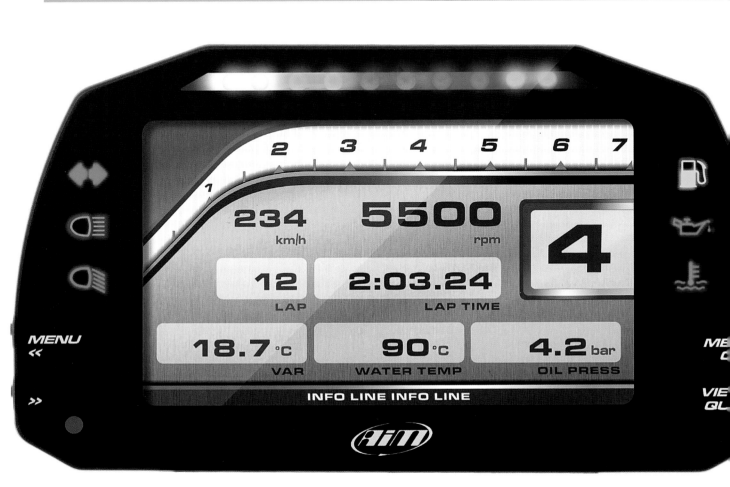

9. UPGRADING DASHBOARDS, INSTRUMENTS AND WARNINGS

Chapter 9

Upgrading dashboards, instruments and warnings

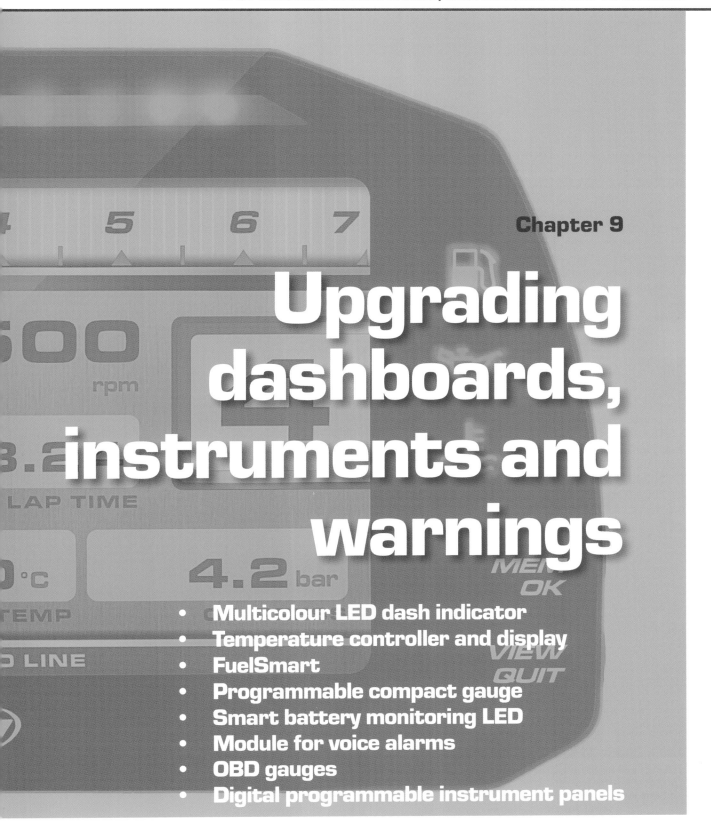

- Multicolour LED dash indicator
- Temperature controller and display
- FuelSmart
- Programmable compact gauge
- Smart battery monitoring LED
- Module for voice alarms
- OBD gauges
- Digital programmable instrument panels

WORKSHOP PRO: MODIFYING THE ELECTRONICS OF MODERN CLASSIC CARS

Gauges, instruments, indicator lights and warning buzzers – they all let you know what the car is doing. The good news is that adding these to a car is now cheaper and easier than ever. And, if you decide you want a really major upgrade, complete new digital instrument panels are also now available. Especially if you're installing programmable engine management, digital panels are also not too difficult to wire into place. So let's have a look, starting with a simple white/red/green LED indicator.

MULTICOLOUR LED DASH INDICATOR

For very low cost it's easy to add a single dashboard LED that can change from green to red to show the status of electrical systems. For example, the LED can be configured so that it's green when the water level in an intercooler spray reservoir tank is high, and red when it's low. No problems with wondering if the LED is broken or a power feed has come adrift.

Or the LED can be configured to be off (white) when the intercooler temperature is low, red when the temperature is high (and the intercooler spray is switched off), and green when the temperature is high (but the spray has been switched on). You can even leave the intercooler spray switch in an 'auto' position where the lighting of the green LED will show when the spray is running! And of course it doesn't need to be an intercooler water spray, You can use the indicator with any switched device.

The use of a single LED also allows you to much more easily integrate the indicator into a busy dashboard or instrument panel.

So all of this must be really complex to do, eh? No, not at all. You will need a cheap three-pin multi-colour LED. These are available in 'common cathode' and 'common anode' designs. The circuit diagrams shown here are for a common cathode configuration, ie the middle pin goes to ground through a current-limiting resistor.

Let's start by looking at how to integrate the LED with a toggle switch, where you want to show via a green or red indication what position the switch is in. You'll need a double pole, double throw (DPDT) switch. All that this means is that two circuits can be switched simultaneously in two directions. These switches are common and cheap; you can recognise them by their six connections. One of the circuits that the switch controls will cause the LED to be either red or green, while the other circuit controls whatever you're turning on or off.

MANUAL SWITCH INDICATOR

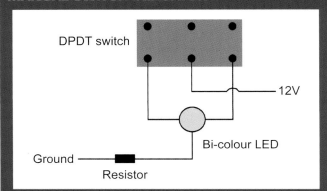

Here a DPDT switch is shown, pictured from underneath so that you can see the terminals. Also shown are the bi-colour LED and a resistor that you'll need (560Ω) to control the current flow through the LED. The 12V input is switched to one or other of the LED leads, while the ground is common for both colours. One with the switch in one position, the LED will show a colour (red or green). With the switch in the other position, the LED will change colour. The upper switch connections aren't used yet.

Test circuit for the bi-colour LED indicator. With the switch in one position the indicator is green, with it in the other position it turns red. The unused contacts of the switch can be used to control the main device being switched.

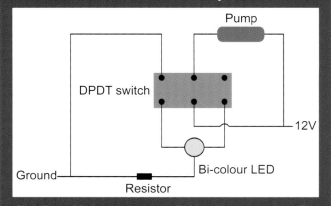

Now I've added a pump, controlled by the same switch. The LED can be red with the pump switched off and green with it switched on. (Reverse the switch-to-LED input connections to reverse the colours.)

9. UPGRADING DASHBOARDS, INSTRUMENTS AND WARNINGS

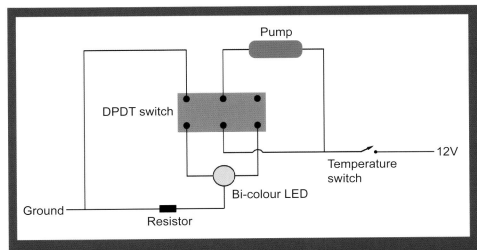

This diagram is the same, but for the addition of a normally-open temperature switch. Let's say that the temperature switch closes at 50°C – the LED now has three conditions, each informing the driver what is happening:

• LED is off (ie white): temperature of intercooler is less than 50°C.
• LED is red: temperature of the intercooler is above 50°C and water spray is switched off.
• LED is green: temperature of the intercooler is above 50°C and water spray is switched on.

You can easily replace the intercooler temperature switch with a boost pressure switch, which, in addition to telling you by a change of colour when you're actually on boost, will also remind you of the spray switch setting each time the LED comes on. (Or you could place the boost pressure switch in series with the temperature switch, so that both boost and temperature have to be high before the spray will trigger.) A suitable temperature switch for this application is available online from a large number of suppliers. One of the cheapest and most effective simple click-action boost pressure switches are the microswitches used in spa baths – again these are cheap and widely available.

WARNING INDICATOR

The above system is primarily for working with a manual switch. But what about that low water level indicator in the intercooler water spray tank that I mentioned earlier? Most switches of this type aren't the DPDT type used above, so how do you make the system work? The answer is to use a relay – a single pole, double throw (SPDT) or double pole, double throw (DPDT), that in turn is controlled by the remote switch. Water level switches of this type are readily available – they're often called float switches.

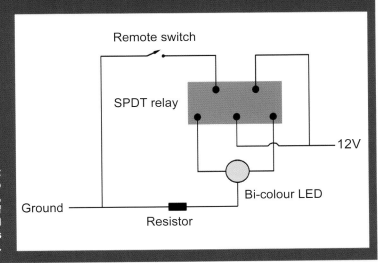

You can see that the circuit is very like the others, except a SPDT relay has replaced the DPDT switch. The two upper connections are for the coil that operates the relay, while the three lower connections are the SPDT part of the relay. Using a relay allows a remote switch to control the LED colour, eg green when water level in a tank is okay, and red when it is low.

TEMPERATURE CONTROLLER AND DISPLAY

You want to control an intercooler fan or water spray, or display coolant temperature, and also switch radiator fans? Or how about monitoring engine oil or gearbox oil temperatures, and sounding an alarm if the temperature gets too high? Well, here's a very low-cost device that can do just that.

The module is widely available online, and can be bought in both Celsius and Fahrenheit versions – here I am covering the Celsius design.

The dimensions of the module are 78 x 71 x 29mm (LxWxH, 3.1 x 2.8 x 1.1in) – and uses a display window that requires a cut-out of 70 x 28mm (2.75 x 1.1in). It uses an LED display that shows temperatures up to 100°C to one decimal place (eg 35.6), and above 100°C in single units (eg 105). The update rate is fast (about three times a second), and the sensor is very responsive to changes in temperature.

In addition to the numerical display, there are two LEDs. One shows when the set-point has been exceeded. (The set-point is the temperature at which you've set the device to activate its output.) This LED has two modes

WORKSHOP PRO: MODIFYING THE ELECTRONICS OF MODERN CLASSIC CARS

The temperature controller/display is a cheap and easy way of monitoring a critical temperature and sounding an alarm or operating a fan or pump. Both Celsius and Fahrenheit versions are available.

– steadily on when the relay is activated, and flashing when the set-point has been passed, but the module is running an inbuilt delay before turning on the output. (You can vary this delay time – more on this in a moment.) The other LED shows that the display is indicating the set-point temperature.

On the face of the instrument are four push buttons – up/down arrows, Set and Reset. Wiring connections are by means of screw terminals on the rear of the module.

Displaying temperature

The simplest use of the instrument is to display just temperature. This requires only four wiring connections and no menu configuration.

Power (12V nominal) connects to pins 3 and 4 – ground to pin 3 and positive to pin 4. The NTC (negative temperature coefficient) sensor that is provided connects to pins 7 and 8 – it doesn't matter which wire goes to which terminal. With these connections made, the display should come alive, and show the temperature at the sensor. (Note: the thermistor wiring can be extended as required.)

The default values programmed into the instrument mean that straight out of the box it will work fine as a digital thermometer.

Controlling an output

The module is fitted with a 5A relay. This means it can be connected directly to buzzers and warning lights. Through an additional heavy-duty automotive relay, it can also control pumps or fans.

To get a feel for how the control system works, it's a good idea to play with it before installation. Let's take a look at how it can be set up.

Pressing the Set button briefly changes the display to show the set-point temperature. This setting can be altered by pressing the up and down keys. When done, press the Set key again or simply wait a few seconds and the display reverts to the current temperature.

Pressing the Set button for 3 seconds brings up a second menu. Different parameters can be selected by pressing the up/down keys. To change the selected parameter, press the Set key a second time, then make the adjustments with the up/down keys. Whatever setting is selected is retained in memory, even if power is lost.

The available parameters are:

HC – this menu configures the module to either turn on its relay when the temperature *exceeds* the set-point ('C' mode), or turns on the relay when the temperature *falls below* the set-point ('H') mode.

d – this sets the difference in temperature between switch-on and switch off. (This is sometimes called the *hysteresis*.) By using the up/down keys, you can set this anywhere from 1°C to 15°C. This is a very powerful control that can make a huge difference to how the system functions.

L5 – this is the minimum temperature the set-point can be configured. Normally, this would not need to be altered from its -50° default.

H5 – this is the maximum temperature the set-point can be configured. Normally, this would not need to be altered from its 110° default.

CA – this function allows you to correct the temperature display by adding or subtracting 1 degree units from the displayed reading.

P7 – this function is used when in C mode you don't want the output cycling on and off at short intervals. The setting can be anything from 0-10 minutes. For example, if it is set to 1 minute, after the relay has activated once, it will not activate again until a minute has passed – even if the temperature set-point has been tripped. In most uses you would set this to zero.

Over-temperature alarm

So let's take a look at how you'd set the module up to turn on a warning buzzer if your turbo'd intake air temperature exceeds 60°C.

Press Set for 3 seconds then select the following:
HC – set to C
d – set to a small value like 2°
L5, H5 and CA left at factory defaults
P7 set to zero
Press Set briefly then use the up/down keys to select 58° as the set-point.

Huh? 58 degrees? We want the alarm to trip at 60 degrees, don't we? This is the only trick in setting up the unit. In 'C' mode the relay will trip when the temperature actually reaches the set-point *plus the hysteresis* – ie 58° plus the 2° we set in 'd' mode.

Whether you use it just as a temperature controller or as a controller/display, this is a very useful and low-cost device to add to your car.

9. UPGRADING DASHBOARDS, INSTRUMENTS AND WARNINGS

FUELSMART – A REVOLUTIONARY FUEL-SAVING DEVICE THAT WORKS

If you have a manual transmission car and the engine uses a MAP sensor, FuelSmart can improve your real-world fuel economy by up to 15 per cent. And there's no shonky science involved! But before we explore the module, a word about how good and bad fuel economy occurs.

Engine fuel economy

It's well known that one of the greatest limiting factors in achieving good fuel economy is the driver. Yes, in real-world use, the person who is steering and operating the brakes and throttle has a dramatic influence on fuel economy. And I'm not just talking about those people who drive everywhere with their foot mashed to the floor. Even gentle driving can give poorer fuel economy than expected. So how can that be? In short, because an engine is much less efficient when it is operating with the throttle only just cracked open.

Looking at the box below, which covers brake-specific fuel consumption, the key point to note is that lowest specific fuel consumption occurs at *high throttle angles and low rpm*. To put this another way, driving around with the throttle only just a little open at high engine revs gives much worse fuel consumption than using lots of throttle and low revs.

But there's more to it than that. When the throttle is floored, nearly all engines add more fuel than usual. That is, instead of adding a 'normal' amount of fuel to the air, at full throttle they squirt in some extra. This allows the engine to develop greater power. Therefore, despite what was said above, running at full throttle (eg when

BRAKE-SPECIFIC FUEL CONSUMPTION (BSFC)

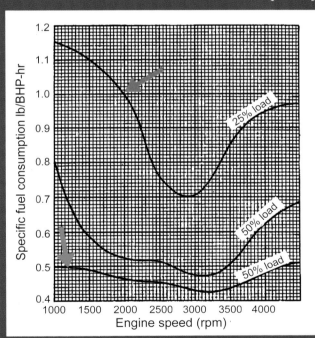

Look at this diagram. What it shows on the vertical axis is the amount of fuel that is used to produce each horsepower for an hour. Obviously, the less fuel used per horsepower/hour, the better the fuel economy of the engine. Across the bottom axis is engine speed in rpm. Look at the line indicated by the green arrow. This shows the amount of fuel used per horsepower/hour at different rpm – measured at full throttle. Now look at the line indicated by the red arrow. The fuel consumption per horsepower/hour is much higher – typically over double. Why? Because the engine is now working at only 25 per cent throttle.

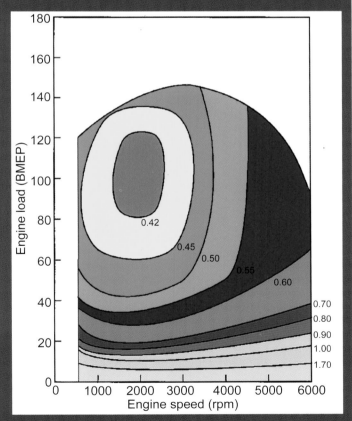

There's another way of showing this characteristic of throttled engines. This diagram shows engine load up the vertical axis, and engine rpm on the horizontal. The red area on the diagram is the 'island' of best fuel economy. As can be seen, the best fuel economy (measured in terms of fuel used per horsepower/hour) occurs at relatively high loads and low rpm.

going up through the gears) will usually give worse fuel consumption. *So a lot of throttle is good – but not full throttle.*

Next, nearly all engines have another important characteristic – they turn off the fuel supply completely (ie switch off the injectors) when the throttle is lifted on the over-run. And of course, if the car is moving and no fuel at all is being injected, then the fuel economy is just great! *So wherever possible, the accelerator pedal should be fully lifted.*

Finally, it's not possible to drive everywhere with the engine operating in its most fuel-efficient area of operation – the trick is to *keep the engine operating fuel-efficiently as much as possible.* In other words, real-world driving is about 'swings and roundabouts' – eg about using more power, produced more fuel-efficiently, to climb a hill so that the car can then roll down the other side, throttle closed and using no fuel.

OK, so let's take a look at what's been covered.
1. Most engines develop their best fuel economy when given lots of throttle (but not full throttle), and engine revs are low.
2. When decelerating, the throttle should be fully lifted.
3. The worst driving style for fuel economy is a slightly open throttle and high revs.

Now hyper-milers (those who chase absolute best fuel economy) have known about these ideas for ages. They train themselves to drive economically by especially avoiding condition #3 above – driving with a slightly open throttle and high revs.

But what happens when you're doing say 60km/h (37mph) in free-flowing traffic? How can you then be in that red island of best fuel efficiency shown in the diagram repeated here? The way hyper-milers do it is to 'pulse and glide' – accelerate in the island of best fuel efficiency by using lots of throttle in as tall a gear as possible (#1 above), and then back right off (#2 above). They then repeat the process. From a fuel efficiency perspective, you can see exactly how it works.

FuelSmart module

So, all well and good. But how does this technical information translate into driving? The trouble is – very often it doesn't. By the time you're observing speed limit signs, being distracted by the screaming kids and watching out for other drivers, the last thing you can concentrate on is changing gears at the right moment – or using the right throttle settings – to gain best fuel economy.

And that's where FuelSmart comes in. FuelSmart is an adjustable electronic module that watches two engine management signals. The first signal it monitors is throttle position – as detected by the standard throttle position sensor (TPS). The module is calibrated so that it can detect

The FuelSmart electronic module.

when the throttle is open, even if it is open only a small amount. The second signal it watches is manifold pressure, as measured by the MAP sensor – and this one needs a little more explanation.

The engine is a pump, drawing air in, mixing it with fuel, combusting the mix and then pushing the exhaust gases out. The throttle blade limits how much air can get into the 'pump' – the engine. If the engine is running with the throttle nearly closed, a low pressure (a 'vacuum') will be developed in the intake system. If the throttle is fully open, the pressure in the intake system will be much the same as the pressure of the atmosphere – there won't be any vacuum. The degree of vacuum in the intake depends on two things – how far the throttle is open, and how fast the engine is spinning. (The faster the engine is turning, the more air it is trying to draw in.) Maximum vacuum occurs on high rpm over-run, when the throttle is shut and the engine is spinning hard. As I said above, minimum vacuum occurs when the throttle is fully open.

Now, here's the exciting bit: *If the throttle is open, but engine vacuum is high, the engine is very likely to be operating with poor fuel efficiency!* That's in fact the reason that fuel efficiency is poor – the engine is working hard to drag air past the nearly closed throttle (in technical terms, 'pumping losses' are high).

What we do with the FuelSmart is light an LED on the dash when the engine vacuum is above a pre-set level *and* the throttle is open. If the throttle is closed but vacuum is high (eg on engine over-run) the LED stays off, and if the throttle is open but the engine load is high (ie vacuum is low) then the LED still stays off.

So how do you use this dashboard indicator? It's easy – if the LED is on, *you should immediately assess your driving style.*

For example, when changing up through the gears, if the

9. UPGRADING DASHBOARDS, INSTRUMENTS AND WARNINGS

LED stays on, you should use more throttle between each gear change. If you are slowing for a set of traffic lights and the LED stays on, you should lift right off – not trail along with the throttle just open. In stop-start urban traffic and hilly areas (among the most difficult of conditions to get good fuel economy), you can use the LED indicator to guide you in variations of 'pulse and glide.' (These conditions lend themselves well to pulse and glide because the throttle is naturally on/off a lot.)

In constant load cruise conditions, the LED will stay on – but that's because in these conditions the engine is actually not working at its fuel-efficient optimum. If you are already in top gear, there's nothing much more you can do. But if you're not in top gear, the LED is a reminder to change up and then use more throttle to maintain speed.

The module

FuelSmart is based on an electronic module produced by eLabtronics. It uses a custom programmed microprocessor, a dash-mounted LED, two set-up pots, one remote-mounted fine-tuning pot, and a pushbutton. It can be bought directly from the company.

FuelSmart needs four connections to the car's engine management system – ground, 5 volt supply *(note: NOT 12V!)*, MAP sensor and Throttle Position Sensor (TPS). The MAP and TPS sensors need to have voltage outputs in the range 0-5V. The TPS must have a voltage output that rises with increasing throttle opening (every TPS I've ever seen does this), and the MAP sensor output must rise with increasing absolute pressure (that is, decrease in voltage output with increasing 'vacuum' – again, they pretty well all do this).

All the ECU connections should preferably be made at the ECU, but they can also be made at the actual TPS and MAP sensors. These sensors nearly always have a 5V supply, and so the 5V connection that powers the FuelSmart can also be made at one or other of the sensors. If no 5V supply is available on any of the ECU sensors (and that would be a very rare car indeed), an external regulated 5V supply will need to be used.

Setup

After it is wired into place, FuelSmart needs two settings calibrated. These calibrations are easily done, and the settings can be changed as often as you like.

The board should by now be wired to ground, 5V supply, TPS signal and MAP signal. The LED should be located where you can

VACUUM GAUGE?

Older readers might be saying to themselves: why not just use a vacuum gauge? In the past, such gauges were often fitted to cars.

As its name suggests, a vacuum gauge constantly displays manifold vacuum. That way, a driver can see what levels of manifold vacuum exist in all driving conditions. Since they also know the position of their right foot, shouldn't this be sufficient to guide the driver?

There are four problems with taking that approach.

Most manifold vacuum gauges have markings on them relating to fuel economy, however these markings are usually incorrect. For example, the lower the manifold vacuum (ie the higher the load), generally the worse the gauge suggests that fuel economy will be. During acceleration and when selecting the correct gear, that's simply not the case.

Second, it is always easier to spot an indicator LED showing that something is wrong, rather than constantly watching the sweep of the gauge needle. In fact, the LED can be sensed without even directly looking at it.

Third, it's often quite hard to tell whether your foot is actually off the throttle or just pushing it down a fraction. In the extensive testing that I did of the FuelSmart, I was often caught using trailing throttle.

Finally, the FuelSmart is cheaper than most manifold vacuum gauges and the single LED is much easier than a gauge to locate on the dashboard.

easily see it while driving, and the remote mount pot should be within reach. In addition, you should have available a small screwdriver to adjust the onboard pots and an assistant to do this while the car is being driven.

Note that the underside of the bare board should not be allowed to come into contact with any metal surfaces – it's fine resting on the lap of a passenger or on a seat.

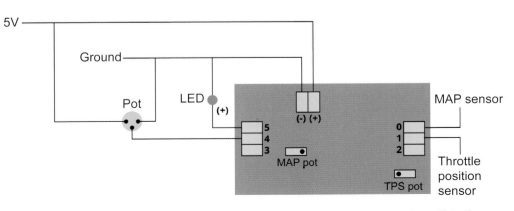

FuelSmart needs power (5V), ground and MAP sensor and TPS sensor connections. Note the three adjustment pots – two on the module and one (a fine-tuning tool) on a remote lead.

WORKSHOP PRO: MODIFYING THE ELECTRONICS OF MODERN CLASSIC CARS

Setting the TPS
The TPS input is set so that it trips as soon as the throttle is opened. Here's how to do this:
1. With the car stationary and idling, press the button on the FuelSmart board. Keep the button pressed until the dashboard LED flashes an acknowledgement. Release the button. The FuelSmart is now in 'TPS Setting' mode.
2. Press the throttle down a tiny amount, until engine revs just start to rise.
3. Turn the TPS pot (use a small screwdriver – it may take lots of turns) until the dash LED just comes on. Adjust the pot until when you release the throttle, the LED goes off, and when you just press the accelerator pedal (even a tiny amount), the LED goes on.
4. Turn off the ignition for at least 2 seconds – this step switches off the 'TPS Setting' mode and reverts the module to standard mode. Note that the module needs to lose power to reset it to standard mode – some cars keep their ECUs energised for a short time after key-off.

Setting the MAP
1. Set the remote mount pot in the middle of its travel.
2. Drive the car in second gear at around 2000rpm, and have the assistant adjust the MAP pot until the dashboard LED stays on at anything less than about half full throttle. (Again, lots of turns of the pot may need to be made.)
3. Test this setting by accelerating through the gears. At low revs, the dashboard LED should stay on until about half throttle is reached.

Testing
The next step is to drive the car in a variety of situations – urban, open road, heavy traffic, hilly terrain.

If the dash LED is on and you're not already in top gear, change up a gear. Does this then cause the car to stumble and lug – the gear change is obviously too early? If so, the MAP set-point of the FuelSmart needs to be reset at higher rpm (eg with the car at half throttle, using 2500rpm rather than 2000rpm).

Alternatively, does the LED indicate up-changes that are too late – when you change up a gear, the engine easily pulls the car along with only light throttle? In that case, the MAP set-point needs to redone using lower rpm.

The optimal setting is one that requires at least half throttle travel to be used in each gear as you accelerate up the gears – otherwise the LED comes on.

Fine-tuning
The external pot allows easy fine-tuning of the main MAP pot setting. The external pot is provided for three reasons:

1. It lets you make *easy on-the-move adjustment* of the MAP trigger point.
2. To *cater for different driving styles*. For example, sometimes you might want to have the pot set so that you can drive pretty normally – or at other times, you might want it set to allow optimal pulse-and-glide driving. The pot also allows easy adjustment to suit different drivers.
3. If you drive up or down a large mountain, the *change in atmospheric pressure* is sufficient that the optimal MAP setting point will actually alter. The remote pot allows you to easily cater for this change.

Mounting
When calibration is finished, the FuelSmart module can be placed inside a plastic box, or simply wrapped in tape or heat-shrink, and placed up under the dash. Even if the sensor and power connections are made in the engine bay, the FuelSmart unit should be mounted inside the cabin of the car.

Many modern manual transmission cars have 'up' and 'down' dashboard indicators, suggesting when it is best to change gears to achieve good fuel economy. This is a diesel Mercedes. The FuelSmart indicator provides similar guidance.

Driving with FuelSmart
Aim for a driving style that keeps the FuelSmart LED off as much as possible – especially when accelerating up through the gears or slowing. For example, when I first fitted FuelSmart I lived in a semi-urban area that was fairly hilly. The speed limit was 60km/h (37mph). Driving on these roads, it wasn't difficult to be either accelerating sufficiently hard that the FuelSmart LED was off – or alternatively, have the accelerator pedal fully lifted, and so the LED was again off.

9. UPGRADING DASHBOARDS, INSTRUMENTS AND WARNINGS

In addition to the FuelSmart LED, when driving for economy you should also use the tachometer. Keep the revs as low as possible – in other words, change-up early.

Some of you may be saying – so why not just use the tachometer by itself? In driving reality, the FuelSmart LED lights up in lots of situations where the tacho is of no use – eg slowing for a traffic light where you've left your foot a little on the throttle (so wasting fuel); maintaining station in slow-moving urban traffic, where the LED shows that you're a gear too low for best fuel economy; and accelerating away from a stop, where the LED shows when insufficient throttle is being used.

And, as with any car instrument, the FuelSmart LED needs to be used intelligently. Inevitably there will be situations where you ignore the action of the FuelSmart LED – when it lights, it is an indication that *you should assess your driving style* at that particular moment, not always change it!

To be completely honest, if you set FuelSmart up to provide optimal fuel economy, and then drive absolutely according to the LED indicator, it can be exhausting! That's because it's hard work keeping the engine in its best efficiency area as much of the time as possible. (And, if it wasn't hard work, there wouldn't be the incredible fuel economy gain that is possible.) But even with FuelSmart set up as a gentler indicator of inefficient driving (ie you've got to be doing things really wrong before you get an LED indication), you will still save fuel. And, best of all, in just the most difficult conditions in which to gain good fuel economy – heavy urban traffic and hilly terrain – you will get the greatest fuel economy gains.

WARNINGS FROM ANY VOLTAGE SENSOR

Any sensor in the car that outputs a voltage can easily have a warning alarm added to it. That alarm can be a buzzer, siren, light or even voice warning. This means you can trigger alarms off any conventional instruments that use a voltage reading (eg coolant temperature or fuel level) or any engine management system sensor (throttle position, MAP, airflow meter, intake air temperature, etc). To achieve the alarm or warning, you will need to use a voltage switch that is able to be accurately configured for:
1. Setpoint and hysteresis (the difference between on and off voltages)
2. Tripping on a rising voltage, or tripping on a falling voltage

It also helps if the module has a relay output as it can then be more easily use to switch other devices (like a warning light or buzzer). The good news is that such modules are now readily available, and at very low prices. Let's look in detail at one such module.

Note that while there are a lot of digital voltage

This low-cost voltage switch is able to be configured to sound a buzzer or turn on a warning light when monitoring any voltage-outputting sensor.

switches available, they vary substantially in features. This particular one, identified by YYV-1 on the board, may not be the first that you come across when searching, so ensure the module you buy looks exactly like the one shown here. Search eBay under the search term "Charging Discharge Voltage Monitor Test Relay Switch Control Module DC 12V G9Q3" or similar.

Connections

The module's connections are straightforward. Power and ground (12V in the module I used, but it's also available in 5V and 24V versions), and signal input and signal ground. (In this case, the signal is the voltage that we're monitoring.) The relay is a SPST design so has common, normally open and normally closed connections. All connections are via screw terminals.

Note that the signal input impedance is about 20kΩ, so that any sensor (but a narrow-band oxygen sensor) can be monitored without upsetting the original system.

The module has seven mode configurations.

Mode 1

When the input voltage is below V1, the relay is activated. When the input voltage is higher than V2, the relay is deactivated. This mode is typically used when you wish to monitor a falling voltage and want the relay to trip when the voltage drops below the set-point (V1). The difference between V1 and V2 sets the hysteresis.

Mode 2

When the input voltage is higher than V2, the relay is activated. When the input voltage is lower than V1, the relay is deactivated. This mode is typically used when you are monitoring a rising voltage and want the relay to trip when the voltage rises above the set-point (V2). The difference between V2 and V1 sets the hysteresis.

Mode 3
In this mode the relay is activated only when the monitored voltage is outside the range set by V1 and V2. For example, if V1 is set to 4V and V2 is set to 5V, the relay is off when the measured voltage is between 4 and 5V, and is on when the voltage is less than 4V or greater than 5V.

Mode 4
This mode reverses the logic of Mode 3. That is, the relay is activated only when the monitored voltage is between V1 and V2.

Mode 5
This mode activates the relay when the monitored voltage is above V1 and switches off the relay when the monitored voltage is below V1. The difference between this mode and Mode 1 is that the hysteresis is not able to be adjusted. In use, the hysteresis in this mode appears to be 0.05V. (V2 therefore does nothing, even though it still appears as an item to be set.)

Mode 6
When the monitored voltage exceeds V1, the relay is activated. The relay stays activated until power is lost to the module. This mode is particularly good when you need to monitor a signal that's might go too high only for a very short period (eg turbo over-boost). A normally closed 'reset' button in the module's power supply could be easily added to switch off the alarm. Note that the LED display keeps showing the actual 'live' value, even if the relay has been tripped.

Mode 7
The relay is activated only when the monitored voltage is precisely at the set-point (V1). Even an indicated 0.05V variation away from V1 switches off the relay.

Mode selection
There are four onboard press-buttons, marked as K1-K4. K2 is 'up' and K3 is 'down.' The mode is selected by holding down K1 for more than 1 second. The LED display will then flash, eg P-2, indicating that Mode 2 is selected. To change the mode, press either the up or down buttons until the required mode is selected. Pressing K1 again will then show the V1 that is selected; this can be changed with the up and down buttons. Pressing K1 again will select V2, that can then be set. A final push of K1 returns the display to normal. And K4? This button turns the LED display on and off.

In use
The best way of setting-up the module is to first use a multimeter to monitor the sensor that you wish to connect to. Work out if the voltage rises or falls with an increase in the parameter (eg a MAP sensor will probably have a rising voltage with increasing manifold pressure, while a temperature sensor will probably fall in voltage with increasing temperature) and configure the module switching mode accordingly. Then decide at what voltage you want the alarm to trip. In some cases, you might want to create this condition artificially, eg by pulling out the sensor and heating it, or applying pressure to it until the warning level is reached, and then measuring the voltage at which that occurs. This becomes your module setpoint. Start with a small hysteresis, and if the warning chatters on and off, increase it. With the low price of the module and its ability to monitor existing sensors, this is a very cost-effective and straightforward way of adding warning lights or buzzers.

PROGRAMMABLE COMPACT GAUGE
The SGD 21-B PanelPilot voltmeter from UK company Lascar has got lots of in-car applications. The display uses a dot matrix e-paper design, allowing it to be read in any lighting condition – from full sunlight to very dim indeed. (However, there is no backlight, so it cannot be viewed in complete darkness.) Also, the display is visible from a wide

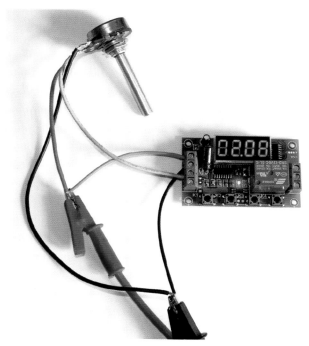

Using a pot to vary the monitored input voltage during testing. The LED display shows the monitored voltage, making setup easy.

9. UPGRADING DASHBOARDS, INSTRUMENTS AND WARNINGS

The PanelPilot voltmeter uses a low-current dot matrix e-paper design that has excellent clarity. The display is fully configurable via free downloadable software. Here it is displaying ride height – but it could be configured (and labelled) to show any voltage-based sensor's output. Note that the output can be scaled to be in engineering units.

range of viewing angles – far wider than a conventional LCD. Use of the e-paper design allows the meter to be powered by a low current supply (just 5mA), and that supply that can be anywhere in the 4-9V range. It can therefore be powered from a 5V supply picked up from the ECU.

The major benefit of the meter is that it is software programmable. By using the accessory cable, the meter can be connected to a PC via the USB port. Using the free Panel Pilot B software, downloadable from the Lascar site, the meter is then able to be configured in a variety of ways. I'll look at these in more detail in a moment, but the configurability includes not only the type of display (eg a bar graph and a numerical display, or a numerical display that includes maximum and minimum, etc) but also two alarms, scaling, the number of displayed decimal places, and the function of a digital input pin.

Specifications

The meter uses a nominally 2.1in display with an actual display area of 48.3 x 23.7mm (1.9 x 0.9in). The meter's measuring voltage range is 0-1.25V, and with the placement of two voltage dividing resistors on the PCB, voltages of up to 30V DC can be read. Suitable resistor placement will also allow the meter to directly read up to 200mA. The display updates twice a second and accuracy is said by the maker to be typically 0.05 per cent with a maximum of 0.1 per cent.

Three header pin connectors are located on the back – these are designated PL1, PL2 and PL3. PL1 is for the USB cable's connector that is used when configuring the meter. This cable must be bought separately, and note that the connector needs to be inserted with the projections facing downwards – nothing stops the connector being inserted upside-down. PL2 is for 'simple' meter use – the pins comprise power supply and ground for the meter, and input signal and ground for the measured voltage. PL3 is for more complex meter set-up. In addition to power supply and signal connections, this connector also includes pins for digital input, an alarm output, and a regulated 3.3V output that can be used to power other devices (max of 145mA). The alarm pin is an open collector output (ie connects to ground when activated) and can sink only 10mA. (Therefore, you will need to use an electronic – not mechanical – relay to use the alarm output to trigger a buzzer or light.)

The function of the digital pin depends on the meter configuration that has been set. For example, if the meter is configured to sound an alarm, this pin can be configured as an alarm reset button. In other configurations it can be used to set tare (ie set a zero point), reset the maximum and minimum that is displayed, hold the display, and so on.

Six different pre-formatted displays are available. Each one can be custom configured for label, units and scaling.

Displays

The main advantage of the meter is the versatility in displays that can be shown. Six different types of display are included in the software, including those that are just numerically based, through to those with graphs and numbers. Each of the displays can also be reversed (ie black on white becomes white on black) but I found the black on white displays easier to read. In addition to the overall type of display, the label and units can both be set. For example, you can select 'Temperature' as the label and then select °C or °F for the units. Furthermore, you can specify your own labels and units, just by highlighting the boxes and typing.

You can also optionally have a splash screen – a screen of text (or an image) that appears for a pre-set time before the display switches to its measuring mode. The display has a 250 x 122 pixel resolution and an image of this size can be uploaded to the meter.

The scaling of the meter is able to be easily set – an example will show how it works. Let's imagine that you are monitoring a throttle position sensor that has a maximum output of 5V. Firstly, set in the software the maximum voltage the meter will see to 5V. The required voltage scaling resistors are then displayed – in this case, 1MΩ and 330KΩ for Ra and Rb respectively. (These resistors then need to be soldered to the meter PCB in the correct places to give the meter the ability to measure up to 5V.)

Let's say the MAP sensor outputs 1.35V at 0 per cent throttle and 4.2V at 100 per cent throttle. Set these values in the digital scaling output – the meter will then read in percentage of throttle. (Note that the sensor output must be linear for this to work correctly.)

We also want an alarm to sound when the throttle position is above 80 per cent, and we want to display the maximum and minimum values. The digital input pin we configure to have a 'reset max and min' function. Set these factors in the software and then upload it – the meter is now configured. When connected to the PC, the meter displays the uploaded configuration (no other connections needed), making it easy to check you've got what you want.

This meter is a very powerful tool to use, especially because it can be scaled to show the measured parameter in engineering units.

LED BATTERY MONITOR

At first appearance this device looks just like a 10mm LED mounted in a bezel. But then when you look closer, you'll see a tiny pushbutton – and if you pull the assembly from the bezel, you'll also see a programmable chip and a few other components. In fact, what we have here is a 6, 12 and 24V battery monitoring LED that is user-programmable to run one of six different in-built maps. The single LED can show green, red, yellow and white (off). The LED can also flash at different rates. The voltage is monitored as a rolling average over 2 seconds, is claimed to be accurate to 1 per cent, and will operate over the range of 3.8-30V. Wiring is as simple as you can get – red to positive and black to negative. The device is made by Gammatronix in the UK and is available via eBay – search for '6v, 12v, 24v programmable LED battery level voltage monitor meter indicator.'

At a glance, the battery monitor just looks like a 10mm white LED. However, its functionality is much greater than it first appears.

Pulling the LED from the bezel reveals an attached circuit board and pushbutton. Programming of the LED's functions is achieved by operating the pushbutton.

The maps

So what are the different in-built battery monitoring maps that are available? There is one for almost every use you can think of.

Here are the different maps:

Map 1: Battery voltage monitor
This is the factory default voltage indicator mode. Low distraction, minimum of colour changes in normal operation, suitable for vehicle use, such as motorcycles,

The rear of the meter. At its simplest, just four connections are needed – 5V power and ground for the meter, and signal input and signal ground. Also able to be seen here is the connector for the USB cable and the pads on which the input voltage divider resistors are added, as required.

9. UPGRADING DASHBOARDS, INSTRUMENTS AND WARNINGS

cars, boats, campers, etc. By flashing and using different colours, it shows eight different voltage ranges from 10.5 to 15+ volts.

Map 2: Vehicle charge indicator
This map illuminates the LED green when under charging conditions, ie when the vehicle alternator is working. Yellow and red will show if the battery is discharging. This mode monitors seven different voltage ranges from 10.5 to 15+ volts.

Map 3: Vehicle monitor, includes fake alarm
This is great for motorcycles, and vehicles stored long-term. When riding/driving and charging, the LED is steady green. 30 seconds after charging stops (ie the vehicle is parked), the unit will enter low current mode to show battery status while the vehicle is in storage. The LED will blink green, yellow or red to show stored state battery condition. An added benefit is that LED blinking looks like a vehicle alarm. This mode has only a very small current draw (0.5mA) from battery.

Map 4: Charge indicator (stealth mode)
This is similar to mode (2) but the LED is not illuminated under normal charging conditions. That is, the LED is blank in normal operation. Yellow and red illuminations signal charging faults or discharging battery.

Map 5: High res 10 step voltage monitor
This mode is a high-resolution mode where maximum resolution is important and colour changes and flashing are not distracting. This mode monitors 10 different voltage ranges from 10.5 to 15+ volts.

Map 6: Minimal monitor
This mode uses a simple low current (less than 0.5mA), three colour battery status monitoring. A short flash every 2 seconds indicates current state, with 5 different voltage ranges from 10.5 to 15+ volts.

Using it
As delivered, the LED is set to Mode 1. To move to the next mode (ie in this case Mode 2) you do the following:
1. Power-up the LED
2. Hold in the pushbutton
3. Wait until the LED flashes green
4. Release the button

To confirm what mode you have now set, turn off power and then reconnect it.

In this case you would expect to see three green flashes (indicates the LED is still in its 12V battery monitoring setting), a pause, and then green flashes (indicates Map 2).

VOLTMETER
Digital LED voltmeters suitable for mounting in cars are now very cheap. So if you'd prefer to have more detailed information on battery voltage than the single LED described on these pages, it's easy to achieve this.

This voltmeter is completely sealed and so is waterproof. It's ideal for mounting under the bonnet near the battery, where it can be triggered by a manual pushbutton.

You can mount the voltmeter on the dash and run it from an ignition-switched 12V source, or mount it in the engine bay or boot of a car, and operate it by a momentary pushbutton. The dash display will show you if the alternator is charging, as well as the voltage with the engine off (but the ignition on). The display mounted elsewhere and operated by a pushbutton is the approach I take: it works well, especially if the car isn't driven every day. If the car has been parked for a while, and you're concerned the battery might be getting low, just access the display and press the button.

A voltmeter mounted near to a rear-mounted battery. The meter is triggered by the pushbutton being pressed. A 100A circuit breaker is located close by – putting everything into a box gives a neat job.

The maps settings (the *second lot of flashes*) are as follows:
Map 1 – red flashes
Map 2 – green flashes

WORKSHOP PRO MODIFYING THE ELECTRONICS OF MODERN CLASSIC CARS

Map 3 – yellow flashes
Map 4 – red and then green flashes
Map 5 – yellow and then red flashes
Map 6 – yellow and then red flashes

Once you've sorted this aspect out, changing the voltage to 24V or 6V monitoring is straightforward. The procedure is:
1. Power-up the LED
2. Hold in the pushbutton
3. Wait until the LED flashes green, then red
4. Release the button

The voltage mode will switch to the next, so 12V to 24V to 6V to 12V – and so on. Then examine the *initial* flashes after switch-on on the basis of the following:

6V – red flashes
12V – green flashes
24V – yellow flashes

If you want the simplicity of mounting a single LED that can effectively monitor battery voltage, this little gadget is hard to beat.

VOICE ALARM MODULE

Here is a small and cheap module that is easily configured to give a voice warning when triggered. When matched with a speaker, the sound quality is high, and you can record whatever message you wish. To find the board online, search for 'ISD1820 module.' Note that some suppliers include a speaker, but the one I used did not come with a speaker and so I added my own.

So what are the specs on this little module? Measuring only 43 x 38mm (1.7 x 1.5in), the module is equipped with the ISD1820 chip, three pushbuttons, power connections via header pins (note: 3-5V, *not* 12V), a connector for an external speaker, and an onboard microphone. The message can be up to 10 seconds long.

At its simplest, connect power and a speaker (nominally 8Ω). Press the 'REC' button and speak into the microphone. Best quality occurs if your mouth is close to the microphone. Release the 'REC' button and press the 'PLAYE' button. The message you just recorded will then be played through the speaker. If you press the 'PLAYL' button, the message will be played for only as long as you have your finger on the button – if you release the button, the message will cease playing at that point. Note that in these modes, the message will be played only once. Alternatively to pressing the buttons, these functions can be triggered by connecting the 'REC,' 'P-E' or 'P-L' header pins to the positive supply as required.

The message is retained even if power is removed – an important aspect considering the next function. Moving the provided link so that the 'P-E' pins are bridged changes the way in which the module operates. In this configuration, the message is played on a continuous loop whenever power is connected to the board. This is a very useful function because it means that the board is drawing no power at all until the message is required – this is the mode that is perfect for that fault condition warning function. Just use a small relay (eg with its coil wired parallel with a warning light) to supply 5V power to the module when the fault conditioning occurs. The module will then 'speak' your pre-recorded warning. The level of audio output provided straight from the module is adequate for normal in-cabin warning functions – if you need it louder, add a small amplifier.

OBD GAUGES

If your car is equipped with OBD functionality, gauges are available that plug into the OBD port and can display a wide variety of parameters. These include car speed, engine rpm, coolant temperature, intake air temperature, long- and short-term fuel trims, and many others.

These gauges are available as colour or B&W dashboard displays, including head-up displays that show the readings reflected on the inside of the windscreen.

While not as flashy as some of the readers that are available, ScanGauge has a good reputation as a general-

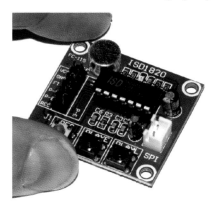

This tiny module can record up to 10 seconds of audio and then play it back on demand. It's perfect for warning messages, and can be configured so that it plays the message on a continuous loop when power is applied. The board needs 5V, so supply it from an ECU 5V regulated feed or add a 5V supply.

The ScanGauge plugs straight into the OBD port and can display parameters like intake air temperature, throttle position, manifold pressure and engine coolant temperature.

9. UPGRADING DASHBOARDS, INSTRUMENTS AND WARNINGS

More complex OBD gauges are available as head-up displays, where the module lies on the dash and the information is reflected by a sticker attached to the inside of the windscreen. However, these displays can be quite ineffective in bright sunshine, so try before you buy. (Courtesy Yalla Savers)

purpose OBD display (and it can also clear fault codes). I have used a ScanGauge, and it worked just as advertised. ScanGauge is also small enough to be positioned easily on the dash: for example, on top of the steering column.

DIGITAL PROGRAMMABLE INSTRUMENT PANELS

A major upgrade in gauges and instruments can be achieved if you fit a new LCD digital dash panel. These are typically made for motorsport, kit-cars and motorcycles. The cost varies from quite low to very high, so carefully consider your requirements before spending any money. Aspects to consider are:

- **Colour or B&W:** older programmable panels, and cheaper current panels, are B&W, not colour. Colour aids clarity and allows you to emphasise key aspects, such as displaying warnings in red.
- **Road car or race car:** road car panels incorporate an odometer and warning lights for indicators, hand-brake (e-brake), brake failure and high beam. Race car dashes typically don't have any of these, though they may have LEDs that can be programmed to perform these functions. Some dashboards are illegal for road cars – check local legal aspects carefully before buying.
- **Inputs and outputs:** in many dashes, the number of parameters that can be displayed is fixed by the number of digital and analog inputs the panel has. For example, a dash may have six analog inputs and one digital input, so limiting it to displaying (say) speed, coolant temperature, intake air temperature, fuel level, battery voltage, oil pressure and oil temperature. On the other hand, a dash that can accept CAN or RS232 data has the potential to display a lot more parameters – that is, if you have this information available from another source (eg a programmable engine management system). If you are using the standard engine management and your car is OBD compliant, some digital dashes can display OBD data (although it may be slow to update).
- **Programmability:** some dashes are sufficiently versatile that they can operate with almost any type of sensor, and display that sensor's output in engineering units (eg oil pressure in kPa or psi). Other, cheaper, dashes will work only with the provided sensors. Some dashboards are programmable in appearance, location and scaling of gauges, colours and fonts. Other dashboards are fixed in display appearance, but the function of each numerical display can be altered (eg the top-right numbers can be set to display road speed or oil pressure, etc). Some dashes have multiple screen displays that the driver can switch through – eg one display set up for normal driving; another for starting, and another for track use. Some dashes have programmable outputs to run an external shift-light or alarm buzzer. Finally, user-programmable alarms are an important feature to have available, and are even better if their logic can be programmed (eg alarm sounds when oil pressure is low *and* revs are above idle).
- **Logging:** dashes designed with more of a race focus may have internal logging, GPS and g-sensor functions. In a road car, these can also be useful.

This low cost digital dash is designed to work only with the provided interface box and sensors. (Courtesy Sinco Tech)

225

WORKSHOP PRO: MODIFYING THE ELECTRONICS OF MODERN CLASSIC CARS

- **Backlight:** some older B&W LCD dashes are not backlit. It may be possible to organise lights forward of the dash to illuminate the panel at night, however a backlit dash is much nicer to use.

This Holley dash is plug-in compatible with Holley EFI systems via a CAN bus connection. It is fully customisable in its display, can log data internally or to an external USB stick, and has user-programmable alarms. (Courtesy Holley)

The Aim MXS 1.2 Strada dash is available with road car indicators (fuel, oil, water, high-beam, etc). The dash display and top warning LEDs are fully configurable. The dash has two CAN inputs, two digital and four analog inputs, and one digital output. Through expansion boxes, it can also have additional inputs for GPS and thermocouples, and it can control a wideband air/fuel ratio sensor. Dimensions are 166 x 97mm (6.5 x 3.8in) and the display resolution is 800 x 480 pixels. (Courtesy Aim)

This MoTeC dash is large (341 x 148mm, 13.4 x 5.8in) and is fully programmable. It has two CAN bus inputs and two RS232 inputs. Only two digital inputs and three speed inputs are provided as standard – most purchasers would use the communication buses to send information to the dash. (Courtesy MoTeC)

- **Physical size and shape:** if you are intending to fit the new panel in the location where the original instruments sat, measure carefully. Digital dashes are available in a wide variety of sizes and shapes. As with the original instruments, the panel needs to be shaded so that it less affected by sunlight and does not reflect in the windscreen at night.

I use a MoTeC dash in my Honda Insight, and I'd like to cover its installation now. The MoTeC CDL3 is one of the older range of MoTeC dashes – these are able to be identified by their B&W LCD and arched upper shape. These dashes are less configurable than the current colour square and rectangular MoTeC dashes but from my perspective, have a major advantage – the shape of the dash means it can be fitted within the standard Honda dash binnacle.

The backlit CDL3 has a 70-segment bar graph (normally used to show engine revs), central large number

Testing the CDL3 dash on the bench. Simulating an analog input (eg by using a pot) and then playing with the programming is a good way of becoming familiar with how the dash works.

The MoTeC SLM is shown mounted directly below the main panel. The LEDs in the SLM are multicoloured and can be individually addressed, allowing a wide variety of warnings to be shown. Here the 'all red' is a major warning, supported by the text that is brought up on the dash panel.

9. UPGRADING DASHBOARDS, INSTRUMENTS AND WARNINGS

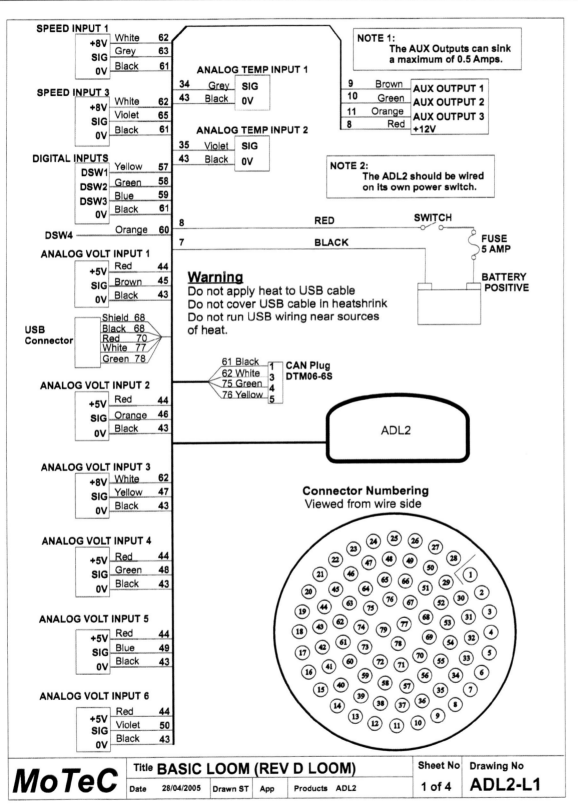

Wiring diagram for MoTeC ADL2 digital dashboard. This shows optional inputs and outputs, as well as those provided as standard. To install a digital dash, you need a wiring diagram that is as clear as this one! (Courtesy MoTeC)

WORKSHOP PRO: MODIFYING THE ELECTRONICS OF MODERN CLASSIC CARS

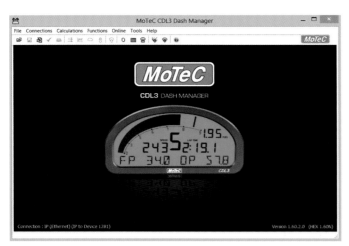

As with any software, expect to have a learning curve when configuring a dash. Most companies will allow you to download and try before you buy.

(normally used to show selected gear), three numerical displays, and a programmable lower line that can be switched between many different inputs to show other data on demand. This lower line is also used to display warning messages.

There are also three dash 'pages' that can be chosen – that is, the dash as a whole has three different screens, with each able to be configured separately.

The CDL3 (optionally) comes with four analog voltage inputs, two analog temperature inputs, two digital inputs, three speed inputs and four PWM or switched outputs. Internally, it runs a three-axis accelerometer. It is able to accept RS232 and CAN communications. Laptop communication with the dash is via an ethernet cable.

The dash is able to receive data from the ECU via the CAN bus. In my case, where a MoTeC M400 engine management system is being used, this substantially reduced the number of inputs the dash itself needed to have. This is because all the inputs to the ECU (MAP, coolant temp, throttle position, etc) can be displayed on the dash via the CAN bus.

The CDL3 I bought was secondhand and came with the optional input/output upgrade and a GPS sensor. I purchased the CDL3 track logging kit upgrade loom, the two-pushbutton loom (that includes the buttons), and the input/output upgrade loom. This allowed me to plug in the GPS sensor and the two pushbuttons (used to control dash functions). I subsequently bought an SLM multi-function LED display (covered in a moment) that also plugged straight into this new loom.

Like much MoTeC equipment, all the information about these upgrades is available – but you need to search hard to find it. I found using Google better than trying to find it on the MoTeC site itself. Note that with some digging, MoTeC support for older products is very good – helping to explain why secondhand MoTeC gear still commands good prices.

The MoTeC Shift Light Module (SLM) comprises a small unit housing eight LEDs. Much more than a shift light, each of these LEDs can be separately configured via the CAN bus for colour and intensity. (Note that there is also a cheaper SLM-C version that has single colour LEDs.)

The SLM allows different driver warnings to be indicated with different colours and LED positions. Static and flashing modes can be configured.

The main advantage of the SLM is that it doesn't use up any outputs from the ECU or dash. In my application, where the dash has four switched outputs, these would have all been used in triggering four warning lights – eg fuel, shift, oil pressure and lean condition. With the SLM, these warnings – and dozens of others – can be configured as required, using the CAN bus and so using no dash or ECU outputs. While initially I thought the SLM a bit of an extravagance, it has become an integral part of how I use the dash.

INSTALLING THE DASH

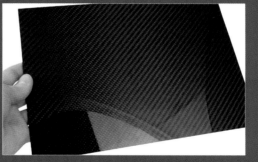

The CDL3 was mounted on a carbon fibre sheet cut to shape. The carbon fibre panel was sourced online. It was an A3-sized, 3mm thick single-layer sheet. By 'single layer,' the description means that most of the sheet comprises glass fibre; the carbon fibre is just the top layer.

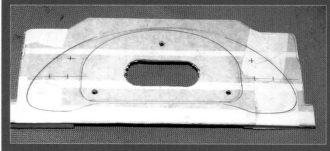

The panel was masked and marked out for the required holes. Holes were drilled using a drill press, starting with small drill bits and working upwards in size. A hole-saw was used to make larger diameter holes. Cutting of the panel was achieved with a thin 1mm cutting disc in an angle grinder.

9. UPGRADING DASHBOARDS, INSTRUMENTS AND WARNINGS

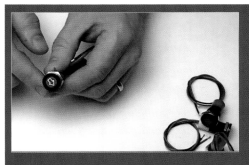

While the SLM can provide multiple warning light displays, some LED warning lights were also needed on the dash panel itself. These include left/right indicator, low/high beam headlights, handbrake/brake system failure, and a red warning that will become the alarm flasher. I chose to use Black Bezel LED warning lights from Car Builder Solutions in the UK. These LEDs incorporate a dropping resistor, so can be connected straight to 12V.

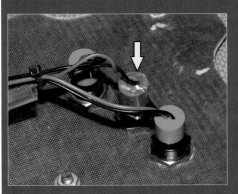

Note that despite being quite expensive, the warning lights aren't perfect. In the high beam warning light, the internal display would rotate unless the rear was held in position with some hot melt glue (arrow).

The high beam light proved to be too bright, so a 20kΩ pot was wired in series as a variable resistor and adjusted until the required intensity was achieved.

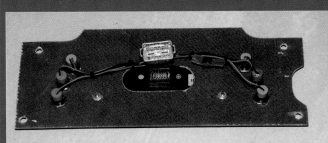

This view of the rear of the panel shows the wiring for the warning lights connects to the car loom via a 12-pin Deutsch connector. Also visible is the multi-pin CDL3 connector.

Six millimetre threaded rod was used to provide the mounting pots for the new panel. These rods screwed (via nuts and washers) into the original mounting holes of the dash, but allowed the location and angle of the new panel to be adjusted as required.

The surface finish of most carbon fibre sheet is very glossy: often too glossy for a dash panel. To reduce the gloss, I buffed the dash panel gently with a Scotchbrite belt in a belt sander, then followed this up with hand work in a circular motion with fine steel wool, three applications of 'rejuvenating' car polish, and a final buff with a soft cloth. This gave a relatively dull finish, but one where the carbon fibre weave was still quite prominent.

The plastic dash binnacle required extensive reshaping of its shroud to fit flush against the newly positioned panel. Not able to be seen here are the two pushbuttons that were mounted low on the dash (these select 'page' and 'alarm acknowledge') while I added another input that utilises the standard Honda FCD (Fuel Consumption Display) button mounted high on the dash. This FCD button now scrolls through the lower line of text on the dash, allowing the showing of driver-selected data.

WORKSHOP PRO: MODIFYING THE ELECTRONICS OF MODERN CLASSIC CARS

As with the MoTeC engine management software, starting from scratch and becoming proficient in the dash software took me many, many hours. That said, in many ways the dash software is easier than the ECU software, so it's a good place to start! I initially set the dash up on the bench and, using just a few inputs (GPS and a pot on an analog input), played with the software until I was able to get these signals displayed on the dash. That successful, I was then confident enough to install the dash in the car.

The dash can be configured in literally hundreds of ways. During tuning of the engine management system, I often changed the data appearing on the dash. For example, when tuning the action of the intercooler water pump speed, I brought this aux output duty cycle up on the dash so I could see how well my programming of pump speed was working on the road.

So rather than describe a final configuration (there probably will never be one!), here are some of the ways in which the dash is being used:

Engine RPM
Engine rpm is displayed on the bar graph, with the signal coming from the ECU via the CAN bus.

Speed
Road speed is displayed in two different ways. The speed input from the gearbox to the ECU is shown on the dash (via the CAN bus), and the speed sensed by the GPS plugged into the dash (RS232 communication) is also shown. Initially, when calibrating the gearbox speed input, I brought up both displays side-by-side and then changed the calibration number until the two matched.

Gear
The selected gear (1 through to 5 plus neutral) is calculated by the dash using speed and rpm inputs (both sent from the ECU). This works surprisingly well on the road – with the settings carefully tuned, the displayed number changes as soon as the gear is selected and the clutch pulled out. (Note, though, that reverse gear is shown as '1' on the dash!)

Fuel level
Fuel level is displayed using the input of the factory Honda sensor. The resistive sensor connects to an AT (normally analogue temperature) input on the dash. Fuel level is displayed as 0-100 per cent, with the relationship between sensor resistance and percentage display calibrated in the dash software.

Engine temperature
Engine coolant temperature is displayed in degrees C, sent from the ECU via the CAN bus.

When the turbo boost trim is changed by the driver, the SLM displays two green LEDs, the warning BOOST CHANGE appears, and the number at middle-left shows the trim change – in this case, -46 per cent. Incidentally, the larger figure in the centre of the display represents 'n' for neutral gear.

Air/fuel ratio
Air/fuel ratio is shown, with this sent from the ECU via the CAN bus.

Others
Also available by pressing the FCD button are the following data, all sent from the ECU and shown on the bottom line of the display:
- Manifold pressure
- Inlet air temperature
- Fuel injector duty cycle
- Engine oil pressure
- Engine oil temperature
- Ignition advance
- EGR valve duty cycle
- Water/air intercooler pump duty cycle
- VTEC on/off
- Lambda short term trim
- Lambda long-term trim

> **So what doesn't the CDL3 do?**
> Maths functions!
> And the more I wanted the dash to perform different functions, the more I wanted access to those maths calculations.
> So when I found a MoTeC ADL3 for sale, brand new but being sold as a used item, I grabbed it. Looks the same as the CDL3, but can do vastly more …

Warnings
There are warnings (text displayed on the dash and specific SLM LED configuration activated) for over 10 different conditions – from low fuel to battery level to seatbelt left unfastened. These are shown on the table opposite.

While the learning curve required to program the dash was immense, the end result looks great and works very well.

9. UPGRADING DASHBOARDS, INSTRUMENTS AND WARNINGS

Displayed warning	SLM configuration	Notes
–	8 yellow/black flashing at fast rate	Gear change redline indicator
Seatbelt	8 red	Uses standard seatbelt buckle switch + speed over 5km/h
Door open	8 red	Door switch + speed over 5km/h (3mph)
Engine hot	8 red	Engine over 100°C (212°F) for more than 5 seconds
Oil pressure	8 red	Configured on minimum pressure at 1000+ rpm
Lean	8 red	Manifold pressure above 100kPa (ie in boost) and air fuel ratio is 14.7:1 or higher
Battery level	2 yellow	Less than 12V for more than 5 seconds, uses internal dash battery voltage monitoring
Inlet air temp	2 yellow	Over 50°C (122°F)
Fuel level	2 yellow	Less than 20 per cent for more than 10 seconds
Change up	2 blue	A higher gear should be selected for best fuel economy. Based on throttle position, manifold pressure, gear and engine rpm.
Change down	2 purple	A lower gear should be selected. Based on throttle position and engine rpm.
ECU hot	–	Uses existing internal temperature sensor
Dash hot	–	Uses existing internal temperature sensor

The warnings configured to appear on the digital dash and multi-LED SLM. In use, these work very well.

WORKSHOP PRO — MODIFYING THE ELECTRONICS OF MODERN CLASSIC CARS

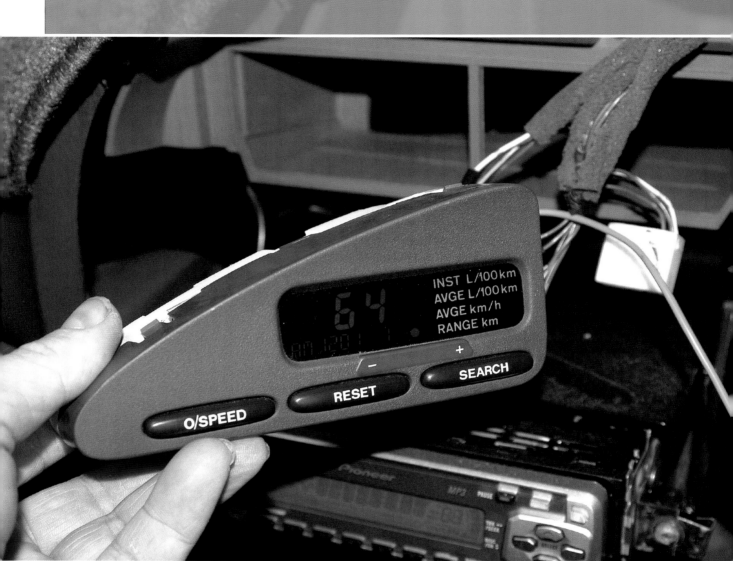

Chapter 10
Improving convenience and security

- **Adding an outside beep to your central locking**
- **Upgrading the horn**
- **Adding a 12V (or USB) power socket**
- **Adding electrical items from higher trim level models**
- **Extending the washer spray time**
- **Headlights-on warning**
- **Stealth CB radio installation**
- **Car alarms and immobilisers**
- **Interior light delay**
- **Installing an auto-dimming rear vision mirror**

WORKSHOP PRO: MODIFYING THE ELECTRONICS OF MODERN CLASSIC CARS

One area where older cars are often lacking compared with current models is in electronic convenience and security features. However, many of these features are easily added. Whether it's a delayed 'on' period for the interior light when you shut the door, extra 12V (or USB) power sockets built into the car, an alarm/immobiliser, or even an auto-dimming interior rear vision mirror, these features are relatively cheap and easy to add. They're also good starting points if you're new to modifying car electrical and electronic systems.

ADDING A BEEP TO YOUR CENTRAL LOCKING

Many cars of the 1990s and 2000s have remote central locking as standard. You press the button on the key and the locks close. Sometimes you're notified of what has happened by a beep or flash of the indicators, but in other cars there's no indication at all. The latter can be a bit annoying – until you test a handle, you're never quite sure that the car is locked. What this little project does is add a beep to the system when the doors lock. There's no indication when the doors are unlocked (none is usually needed) and the modification will work with all systems – including those that use a single button on the key to both lock and unlock the doors.

The basic idea is to wire a polarised electronic buzzer in parallel with the door lock solenoid or motor. These motors (or solenoids) are fed current with one polarity when unlocking the doors, and the opposite polarity when locking the doors. If you put the buzzer the right way across the wiring, it will sound only when the doors are being locked.

To make the sound, you'll need a 12V electronic buzzer (cheaply available online). The buzzer should be fairly loud, as in use it will be buried inside the door, muffling it. You might also need a diode, but I'll come back to that later.

Be careful when removing the trim. In addition to clips, there are often a few screws, usually hidden. In this car (a Toyota Prius), there was a screw hidden behind a flap inside the door handle …

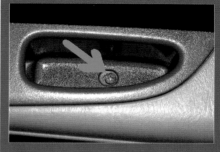

… and another inside the armrest. Remove these screws before gently prying the door trim loose.

The inners of the door cavity are usually protected by a plastic liner, held in place with adhesive black mastic. Gently pull this loose, starting in the area where the door lock motor is likely to be.

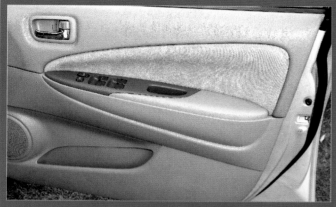

The first step in the installation is to remove the door trim. I chose the driver's door, but any door with electric locking can be used.

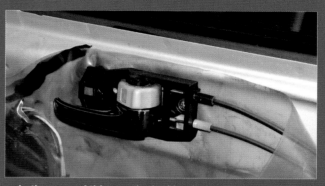

In the case of this car, the locking button is connected to the door lock mechanism by Bowden cables. By following these I was able to find the door lock motor itself.

10. IMPROVING CONVENIENCE AND SECURITY

Here is the lock motor, almost completely hidden inside the door. But there's no problem – we don't want to access the motor itself, just the wires leading to it.

And here are those wires. The bundle is wrapped in loom tape but …

… this is easily sliced open to reveal a bunch of wires. But which ones are the right ones to connect the buzzer to?

The answer is easier than you'd first think. Just carefully bare a small portion of each of the wires, making sure that they cannot touch each other or ground (car metalwork). Then try connecting the buzzer to the different wires until it sounds briefly when the doors are locked. Yes, that easy. If you are using an electronic buzzer, you won't do any harm if you connect it across the wrong wires first.

Once you have found the right wires for the buzzer, try unlocking and locking the car a few times. In some cases you might find that the door locking mechanism is now a bit erratic. If this is the case, a diode is likely to be able to solve the problem. (Small diodes are available online – a 1N4004 is fine.) Wire it into one wire going to the buzzer (try each way around until the buzzer still works) and then check that all is again fine.

Insulate the wires that aren't required and then solder the pair of wires from the buzzer to the correct leads. Thoroughly insulate these connections with tape.

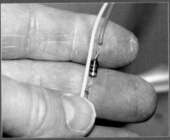

In the case of the Prius, the diode was needed. One wire to the buzzer was cut and the diode inserted, these joins were then insulated.

The new cable was wrapped in tape, and then the system again tested.

WORKSHOP PRO: MODIFYING THE ELECTRONICS OF MODERN CLASSIC CARS

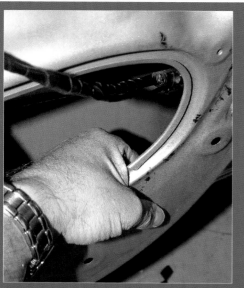

The buzzer was mounted inside the door using double-sided tape – out of view in this photo, but that's what my fingers are doing! Be careful when picking the mounting position to ensure the window can still move up and down freely.

Again, to avoid fouling the window mechanism, cable-tie the new wiring into place.

After that it's just a case of reassembling the door trim and you're finished! Now, no more wondering if you really did lock the car!

UPGRADING THE HORN

Having a loud and penetrating horn is important in any car. It might be to stop someone reversing into you in a car park, warn someone swapping lanes when you're passing

Secondhand horns from a Nissan Patrol. These are loud and were bought quite cheaply.

through their blind spot, or – on the open road – the need to scare away birds or wildlife. But many cars have feeble, ineffective horns that can barely be heard. However, upgrading the horn is easy, cheap and effective.

Selection

There are some excellent aftermarket horns now available. Electric horns, where a diaphragm is vibrated, are used universally as standard equipment – and you can also buy more powerful aftermarket horns that use this design. An alternative approach is to use air horns, where a compressor is matched to one or more trumpets. These are loud – in fact very loud. They are also penetrating (not the same thing as loud) and new-design units that integrate the compressor with the horn assembly are more compact than the designs of yesteryear.

However, in this case, I decided that a pair of electric horns from another car would be fine. I settled on a pair of secondhand horns from a Nissan Patrol – and since almost no one wants horns from dismantled cars, I got them very cheaply. The upgrade was occurring on a Skoda Roomster.

Installation

The first step in upgrading the horn was to find the standard one. It proved to be behind a pull-out grille in the bumper. Removing the grille revealed a single horn with a 400Hz output frequency.

Unplugging the horn and operating the horn switch showed, by the presence of a clicking sound, that the horn was operated by a relay. It's important that you check whether a relay is used – if it is (as will be the case in nearly all cars of the era covered in this book) then you can upgrade the horns without having to fit a relay. If there is no relay fitted, you'd better add one. The two Patrol horns were each marked as drawing 3.5A – so 7 amps

10. IMPROVING CONVENIENCE AND SECURITY

Measuring the Roomster's horn wiring, to work out which was the ground wire and which was the 12V supply from the horn relay.

The two new horns in place. With the grille replaced, they're invisible.

total. That's actually low for horns, so I was confident the standard relay would be fine driving that load.

The Patrol horn each has a chassis connection (ground) for one of its two connections. That makes it easy – the chassis connection goes to ground and the other is the power feed. On the other hand, the Roomster horn is plastic-bodied and has two connections – a ground and a 12V supply. But which of the supply wires was which? By connecting a multimeter and probing the car loom plug I could find out which wire was 12V and which was ground. A check was then made between the ground wire and the negative terminal of the battery to ensure they had continuity (that is, they were connected together).

The plug was then cut off, and a spade terminal attached to the ground wire, which would be in turn attached to the bolt that would hold the horns in place.

The other wire was attached to a female terminal that also ran to a second female terminal, allowing both horns to be powered. The horns were bolted into place, with care being taken to ensure they could still vibrate freely on their laminated mounting strips. (Never solid-mount a horn – it will lose a lot of output.)

Results

Testing with a sound pressure level meter showed that at a distance of about 150mm from the bumper, the standard horn had a maximum output of 112dB(A). The new horns? Well they output 120db(A) – with the logarithmic dB scale, that's massively louder. Furthermore, the two discordant frequencies give much improved penetration. For so little money and an hour of my time, the upgrade was well worth it!

The new wiring – the black spade terminal is the ground connection to the horn bracket, and the two new blue terminals feed power to the twin horns.

Using a sound pressure meter to measure the increased output. The standard horns had an SPL of 112dB(A), while the two new horns were much louder at 120dB(A).

WORKSHOP PRO — MODIFYING THE ELECTRONICS OF MODERN CLASSIC CARS

ADDING A 12V POWER SOCKET

Here's a really easy project that is also very useful – no matter what car you drive.

These days, plenty of cars have extra accessory 12V power outlets. They might be located in the back (eg the boot or the load area of a wagon) or they might be in the centre console. You can plug in phones, an accessory air compressor, fluoro lights – you name it. They're also good when camping, to run lights or a portable fridge. And they're much safer than using alligator clips to connect directly to the battery or running some half-baked wiring around the car.

(If you'd prefer to have extra USB power outputs, that's easy as well. Just buy a suitable USB car power socket online and then install it in the same way as shown here for the 12V socket.)

The most important part is the power socket. While you could use any old cigarette lighter socket, it's much neater to use a socket designed for use as an external power socket. This one came from a car dismantler, and was very cheap.

Test-fit the power outlet – this one was placed under the bonnet near the battery. Its primary purpose will be to power camping lights.

In addition, you'll need a fuse holder and appropriate fuse (10A if the outlet is rated for 120W), and spade and eye terminals. You'll also need some cable and a small piece of metal to make the mounting bracket (I used an off-cut of aluminium).

Remove the power outlet from the bracket and then make a wiring loom to suit. At this stage leave out the fuse. I suggest both positive and negative wiring be run right back to the battery.] Place the fuse holder close to the positive terminal of the battery, and solder or crimp all the connections. The positive from the battery goes to the tip connector of the power outlet, and the other connection goes to the negative of the battery.

Decide where you're going to mount the power outlet. Keep in mind that there needs to be clearance behind and in front of it. Cut a hole in a bracket to suit the power outlet – this one clicked into place in the right sized hole and its security was aided with some silicone sealant. Make more holes to allow the bracket to be bolted in place.

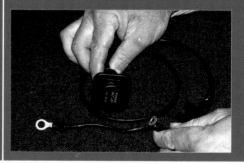

To neaten and help weatherproof the loom, I wrapped it in black insulating tape.

10. IMPROVING CONVENIENCE AND SECURITY

Here's the finished product (arrowed). There's nothing to stop you running multiple outlets (just wire them in parallel) – you can be sure that they'll prove very useful.

ADDING ELECTRICAL ITEMS FROM HIGHER TRIM LEVEL MODELS

If you have the base model, or one that lacks desirable options, it's very likely that your car is missing bits and pieces. Things like oil pressure and volt gauges, a trip computer, climate control, cruise control, electric windows and electric seats. Ah, you think, that's easy enough to solve – just visit a car dismantler and pick up all these good bits. Trouble is, what you see is sometimes only the tip of the iceberg, and getting the system working might require a lot more than just what you can see.

What's easy or hard to add depends a lot on the car. For example, on one car that uses electronic throttle, adding cruise control is as easy as buying the factory cruise control stalk, buying a new brake light switch, and installing both. That's pretty well it – the cruise control light is already built into the dash, the wiring loom is mostly already there, and the ECU logic is all built in, ready to be accessed. For just a few hours of time, it doesn't get much easier than that. But installing factory cruise on another car can be much harder. In some trim levels the loom isn't present, so that part of the system needs to be bought. Then you'll need to add the throttle actuator, control stalk and additional brake light switch.

Standard dashboards that have extra instruments are another case. A dash that integrates a voltmeter and oil pressure gauge into the display sounds easy to fit – just swap-in the whole new instrument panel. And the voltmeter part won't cause any problems – the panel already gets a battery voltage feed, and so the new voltmeter will just monitor this. But changing from a dash that uses an oil pressure warning light to one with an oil pressure gauge will require a new engine sender, as the signal output is no longer just on/off but variable.

And it might look easy to pull out the original heater/ventilation controls, and replace them with digital climate control. In most cases the panels are physically the same size and the climate control panel might not cost much. But stop! A climate control system uses output actuators to vary the position of heater valves and ventilation flaps; and it uses external sensors that include sun, outside temperature and cabin temperature. It also has its own ECU. Basically, the control panel is perhaps only 25 per cent of the system. And installing the other 75 per cent is likely to be a complete dashboard-out job!

What you need to do before you spend anything is to get information. Is the upgrade you're considering possible, and what *exactly* needs to be done to achieve it?

In these sorts of the swaps, web discussion groups are really useful: chances are that someone before you has tried to do the same upgrade. So, pose the question: what else needs to be done to swap in the instrument panel that contains the oil pressure and volt gauges? Or, has anyone fitted the optional cruise control? Don't necessarily rely 100 per cent on the answers, but take them as generically valid. So if someone says that the upgrade instrument panel swap has been done many times, you know it's possible. But that answer doesn't tell you all the bits that you might need …

Another excellent resource to use is the workshop

WORKSHOP PRO: MODIFYING THE ELECTRONICS OF MODERN CLASSIC CARS

manual for the car. For example, the wiring diagrams of the base and upper trim models will show you some of the differences in instrument panels, climate control and cruise. The manuals might even have two clearly different wiring diagrams: those for the lower trim models and those for the upper trim models. If the upper trim model runs a different body ECU, instrument panel and trip computer – and they're all linked together – it's obviously going to be a lot harder than if the basics are all the same and the good bits just plug into the same system.

Finally, a dismantling yard can tell you lots of stories.

For example, if there's the top-line model (complete with all the good bits you want) and right next door there's the base model you own, and both have their dashes out, bumpers off and door-trims removed, you can quickly and easily compare wiring looms. Is it the case that both cars have the same looms, but in the base model the plugs are just unused? If so, great. Or are the two looms completely different? If so, well, not so good. If you want to install the topline model's fog lights and fog light switch, are there wires running to the front and back of the car for the lights? A quick comparison of the two cars will soon show you.

TRIP COMPUTER UPGRADE

Any electrics upgrade will be specific to the make and model of car, so let's take a look at one. It's unlikely that you will be doing this upgrade, but it shows you the sort of complexity that may be involved. The addition is of the Fairmont trip computer to an Australian EF Falcon Futura, which normally fills the same space with a clock. In this case, the trip computer was purchased online complete with the trip computer wiring plug, a new female plug connection, and specific instructions on how to do the swap.

This example shows how tricky these upgrades can be – if I'd bought just the trip computer and had no instructions, I'd never have got it working.

Here's the standard Futura clock.

But the tricky bit is the wiring. The trip computer gets its information (fuel remaining, speed, injector opening percentages) via a single wire data feed from the dashboard. To gain this info, a new terminal needs to be inserted into the right-hand instrument cluster connector, seen here.

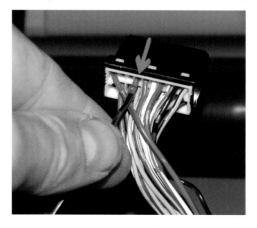

This new female pin goes into the connector in this terminal position.

With the dash trim removed, the clock simply unclips, with the trip computer able to be pushed straight into the same space.

10. IMPROVING CONVENIENCE AND SECURITY

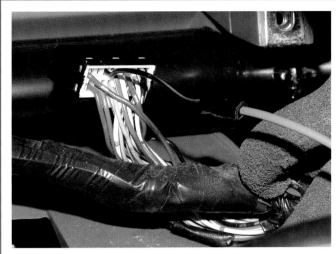

This pin is then connected to a new wire that goes to the trip computer connector. The original clock connector can be used for the trip computer, so long as you add some extra pins. However, it's easier to take the trip computer plug and a short length of loom from the donor car. In this case, in addition to the pin for the dashboard connector being supplied, the trip computer plug was also supplied. The changeover is as follows:

Pin no	Pin function	Clock connector	Trip computer connector	Action when fitting trip computer
1	Trip computer data signal	No connection	Blue	New wire to instrument cluster plug
2	Clock ignition power	Red/dark green	Not connected	Cut wire
3	Battery positive	Yellow/black	Yellow/black	No change
4	Earth	Black	Black	No change
5	LCD dimmer signal	White	Black/white	No change
6	Trip computer buttons light	No connection	Brown	Connect to brown wire on rear demister switch connector
7	Dash lights dimmer signal	No connection	Dark blue/red	Connect to dark blue/red wire on rear demister switch connector
8	None	No connection	No connection	No change

With the pin numbers as follows, and the plug viewed from the rear with the plastic spring clip on the left:

1	5
2	6
3	7
4	8

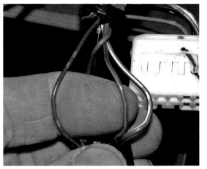

As indicated in the table on the left, the brown and dark blue/red wires from the trip computer plug need to tap into the 'like' colour wires on the adjacent rear demister switch. The two rear demister switch wires that need to be accessed can be seen here.

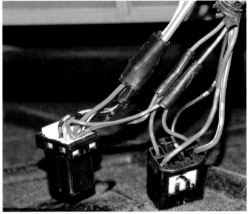

The wired-in trip computer plug on the left, with the rear demister switch plug at right.

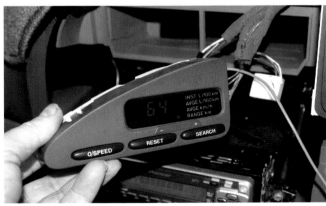

The trip computer temporarily plugged in to make sure everything is working …

… and then working with the dash put back together.

241

WORKSHOP PRO: MODIFYING THE ELECTRONICS OF MODERN CLASSIC CARS

EXTENDING THE WASHER SPRAY TIME

In some cars, you flick the windscreen washer stalk and the washer sprays for three or four seconds, wipers automatically activated. On others, you do the same and the wipers stay still but the spray washes for a short, pre-timed period. And in yet other cars, the washer is on only when the stalk is pulled (or button pressed), and the wipers are controlled quite separately.

So? Well if you're swapping back and forth between a car with a pre-timed wash and one without, nearly every time you get into the car with the simpler washer system, you'll scratch the windscreen by running the wipers dry. So here's a simple fix. What it does is run the washers for a pre-timed period each time you flick the lever or press the button. Furthermore, if you wish, you can configure the system so that the washers will run for twice as long if you flick the lever twice (or press the button twice), three times as long if you operate the button three times, and so on.

If this was a difficult project that involved your delving into complex wiring looms for hours, upside down under the dash and juggling wiring diagram and multimeter – well, I honestly wouldn't bother with this project. But since it takes literally less than an hour to install the pre-built module, the effort/benefit trade-off is pretty good!

This section is also a good introduction to a configurable, smart timer module that can be used for other purposes as well. The windscreen washer timer uses as its control system a pre-built electronic module – the eLabtronics Timer available from www.elabtronics.com. The module has only four wiring connections. Since two of those are power and ground, it's very easy to wire into place. The timer module can directly drive the washer pump, so no relay is needed.

THE ELABTRONICS TIMER MODULE

An electronic timer sounds like a pretty ho-hum kind of module, right? But that's only until you try to come up with solutions to problems! So, what sort of problems?

Okay, how about adding a delayed 'on' time to a press-button? That's so when you push the button and release it, the device stays operating for a preset time. That could be as useful as keeping your car headlights illuminated after you've arrived home in the dark, giving you time to walk to your door.

Or, how about adding an automatic 'on' time that keeps your car's electric windows working for 1 minute after you've turned off the ignition? Many modern cars already have this feature – but if you live in a household where there are two cars (one with and one without delayed electric window operation) you're sure to be driven nuts whenever you're in the 'wrong' car. I know I am!

Or what about a timed turbo over-boost that lets you automatically run higher than normal boost for (say) 10 seconds? You put your foot down and the boost is allowed to rise to its new level – but only for a limited period. With such a short period of over-boost, even a standard intercooler will cope, and yet in passing and short-term acceleration, you'll have plenty more performance than normal.

Each of these scenarios needs a timer with a slightly different requirement – not only in the length of the timed period but also in the *way the timer is triggered*. In one situation, the timer triggered with a button push, in another it automatically turned on when the ignition turned off, and in another it needed to be triggered when a sensor (throttle position) exceeded a certain voltage level.

As you can see, when it's required to perform lots of real world functions, what sounds like something rather simple – a universal timer – actually requires a sophisticated design. But with the eLabtronics Timer, all the work has been done for you. Let's take a look at it.

Connections

The eLabtronics Universal Timer's wiring connections are:
- Power – marked on the board as '+.' Power is nominally 12V.
- Ground – marked as '-'. You would normally connect this to chassis ground or the negative terminal of the battery.
- Input – marked as 'in.'
- Output – marked as 'out'.

The intelligent eLabtronics timer module. It requires only four connections and can be configured in multiple ways.

10. IMPROVING CONVENIENCE AND SECURITY

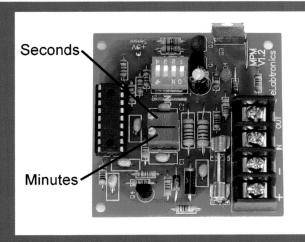

The timed period is set with these two pots.

When the eLabtronics Universal Timer output MOSFET power transistor is activated, battery power is available at the output terminal. So all you need to do is to wire your load (lights, buzzers, horns, solenoid, fans, pumps, etc) between the output terminal and chassis ground.

If the load has a polarity, the positive terminal goes to the Universal Timer. (Note that as with all MOSFETs, there is a slight voltage drop across it, so at high loads, a little less than full battery voltage will be available at the output at high loads.)

Pots

The two onboard pots set the period the Timer runs. One pot sets seconds (0-60) and the other sets minutes (0-60). Because these pots are multi-turn, you can accurately set a timed period from 1 second to just over 1 hour – and everything in between.

For example, if you want the timed period to be five seconds, set both pots fully anticlockwise. Then rotate the 'seconds' pot clockwise a little and test the timed period. (It's easy to see what the timed period is, because, when the timer is activated, the onboard LED flashes twice per second.)

DIP Option Switch positions

The real smarts in the eLabtronics Universal Timer are in the four-position DIP Option Switch. Position the board so that the terminal strip is on the right and then the following switch positions give the listed behaviour.

Mode 1

X	X	X	X

In this mode, the output is switched on for the timed period when the input signal rises above 2.6V. After the timed period has elapsed, the timer switches off – even if the input signal is still above 2.6V. This means it's a one-shot timer – the input signal needs to fall and then rise again to re-trigger.

This is the mode that would be used to run a timed turbo over-boost. The input would be connected to the throttle position sensor, and the output would run a solenoid bleed allowing higher boost. When the throttle position sensor output rises *above* 2.6V, the output is triggered – but it won't let boost remain high for more than the timed period, even if your foot stays flat to the floor.

If all that sounds complex, here's another example of this mode's use. Just connect the input to power via a normally-open push-button. When you press the button, the timer activates for the set period and then switches off – even if you leave your finger on the button.

Mode 2

X	X	X	
			X

This mode is the same as the one above, except the module triggers for its timed period when the input voltage falls *below* 2.6V. This mode is ideal for triggering things when the ignition is switched off – simply connect an ignition-switched power supply to the input and power the timer module from an always-on power source.

For example, the module can be used to feed power to the electric windows for a minute each time the ignition is switched off. Or, you could run a turbo cooling fan for 2 minutes every time the car is turned off.

Mode 3

X	X		X
		X	

This mode is easiest thought of as adding an extended 'on' time. The output switches on when the input rises above 2.6V, and then stays on for the timed period after the input drops below 2.6V.

For example, you might have a radiator fan triggered by a temperature switch. You'd like the fan to say running for two minutes after the temperature switch has turned off, so you use the temperature switch to feed power to the input terminal of the timer. The timer's output powers the fan. When the temp switch says 'run that fan' the timer will do just that. But when the temperature switch says 'stop that fan' the timer will keep it running for the selected period.

243

WORKSHOP PRO MODIFYING THE ELECTRONICS OF MODERN CLASSIC CARS

Mode 4

x	x		
		x	x

This mode is the same as the one above, but the timer triggers when the input falls below 2.6V and then stays on for the timed period after the input rises above 2.6V.

It's ideal for triggering from sensors that have an output voltage that decreases when the variable (eg temperature or airflow) is increasing. For example, you could use this mode to trigger an intercooler water spray, using the output of an airflow meter that falls with increasing load. At high loads, the water spray pump turns on, then its stays on for the pre-set time after the load falls.

Mode 5

x		x	x
	x		

This mode is different to the others in that it is designed to be used with a normally-open pushbutton that connects the input to power. In short, it makes the module a manually-controlled timer. But there's a trick in it – and it's a very good trick.

What the timer does is count the number of times you press the button. Let's say you use the onboard pots to set the timer for a 1-minute period. If you press the external pushbutton once, the timer will turn on for a minute. But if you press the button twice, it will turn on for 2 minutes! And so on – the number of button presses multiplied by the timed period sets the total output time. (This is the mode that we will use in a moment with the windscreen washer.)

This is the perfect mode when you want manual control over when the timer operates, and for how long it operates. Almost anything requiring a manually timed period can be run in this way. One example is a manually controlled intercooler spray – all you need is a pushbutton on the dash and you can give the intercooler a short squirt, a medium squirt or a long squirt – even with your hands back on the steering wheel.

Incidentally, you can also cancel the output at any time, just by keeping your finger on the button for a few seconds.

This mode (and mode #1) is also useful when setting the timed period. After setting the DIP switches correctly, trigger the timer by momentarily connecting the input terminal to power.

The DIP switch that sets the different configurations.

Mode summary

Here is a summary of the five different timer modes.

Mode	Output	Input trigger
Mode 1	One shot then off	Activates when input rises above 2.6V
Mode 2	One shot then off	Activates when input falls below 2.6V
Mode 3	Extended 'on' time	Activates when input rises above 2.6V
Mode 4	Extended 'on' time	Activates when input falls below 2.6V
Mode 5	One shot then off	Number of press-button pushes

Universal timer specifications

Operating power: 10-40V DC
Output current: up to 10A (over 2A will need added heatsink on MOSFET)
Timed period: User-selectable from 1 second to 61 minutes
Timed outputs: one shot, delayed 'on' or manual pushbutton
Timer trigger: User-selectable from rising input voltage, falling input voltage or push-button
Onboard fuse: 15A

INSTALLATION

There are two approaches that can be taken to installing the windscreen washer timer module – the best approach depends on the car. The first step is to look under the bonnet and find the windscreen washer pump. If the wiring to the pump is easily accessible, you'll probably want to mount the controller nearby.

On the other hand, if the washer wiring is buried (eg the washer bottle and pump are mounted within an inner guard (fender)), perhaps the washer wiring is easier to access

10. IMPROVING CONVENIENCE AND SECURITY

inside the cabin. However, in this case you'll probably need a wiring diagram for your car – but even a simple generic workshop manual should be sufficient.

Let's do the under-bonnet approach first.

Under-bonnet

Invariably, the washer pump is grounded on one side and is fed 12V when it needs to operate. That is, the positive side of the circuit is switched, with one side of the pump grounded (maybe not at the actual pump, but somewhere).

If you ground the negative terminal of a multimeter, unplug the washer and probe the wiring going to the washer pump, one side will always have 0 volts on it (ground) and the other will have battery voltage on it when the washer pump is turned on. (Don't forget you need to unplug the washer pump to do this test!) From this test it is easy to work out which is the ground wire and which is the 12V supply wire.

To install the windscreen washer timer, cut the power supply wire to the washer pump. Feed the supply wire from the switch to the 'in' terminal of the module, and connect the 'out' terminal of the module to the washer. Then all you need to do is connect to the module 12V (use an ignition-switched source) and ground. And that's it! The module is triggered by the standard windscreen washer pump switch momentarily feeding power to the 'In' terminal.

The module has sufficient power handling capabilities to drive all normal windscreen washer motors – but ensure that the output MOSFET doesn't get hot when driving the pump. If it does, add a suitable heatsink.

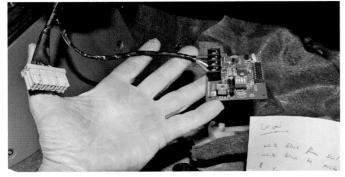

The module wired into place. Note the 'cheat sheet' of wiring colour codes at bottom-right – a good approach so that you don't have to try to remember the different colours.

In-cabin

The in-cabin approach can be even easier – but only if you have a wiring diagram for the car. That's because you can normally connect ground, ignition-switched 12V and the 'in' and 'out' wiring in just a few minutes – and all to the same cluster of wires.

I chose the 'in-cabin' approach on the Honda Insight to which I fitted the system. The wiring diagram for the Honda showed the following colour codes:
- Green/black = ignition-switched power
- Black = ground
- White/black = switched power to washer pump

Furthermore, just by taking off the lower steering column shroud, all these wires were very easily accessed near the washer/wiper stalk. The white/black wire was cut, with the wire coming from the switch connected to the 'in' terminal, and the wire going to the pump connected to the 'out' terminal. Power and ground for the module were picked up from the other two wires.

DIP SWITCH AND POTS

On the board there's a DIP switch that's used to configure the module for different functions. Orientate the board so that the terminal strip is on the right, and then use a ballpoint pen or a small screwdriver to set the switches so that they look like this:

X		X	X
	X		

As described in the breakout, this setting configures the module to count the number of times you operate the washer control (other options in a moment).

Now it's time to set the length of time that one button press will operate the windscreen washer timer. You can set this anywhere between 1 second and 1 hour (but I reckon 1 hour might be a bit excessive!). Again orientate the board so that the terminal strip is on the right. Use small flat-bladed screwdriver to rotate the bottom pot anti-clockwise at least 15 turns, or until it can be heard clicking. (Why so many turns? Multi-turn pots like the ones fitted don't have clear 'end stops,' so to make sure you've adjusted the pot as far as possible to the minimum value, turn it lots of times!)

Then do the same for the upper pot – rotate it anti-clockwise at least 15 turns, or until it can be heard clicking. Okay, now both pots have been adjusted to give the shortest possible time. Now turn the upper pot clockwise four full turns. The spray time with one push of the button will now be about 5 seconds.

You can now test it. Pull and release the washer stalk or button once and washer should start spraying. (The onboard LED will also flash twice per second.) After the timed period has elapsed, the LED and pump will stop operating. Pull the washer stalk twice in quick succession and the spray should work for twice as long. Three times – three times as long.

If you want a shorter time for a single wash, rotate the upper pot anti-clockwise. If you want a longer time, rotate the upper pot clockwise. If at any stage you want to cancel the spray, keep your finger on the stalk or button for a few seconds.

Now what if you decide that you don't want the spray to count the number of times you operate the control? Instead, you just want a simple time extension? All that you need to do is to orientate the board so that the terminal strip is on the right and set the DIP switch like this:

X	X		X
		X	

Now the spray will start as soon as the control is operated and will add the pre-set extension time to the spray operation when the control is released. Again, you can adjust the timed period by altering the pot position.

With the setup complete, install the board in a box and then mount the box somewhere convenient – eg up under the dash. Insulate all wiring joins and keep the wiring neat and tidy with cable ties.

ADDING A HEADLIGHTS-ON ALARM

If you have a car that has no indication that you've left your headlights on as you exit, you'll likely be familiar with that sinking feeling you get when you find your battery has gone flat. However, adding a buzzer that sounds when you open the door with the lights still on is very easy. You need to connect only two wires, and you need either a polarised 12V buzzer (ie, it will sound when connected only one way around) or a normal buzzer and diode.

The diagram shows how easy it is. Connect the positive of the buzzer to the headlight (or parking/side lights) power feed. Connect the negative of the buzzer to the door switch wire. When the lights are on and a door opens (so closing the door switch), the buzzer sounds. You can still leave your lights on if you want, because the buzzer turns off when you close the door.

If you want to use a buzzer that is not polarised, simply insert the diode in the feed wire to the buzzer band of the diode nearest to the buzzer.

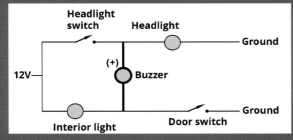

A lights-on buzzer (shown in bold) is easily added to most cars. Only two connections are needed – to the lights and door switch. Use a polarised buzzer, or add a diode.

STEALTH CB RADIO INSTALL

Over the years, and in different countries around the world, CB radio in cars has varied enormously in popularity. Thirty years ago, it was an urban craze, with cars sporting huge antennas, and terms like '10-4' part of the popular lexicon. Then things started to go quiet – nowadays, unless you're into serious off-road driving or you're a dedicated mobile radio enthusiast, it's pretty rare to find CB radio gear in cars.

After all, who wants a big whip antenna (and how are you going to mount it in cars that no longer have gutters?), and where's the space in a car's cabin to integrate another piece of gear? And anyway, aren't CB radio frequencies just full of foul-mouthed idiots with the social skills of brain-dead parrots?

Well, let me tell you that while there are plenty of parrots out there, there's also a huge range of people who every day effectively use UHF CB radio. They're the people on the road – not so much in the cities but on the highway. Every long-distance truck has a CB. Immediately, anyone listening on that channel will be warned of accidents, road hazards, road works, unusual weather conditions – the lot.

On a long journey, a CB radio allows trucks to become your friends – there's nothing like the improvement in safety that comes from trying to overtake a big truck on a two-lane country road when you can actually talk to the truck driver about approaching traffic and upcoming bends. In an emergency, a CB radio can be simply invaluable. Over the years, I've also been warned of many road hazards – one example was when a small mountain of sediment had been washed across the road by a flash flood (at night it was completely invisible). I've hitched a ride with a truck driver after a breakdown (again at night – and who would pick up a hitchhiker without first being able to talk to them on the radio?) and I've had plenty of conversations that dispelled the boredom of long drives.

So when I bought a car in which I intended to do lots of long drives, one of the first steps was to fit a UHF CB. But (a) it's a very small car with little spare dashboard space, (b) for aesthetic reasons I didn't want a big antenna.

The radio is a GME Electrophone TX3400. The beauty of this design is that the front panel (containing the operation knobs, buttons LCD and microphone cable socket) can be removed from the main body and mounted separately, being joined to the rest of the radio by a 1.8m (6ft) cable. Even with the jam-packed nature of modern dashboards, there's almost always room for a piece of equipment only 128mm wide x 31mm long x 29mm deep, (5 x 1.2 x 1.1in) which is the size of the remote mount head unit. The main body can be mounted anywhere under the dash, within the centre console or even under a seat. It's also quite small at 128mm wide x 117mm long x 29mm deep (5 x 4.6 x 1.1in).

10. IMPROVING CONVENIENCE AND SECURITY

The installation was easy – the head unit went to the right of the steering wheel on a part of the dashboard where three small switch blanks were situated. Rather than using the provided bracket, the head unit was stuck into space with double-sided tape – it's easily light enough to be securely mounted in this way. (But if you do the same, don't forget to first clean both surfaces with an automotive grease and wax remover.) An opening compartment directly beneath this part of the dash was also big enough to swallow the microphone and its cord. (If you're going to be listening far more than talking, the microphone can be unplugged and placed in the glove-box. You can then plug it in just before heading off on a trip, or if you need help in an emergency.) The main body of the radio was slim enough to be positioned vertically against the inside of one wall of the centre console.

An external speaker (one originally used for an in-car phone) was plugged straight into the speaker extension jack of the radio – this switches off the small (and tinny) internal speaker. The extension speaker was mounted under the dash.

Now, what about an antenna? Nothing much has changed in mobile antenna design – for best reception, drill a hole in the middle of the roof and screw in a big one! But who wants to do that?

These days, nearly invisible glass mount antennas are available that stick to the inside of the front or rear glass, looking rather like abbreviated demisters. Note that these antennas, which have a gain of 2.5-4.5dB, are not suitable for cars with metallised film tinting.

An alternative stealth antenna is an external glass-mount whip. These look much like the antennas that were previously widely used with in-car mobile phones and

This CB radio has a small, separate front panel linked to the main unit by a cable. This allows easy mounting. (Courtesy GME)

The small glass-mount aerial at the top of the windscreen. I slipped some tube over it to reduce the whistle that otherwise occurred.

The mounted radio. The microphone is stored in the opening compartment beneath it, and the main unit of the radio is remote mounted.

mount in the same way. They have a gain of 3dB (longer whip) and unity (shorter whip). It's easy enough to store the longer antenna somewhere in the car (eg under the carpet in the boot) and have the shorter whip fitted for normal use. When you need the increased range, unscrew the short whip and screw on the long one.

I chose to locate a unity-gain glass-mount antenna at the top centre of the windscreen. To prevent wind noise, I enlarged the diameter by slipping some plastic tube over the aerial and then placing heat-shrink over the whole antenna.

The result? An unobtrusive radio that is there when you need it.

ALARMS AND IMMOBILISERS

I am in two minds about alarms and immobilisers.

My first issue is that nearly all alarms and immobilisers use a remote fob that is added to the car's keys. But then of course, if the thief steals your keys, the alarm/immobiliser is completely useless. With your keys in hand, the thief can disable the alarm or immobiliser, just as easily as you can.

My second problem is that I do not know of any alarm or immobiliser that will stop a determined thief stealing your car. As in – a really determined thief! After all, if the thief winches your car onto a flatbed truck and drives off, there goes your car – purportedly immobilised or not, siren wailing or not. (You could install a GPS tracker, but then that creates a whole lot of issues of their own. Issues like – who is going to track the car down? You and your (suddenly unavailable) friends? Uninterested police?)

But having said that, alarms or immobilisers have stopped my cars being stolen twice. Once, the siren going off alerted me, and when I'd sprinted out to the car, I found the door lock on the ground and the ignition lock broken. On another occasion, where the car had been left in a darkened and deserted car park, the car was still there when I came back to it – but again the door lock had been broken.

So I don't think it is worth paying a lot of money for a brand-name alarm/immobiliser (not unless your insurance requires that you must have a specific make and model fitted). Instead, I think that you should consider these three approaches:

1. Fitting a low-cost remote-control alarm with a siren, so that it will attract your attention if you are nearby and a theft is attempted.
2. Fitting a device that will reduce the speed with which a thief can steal the car. This is based on the idea that if things don't go according to plan for the thief (what – the engine won't start?), the thief is more likely to abandon the heist.
3. Fitting a device that prevents the thief driving off in your car, *even if they have your keys*.

If you wish, you can do all three.

The first of these requirements is easily addressed by fitting a cheap alarm with a remote. Buy one that has a battery back-up siren, so it keeps sounding if the thief accesses the car battery and pulls off a cable. Furthermore, since many of these alarms also have an output for an immobiliser relay, you can easily disable a starting circuit. If you wish to add an immobiliser only, a low-cost remote electronic switch can be used – an approach I cover (with an added wrinkle) in a moment.

Typical immobilising circuits include disabling the:
- fuel pump (but remember that some cars will run for a minute or two without the pump operating, and that might be just long enough for the thief to get to the middle of a busy intersection!)
- ignition-switched feed to the ECU
- power feed to the injectors
- low-current wire to the starter
- ignition coil (often the easiest approach in cars with single coil ignition)

And what if the thief has your keys? The simplest and cheapest approach is to add a hidden switch in the cabin that disables the starting of the engine. It's one of the oldest immobilisers in existence – and now has renewed relevance with the problem of key theft.

Don't place the hidden immobilising switch under the dashboard; instead locate it somewhere you can reach without looking awkward. (The thief might have you under regular observation prior to attempting the theft.) For example, in a car with low seats, the base of the B-pillar is likely to be within easy reach – but it's not somewhere a thief would look for a switch. If you use a large rocker switch, the switch can even be mounted behind carpet or

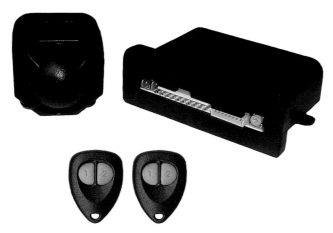

This low-cost remote alarm and immobiliser disables one car starting circuit. It senses thief entry by a voltage drop in the car's electrical system (eg caused by the interior light coming on as a door is opened) and uses a battery back-up siren.
(Courtesy Rhino)

10. IMPROVING CONVENIENCE AND SECURITY

other flexible trim material – and only you know where to press.

The switch can disable as many circuits as you wish, either by the use of a multi-pole switch (eg a DPST switch can turn off two circuits), or by switching multiple relays with the one switch. As with any immobiliser using a relay, the circuit should be configured so that the relay is open-circuit when power is cut.

SIMPLE REMOTE IMMOBILISER

With the availability of cheap 12V remote switching modules, it's easy to add an immobiliser. Search online and you'll find plenty of these switches available.

Here is the remote switch. This unit comprises a small box and a remote fob. The action of the 'on' and 'off' keys can be configured by moving a jumper on the circuit board inside the box.

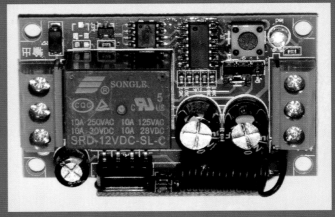

Inside the box. In this immobiliser application, a 'latching' mode needs to be selected. This causes the onboard relay to engage-and-hold with one press of the remote's button, and disengage with another button press. The relay is rated at 10A – that's enough for some disabling circuits, but not others. If in doubt, add a heavy-duty automotive relay.

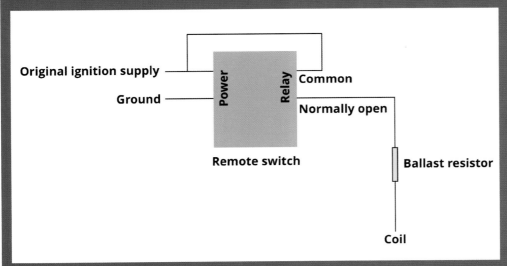

In this case, the immobiliser was being installed on a car with a single ignition coil. Here's the wiring approach I used. I chose to disable the ignition coil, intercepting the power feed before the ballast resistor. Furthermore, I mounted the receiver module under the bonnet and powered it from the wire I was disabling – an approach that made the wiring very simple. Wired in this way, the steps in starting the car were as follows:

1. Turn the ignition key until dash warning lights are lit
2. Press the remote fob button
3. Turn the key further to start engine

Note that pressing the remote with the car ignition off doesn't switch off the immobiliser – rather deceptive for a thief armed with your keys! Note also that you don't need to press the remote button to activate the immobiliser when you leave the car – as soon as you turn the ignition off, the relay automatically opens.

WORKSHOP PRO — MODIFYING THE ELECTRONICS OF MODERN CLASSIC CARS

Here are the coil (black arrow) and ballast resistor (yellow arrow).

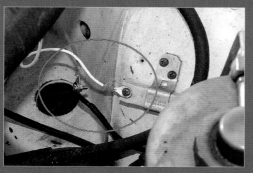

The power supply lead to the ballast resistor was parted at the terminal and extensions put in place to connect to the immobiliser module's relay.

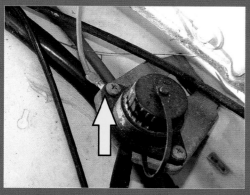

A ground connection for the immobiliser module was made at a convenient screw.

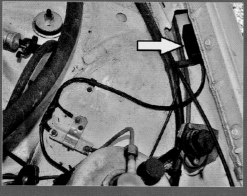

The fitted system, with the receiver module arrowed. Seal the module's box with some neutral-cure silicone.

ULTRA-LOW CURRENT LED FLASHER

A flashing LED may dissuade a casual thief. However, if it stops flashing, it's also an indicator to the thief that they've managed to disconnect power – eg by reaching up to the car battery from under the car. One way around this – and it also works well if you leave a car for a long period and the battery may go flat – is to use an ultra-low current flasher that can be operated off two 1.5V cells. Just use the normally-closed contacts of a relay to switch the flasher on whenever the ignition is off. If you fit a hidden DPST immobiliser switch, you can use one set of contacts to immobilise the car and the other set to turn on the flasher. And how do you make such a flasher? It's really easy.

Any small battery-powered clock can be used as the basis of the flasher. You can buy them new very cheaply or salvage the workings from a discarded clock.

Disassemble the clock until you can remove the circuit board (arrowed). Before you do so, take special note of the polarity of the battery connections, and which solder pads connect to the solenoid coil.

10. IMPROVING CONVENIENCE AND SECURITY

Remove the electronics. In this case, the clock had an alarm – you don't want it in this application, so the buzzer wiring can be cut off. This PCB design uses battery clips that can be slid off the board – other designs will require the battery wiring to be unsoldered.

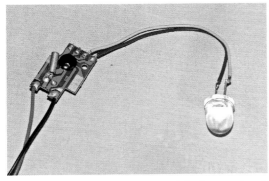

Connect 3V of battery power to the original power terminals (use the correct polarity) and solder an LED directly across the solenoid coil outputs. Use flexible thin insulated wire for these connections. Here a 10mm red LED has been used. You can use whatever size or colour low-current LED you wish.

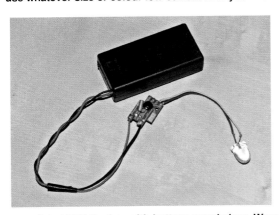

The completed LED flasher with battery supply box. Wrap the circuit board in tape and insulate its connections. The flasher will run continuously for a year from two AA cells, giving one short flash every two seconds.

INTERIOR LIGHT DELAY

If your car does not have an extended 'on' time for the interior light when you get in and close the door, you can easily add one. Modules that connect in parallel with one of the door switches are cheaply available and work well. However, there are two potential traps.

The first is that if you have fitted LED lights instead of filament lamps, the lights may not go completely off after the delay period is meant to have finished. You can counteract this by either selecting a module designed to work with LED lights, or adding a high-power resistor across the LED lights so that the load remains the same as it was when filament lamps were being used.

The other trap is that most modules have a fixed delay period, and in some situations that might be too long. For example, if you jump into the car and quickly drive off, the interior light will stay on – even when that's not wanted. The circuit shown here, that uses an additional relay, addresses this issue.

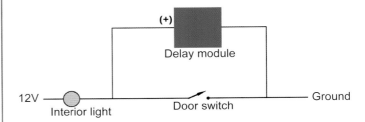

The interior light delay module is wired in parallel with a door switch. It doesn't matter which switch you use, so pick the most accessible. Ensure that you observe the correct polarity when wiring in the module.

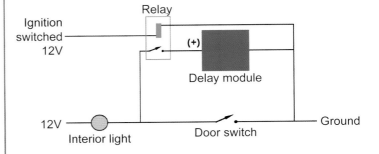

By adding a SPDT relay and using its normally closed contacts, the delay module can be disabled when the ignition is switched on, causing the interior light to go off as soon as the car is started.

INSTALLING AN AUTO-DIMMING REAR VISION MIRROR

One feature that few cars of the decades covered in this book have is an auto-dimming rear vision mirror. However, adding such a mirror is easy. In fact, the hardest part is

WORKSHOP PRO: MODIFYING THE ELECTRONICS OF MODERN CLASSIC CARS

This Saab auto-dimming mirror is widely available secondhand and is easy to wire into place.

attaching the new mirror in place of the old. Typically, it's easiest just to glue the mount to the windscreen – a specific glue is available for this purpose.

The Saab 9-5 has an auto-dimming mirror (part no 12803374) that's available cheaply secondhand. Its dimensions are 240 x 75mm (9.4 x 3in) and it has a depth of 50mm (2in). The wiring is very easy – just cut off the plug and connect the red wire to ignition-switched 12V and the black wire to ground. You can ignore the other wires. When you are testing it, don't forget to press the auto button on the mirror!

NAVIGATION

Many of the premium cars of the 1990s/2000s had inbuilt navigation systems. Unfortunately, all of these used DVDs to provide the maps, and the data was encoded in a way specific to that manufacturer. As these cars have got older, the availability of up-to-date map DVDs has decreased, and where they are still available, the DVDs have remained expensive.

If you're in that situation, you have two main options:
- replace the original head unit with an aftermarket head unit that incorporates navigation
- use a standalone navigation system, that typically sticks to the windscreen

I like in-dash navigation, so I've stayed as long as possible with the original system. However, when the maps are becoming increasingly out of date, and no further upgrades are ever going to be produced, there comes a point where effectively your car no longer has navigation!

In that situation, I've moved to the stick-on standalone screens – and I must say, have been pleasantly surprised. This is an area of technology that will continue to change rapidly, but at the time of writing, I am using a Garmin DriveSmart™ 51 LMT-S that has excellent navigation, and incorporates lane keeping and rear-end collision avoidance warnings, among other features.

MORE FROM JULIAN EDGAR:

Want to restore, modify or repair your car's electrical and/or electronic systems? This handbook is a must-read that takes you from the basics of circuits right through to diagnosing and repairing complex electronic car systems.

ISBN: 978-1-787112-81-0
Paperback • 25x20.7cm • 168 pages
• 262 colour pictures

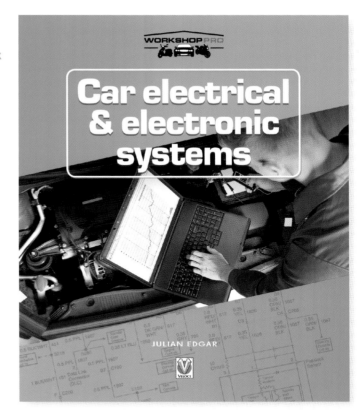

Want to modify, restore or maintain your car at home? This book covers the complete setting-up and use of a home workshop. From small and humble to large and lavish – this book shows you the equipment to buy and build, the best interior workshop layouts, and how to achieve great results.

ISBN: 978-1-787112-08-7
Paperback • 25x20.7cm • 160 pages
• 250 pictures

FOR MORE INFORMATION AND PRICE DETAILS, VISIT OUR WEBSITE AT WWW.VELOCE.CO.UK
EMAIL: INFO@VELOCE.CO.UK • TEL: +44(0)1305 260068

MORE FROM JULIAN EDGAR:

If you want the best ride and handling for your road car, this is the book you need! Julian Edgar shows you how to fit air suspension to your car. It covers both theory and practice, and includes the step-by-step fitting of aftermarket air suspension systems and building your own with parts from other cars.

ISBN: 978-1-787111-79-0
Paperback • 25x20.7cm • 64 pages
• 82 colour and b&w pictures

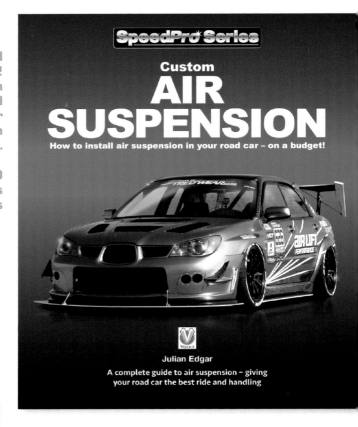

Modifying your car for increased performance? You need this book! It shows you how to easily measure on the road the gains and losses of changing air intakes, exhausts, cams and turbos. Also learn how to test suspension, brakes and car aerodynamics – accurately and at low cost.

ISBN: 978-1-787113-18-3
Paperback • 25x20.7cm • 72 pages • 83 pictures